E. Kreyszig / E. J. Norminton

MATHEMATICA COMPUTER GUIDE

A Self-Contained Introduction

For

Erwin Kreyszig

ADVANCED ENGINEERING MATHEMATICS

Eighth Edition

JOHN WILEY & SONS, INC.

EDITOR Kimberly Murphy
MARKETING MANAGER Julie Lindstrom
ASSOCIATE PRODUCTION DIRECTOR Lucille Buonocore
COVER DESIGNER Madelyn Lesure
COVER PHOTO Chris Rogers / The Stock Market

This book was set in 10/12 Computer Modern
and printed and bound by Von Hoffmann Graphics.
The cover was printed by Brady Palmer Printing Co.

This book is printed on acid-free paper.

The paper in this book was manufactured by a mill whose forest management programs include sustained yield harvesting of its timberlands. Sustained yield harvesting principles ensure that the number of trees cut each year does not exceed the amount of new growth.

To order books or for customer service call 1-800-CALL-WILEY (225-5945).

ISBN 0-471-38669-3 (paperback)

Printed in the United States of America

10 9 8 7 6 5 4 3 2 1

PREFACE

This **MATHEMATICA COMPUTER GUIDE** supplements the new edition (the Eighth Edition of 1999) of **ADVANCED ENGINEERING MATHEMATICS** by Erwin Kreyszig (J. Wiley & Sons, Inc., 605 Third Avenue, New York NY 10016; to order, call 1-800-225-5945 or 212-850-6000; see also

http://www.wiley.com/college/mat/kreyszig154962/).

In this Guide, ADVANCED ENGINEERING MATHEMATICS will be quoted as **AEM**.

CAS's (Computer Algebraic Systems) are useful tools in applied mathematics, and Mathematica is one of the most successful ones, used by many engineers, mathematicians, physicists, and computer scientists. In connection with AEM the algebraic, numerical, and graphical power of Mathematica can help the student in working out class notes, in doing homework and exams, as well as in pursuing self-study. This will generally enhance the student's understanding, skill, and motivation.

This Guide *supplements* **AEM** but, of course, *does not replace* it. Indeed, its use is not needed in studying from AEM itself. Thus, this Guide just invites the student to gain experience with a prominent CAS.

This Guide has two new main features, not shared by its predecessor, the Mathematica Computer Manual of 1995.

1. **It is much simpler than the latter**. It is an elementary introduction that presents Mathematica as simply as possible, in contrast to most other books on Mathematica.

2. **It is self-contained**, so that it can be used without referring to AEM, once the student has learned how to master the subject matter as such. Thus, the reference to AEM given after each example and problem is only for the case that a student needs help with the subject matter or wants to know where to find the same example or problem, or similar problems in AEM.

This Guide contains over 130 carefully selected worked-out examples and about 400 problems for solution, covering all fields discussed in AEM (except for Chap. 21).

This Guide is written in Mathematica Version 4.

Mathematica can do **numerical calculations** (calculation with numbers), **symbolic calculations** (calculations with formulas consisting of letters and other symbols, such as $+$, $-$, $=$, etc.), and **graphing** (plotting of curves and surfaces and other figures). This includes differentiation and integration, solution of differential equations, matrix and vector algebra and calculus, complex analysis, numerical methods, and statistical problems. All this is shown in this Guide.

Commands that you have to type are printed in BLUE, responses by the computer in BLACK.

For efficient help we wish to thank Prof. John Todd (Pasadena), Herbert E. Kreyszig (New York), and the personnel of J. Wiley.

ERWIN KREYSZIG
EDWARD J. NORMINTON

Contents

INTRODUCTION, GENERAL COMMANDS

Familiarity with Mathematica or other CAS's (Computer Algebraic Systems) will not be assumed.

Colors. Commands and data that you have to type are given in BLUE , computer responses in BLACK.

Chapters in this Guide correspond to those in AEM.

Information on the screen. This is often the best way of learning or recalling Mathematica. At the top of the screen you see the words File Edit Cell etc. Click on Help. A list will show up. Click on Master Index. This Index is arranged alphabetically. Type the keyword wanted, for instance, "Matrices". Click on "Go to". This will produce a list of packages and below it a list of keywords, with references to sections in the Mathematica Book [5] in Appendix 1. If you want, for instance, information on eigenvalues, click on 1.8.3 . This will show you the content of Sec. 1.8.3 in the Mathematica Book. Press the down arrow key (or use the mouse) to run through the long list of commands and examples for matrix calculations that will include eigenvalues. When you are done, click on the cross sign in the right upper corner of the sheet of the Master Index. The sheet will disappear and you will be back on the notebook sheet you are working on. In most cases the screen will show you the corresponding content of the Mathematica Book. In some cases the Book will contain keywords not in the Master Index. Hence if your search on the screen did not produce sufficient results, consult the alphabetical Index of the Mathematica Book for further information.

Books are listed in Appendix 1.

Collecting your own experience. Make up your personal manual, recording commands used, typical examples calculated, and positive and negative experience gained by trial and error.

Answers to odd-numbered problems. See Appendix 2.

Instructor's Mathematica Manual is available from the Publisher upon request.

Beginning. Suppose that Mathematica 4.0 has been loaded into your computer and that you are ready to start your first Mathematica session. Then start typing your commands needed for a calculation. A **command** is what you tell the computer by typing it on the keyboard. This can be data or the request to add a few numbers or to solve a differential equation that you typed before, etc. Commands are printed in BLUE . The computer shows something on the screen; this is called the **response** to your command, and is printed in BLACK in this Guide. For instance, to calculate $2 + 4 + 3.5 - 0.6 + 3 \times 2 + (1/8)^4$, type 3×2 as 3*2 or as 3 2 , with space between the two factors. $(1/8)^4$ is typed with ˆ for exponentiation. Thus type the following, and press Shift-Enter .

In[1]:= `s = 2 + 4 + 3.5 - 0.6 + 3*2 + (1/8)^4`

Out[1]= 14.9002

The choice of the notation *s* for this sum is up to you. `s` is not shown in the response. If you want more than 6 significant digits (the usual number shown), say, 10, type

In[2]:= `SetPrecision[s, 10]`

Out[2]= 14.90024414

Similarly if you want fewer digits.

If you do not want the response to appear on the screen, type a **semicolon** at the end.

In[3]:= `s;`   `(* This is the above sum.  Try without ;  *)`

Furthermore,

In[4]:= `1/2 + 1/4`

Out[4]= $\dfrac{3}{4}$

If you want this as a decimal fraction, use `N[...]` ,

In[5]:= `N[%]`

Out[5]= 0.75

Here the very useful symbol `%` avoids retyping.

Square roots are typed as `Sqrt[...]` . For instance,

In[6]:= `Sqrt[2]`

Out[6]= $\sqrt{2}$

In[7]:= `SetPrecision[%, 100]`

Out[7]= 1.414213562373095048801688724209698078569671875376948073176679737990732478462107038850387534327641573

π is typed as

In[8]:= `Pi`

Out[8]= π

In[9]:= `N[%]`   `(* Out: 3.14159 *)`

In[10]:= `SetPrecision[%, 50]`

Out[10]= 3.1415926535897931159979634685441851615905761718750

The natural logarithm ln x is typed as `Log[x]` ; for instance,

In[11]:= `N[Log[2]]`

Out[11]= 0.693147

Sequences, Sums, Series.

Sequences are obtained by the command `Table` . For instance,

In[12]:= `T = Table[1/n^2, {n, 1, 10}]`

Out[12]= $\{1, \frac{1}{4}, \frac{1}{9}, \frac{1}{16}, \frac{1}{25}, \frac{1}{36}, \frac{1}{49}, \frac{1}{81}, \frac{1}{100}\}$

In[13]:= N[%]

Out[13]= 1., 0.25, 0.111111, 0.0625, 0.04, 0.0277778, 0.0204082, 0.015625, 0.0123457, 0.01

You can pick ("**access**") a term, for instance, the fourth, by typing

In[14]:= T[[4]]   (* Out: $\frac{1}{16}$ *)

Sums are obtained by the command Sum[...] . For example,

In[15]:= N[Sum[1/n^2, {n, 1, 10}]]   (* Out: 1.54977 *)

Here you retyped the general term. You can avoid retyping by accessing the terms from the sequence T , namely

In[16]:= N[Sum[T[[n]], {n, 1, 10}]]   (* Out: 1.54977 *)

A **power series** of a function $f(x)$ in powers of $x - a$ can be obtained by the command Series[f, x, a, M] . For instance, the familiar Maclaurin series of the exponential function is obtained by

In[17]:= Series[Exp[x], {x, 0, 10}]

Out[17]= $1 + x + \frac{x^2}{2} + \frac{x^3}{6} + \frac{x^4}{24} + \frac{x^5}{120} + \frac{x^6}{720} + \frac{x^7}{5040} + \frac{x^8}{40320} + \frac{x^9}{362880} + \frac{x^{10}}{362880} + O[x]^{11}$

The last term is the *remainder* (the error term). If you want to use a series for *calculating* values or for *plotting*, you must first drop the error term by typing

In[18]:= S = Normal[%]

Out[18]= $1 + x + \frac{x^2}{2} + \frac{x^3}{6} + \frac{x^4}{24} + \frac{x^5}{120} + \frac{x^6}{720} + \frac{x^7}{5040} + \frac{x^8}{40320} + \frac{x^9}{362880} + \frac{x^{10}}{362880}$

For instance, substituting $x = 1$ gives an approximation for the basis e of the natural logarithm,

In[19]:= N[S /. x -> 1]   (* Out: 2.71828 *)

You see that the command for **substitution** is / . The **arrow** is typed as - followed by > (no space in between!).

Plotting. A **figure** is obtained by the plotting command Plot , in which the command AxesLabel -> {x, y} is "**optional**", that is, you may or may not type it, depending on whether you want the axes to be labeled or not.

In[20]:= Plot[S, {x, 0, 2}, AxesLabel -> {x, y}]

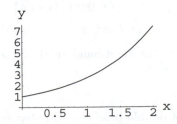

Equal scales on both axes are obtained by the optional AspectRatio -> Automatic .

In[21]:= `Plot[S, {x, 0, 2}, AxesLabel -> {x, y}, AspectRatio -> Automatic]`

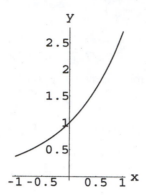

Of course, you may plot e^x by `Plot[Exp[x], {x, -1, 1}]`. Try it.

Removing ("unassigning") assigned values. This is important. The command is `Clear[...]`. For instance,

In[22]:= `x = 2`   `(* This is a value assigned to x. *)`

Out[22]= `2`

In[23]:= `Cos[x]`   `(* Out: Cos[2] *)`

But if you now want $\cos x$ with x arbitrary, type first `Clear[x]` and then `Cos[x]`.

In[24]:= `Clear[x]`

In[25]:= `Cos[x]`   `(* Out: Cos[x] *)`

Equations can be solved by the commands `Solve` or `FindRoot`. For instance,

In[26]:= `sol = Solve[x^2 + 3 x - 4 == 0, x]`   `(* Note ==, not =. *)`

Out[26]= $\{\{x \rightarrow -4\}, \{x \rightarrow 1\}\}$

In[27]:= `FindRoot[Cos[x] == 0.5, {x, 1}]`   `(* Out: {x → 1.0472} *)`

`{x, 1}` means that the search for a root begins at $x = 1$.

In[28]:= `FindRoot[Cos[x] == 0.5, {x, 8}]`   `(* Out: {x → 7.33038} *)`

Use of double brackets for accessing. Suppose that you want to have one of the two roots in `sol`, say -4. Then type

In[29]:= `sol[[1, 1, 2]]`   `(* Out: -4`   `[[1, 1, 2]]` will occur *most often.* `*)`

What do you do if you don't know this? Begin stepwise and see what happens,

In[30]:= `sol[[1]]`   `(* Out: {x → 4}`   Picks the first part. `*)`

In[31]:= `sol[[1, 1]]`   `(* Out: x → 4`   This shakes off the braces. `*)`

Now you want the second part (the right-hand side) of the remaining expression. Hence put a `2` in the previous input

In[32]:= `sol[[1, 1, 2]]`   `(* Out: -4 *)`

This illustrates the handling of double brackets. Go step by step and in each step check whether you are still on the right track. Try this out for the second solution, ending up with

In[33]:= `sol[[2, 1, 2]]` (* Out: 1 *)

Special functions, elementary and higher, are known to Mathematica and will be discussed throughout this Guide. For notations see the Master Index or the Index at the end of this Guide.

Differentiation and integration. Commands are explained at the beginning of Part A and applied in various examples.

Packages are loaded, for instance, by `<< Statistics`DescriptiveStatistics``, that is, preceded by two inequality signs `<<` and with two backquotes `` `...` `` (not primes `'...'`).

PLOTTING GUIDE

Task	Example	Page
Animation	11.1	123
Bar graphs	22.3	241
Complex numbers	12.1, 12.6	136, 146
Conformal mappings	12.4, 12.5	140, 144
Curves given parametrically	3.3	36
Curves plotted jointly	1.5, 3.3, 4.2	13, 36, 45
Curves in space	8.3	92
Curves in the plane	1.1	8
Direction fields	1.2, 3.5	9, 38
Distribution function	22.3	239
Fluid flows	16.3	168
`<<Graphics` (*See* Index)		
Grids in the plane	19.7	222
Histograms	22.2	237
Labeling coordinate axes	1.1	8
Labeling plots	1.3	11
Numbering coordinate axes	1.1	8
`ParametricPlot`	3.1	32
`ParametricPlot3D`	9.5	104
`Plot3D`	11.4	127
Points	12.1, 12.6	136, 146
Polygons	17.2, 19.8	182, 227
`ScatterPlot`	20.1	232
Sequence, plotting terms	12.6, 14.1	146, 155
Series, plot from it	4.1	43
Splines, plotting	17.7	188
Surfaces	11.5, 12.5, 20.1	131, 143, 232
Title of a figure	1.3	11
`ViewPoint`	9.5	104

PART A. ORDINARY DIFFERENTIAL EQUATIONS (ODE's)

Content. First-order ODE's (Chap. 1)
Second and higher order ODE's (Chap. 2)
Systems of ODE's (Chap. 3)
Series solution of ODE's (Chap. 4)
Solution of ODE's by Laplace transforms (Chap. 5)

Derivatives $y', y'', y''', \ldots$ may be typed as `D[y, x]`, `D[y, x, x]`, `D[y, x, x, x]`, ... For instance,

In[1]:= `y = Exp[-x] Sin[2 x]`

Out[1]= $e^{-x} \mathrm{Sin}\,[2\,x]$

In[2]:= `D[y, x]`   (* Out: $2\,e^{-x} \mathrm{Cos}\,[2\,x] - e^{-x} \mathrm{Sin}\,[2\,x]$ *)

In[3]:= `D[y, x, x]`   (* Out: $-4\,e^{-x} \mathrm{Cos}\,[2\,x] - 3\,e^{-x} \mathrm{Sin}\,[2\,x]$ *)

In differential equations you may denote derivatives of unknown functions $y(x)$ by `y'[x]`, `y''[x]`, etc. See the examples.

Integration. `Integrate[f, x]` gives the indefinite integral (the antiderivative) and `Integrate[f, {x, a, b}]` the definite integral of a given function f from a to b. For instance,

In[4]:= `f = 4 x Exp[2 x]`   (* Out: $4\,e^{2\,x}\,x$ *)

In[5]:= `Integrate[f, x]`

Out[5]= $4\,e^{2\,x}\left(-\dfrac{1}{4} + \dfrac{x}{2}\right)$

and the definite integral of f from $-\infty$ to 3 is obtained by typing

In[6]:= `Integrate[f, {x, -Infinity, 3}]`   (* Out: $5\,e^{6}$ *)

In[7]:= `N[%]`   (* Out: 2017.14 *)

If you want more digits, use the command `SetPrecision[...]`. For instance,

In[8]:= `SetPrecision[%, 10]`   (* Out: 2017.143967 *)

Command for solving ODE's and systems. `DSolve` gives general solutions as well as particular solutions of initial value problems. See the various examples.

6

Chapter 1

First-Order ODE's

Content. General solutions (Ex. 1.1)
Direction fields (Ex. 1.2, Prs. 1.1, 1.2)
Separable ODE's (Prs. 1.3-1.8)
Exact ODE's, integrating factors (Ex. 1.4, Prs. 1.9, 1.10)
Linear ODE's, mixing problems, electric circuits (Exs. 1.3, 1.6,
　　Prs. 1.11, 1.13, 1.17, 1.18)
Bernoulli ODE, Verhulst population model (Ex. 1.5, Prs. 1.14, 1.15)
Picard iteration, do-loop (Prs. 1.19, 1.20)

Examples for Chapter 1

EXAMPLE 1.1　GENERAL SOLUTIONS

General solutions are obtained by DSolve . For instance,

In[1]:= Clear[y]　　　　　　　　(* Unassign y that has just been used. *)

Type the given ODE $y' - 3y = 0$, in which $y(x)$ is unknown. Note that the **equality sign** is typed as == , not = .

In[2]:= ode = y'[x] - 3 y[x] == 0
Out[2]= $-3\,y[x] + y'[x] == 0$

In[3]:= DSolve[ode, y[x], x]
Out[3]= $\{\{y[x] \to e^{3\,x}\,C[1]\}\}$

Give a name to the solution by which you can use it further, say, sol . (This name will not be shown in the response; you have to remember what you did.)

In[4]:= sol = DSolve[ode, y[x], x]
Out[4]= $\{\{y[x] \to e^{3\,x}\,C[1]\}\}$

C[1] is the Mathematica notation for an **arbitrary constant** throughout.

Initial value problems are also solved by DSolve . For instance, $y' - 3y = 0$, $y(0) = 2$ is solved by

In[5]:= ypartic = DSolve[{ode, y[0] == 2}, y[x], x]　　(* **Particular solution** *)
Out[5]= $\{\{y[x] \to 2\,e^{3\,x}\}\}$

Carefully distinguish among braces {}, *brackets* [], *and parentheses* (). For instance, replace the braces in ypartic by brackets and see what error message you get.

Checking solutions obtained on the computer is at least as important as it is in working with paper and pencil – the computer will sometimes fool you. Type

In[6]:= `y1 = c Exp[3 x];`   `(* Your general solution just obtained *)`

In[7]:= `D[y1, x] - 3 y1`   `(* Out: 0 *)`

In[8]:= `ypartic /. x -> 0`   `(* This verifies the initial condition. *)`

Out[8]= `{{y[0] → 2}}`

To plot the particular solution, type the following, with the arrow typed by - and then > without space in between.

In[9]:= `Plot[ypartic, {x, 0, 1}, Ticks -> {Automatic, {2, 10, 20, 30, 40}}]`
    `Plot::plnr:`
      `ypartic is not a machine-size real number at x = 4.166666666666666'*^-8. ...`

To understand what went wrong, note that `ypartic` is the whole expression $\{\{y[x] \to 2\,e^{3x}\}\}$. But you want to plot just the right-hand side $2\,e^{3x}$. To 'shake off' all the braces, type

In[10]:= `ypartic[[1, 1]]`   `(* Out: y[x] → 2 e`^{3x}` *)`

To obtain only the right-hand side $2\,e^{3x}$, type

In[11]:= `ypartic[[1, 1, 2]]`   `(* Out: 2 e`^{3x}` *)`

Keep `[[1, 1, 2]]` **in mind**; it will occur frequently. Hence the correct plotting command is the following, where `Automatic` sets tick marks and numbers on the x-interval $0 \le x \le 1$ in a reasonable fashion.

In[12]:= `Plot[ypartic[[1, 1, 2]], {x, 0, 1}, AxesLabel -> {x, y},`
    `Ticks -> {Automatic, {2, 10, 20, 30, 40}}]`

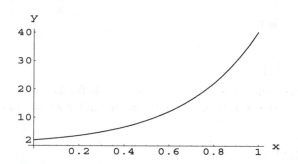

Example 1.1. Particular solution $y(x) = 2\,e^{3x}$

Similar Material in AEM: pp. 6, 9 (#25), 15

EXAMPLE 1.2 | **DIRECTION FIELDS**

Plot a direction field for the ODE $y' = xy$. Include the two solution curves through the points $(0, 1)$ and $(0, 2)$ in order to illustrate that the field gives the correct tangent directions at the points of the curves.

Solution. Type the ODE

In[1]:= `ode = y'[x] == x y[x]`   `(* Space between x and y! *)`

Out[1]= $y'[x] == x\, y[x]$

and the given points (initial conditions for particular solutions represented by those curves)

In[2]:= `inits = {{0, 1}, {0, 2}}`   `(* Out: {{0, 1}, {0, 2}} *)`

Load the package needed for plotting direction fields by typing

In[3]:= `<< Graphics'PlotField'`   `(* Note '...' not '...' *)`

Then give the command for plotting a vector field $\mathbf{v} = [v_1, v_2]$, which is in general `PlotVectorField[...]` and in the brackets $\{v_1, v_2\}$, then an x-range and a y-range. Now for a direction field, $v_1 = 1$ and $v_2 = y'$ because y' is the slope of the solution curves. For our ODE we have $v_2 = y' = xy$. Hence type (choosing the ranges $-3 \le x \le 3$ and $0 \le y \le 3$),

In[4]:= `p1 = PlotVectorField[{1, x y}, {x, -3, 3}, {y, 0, 3},`
         `AxesLabel -> {x, y}, Axes -> True]`

Instead of `x y` you may type `x*y`, (with an **asterisk**) but since this `*` is not necessary, we shall generally omit it from products – but make sure that you always leave ***space*** between factors. Try it without the space.

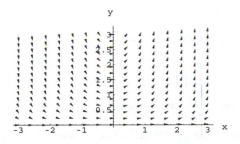

Example 1.2. Direction field for $y' = xy$

To obtain the (exact) solution curves through the points $(0, 1)$ and $(0, 2)$, type

In[5]:= `sol1 = DSolve[{ode, y[0] == 1}, y[x], x]`

Out[5]= $\left\{\left\{ y(x) \rightarrow e^{\frac{x^2}{2}} \right\}\right\}$

In[6]:= `sol2 = DSolve[{ode, y[0] == 2}, y[x], x]`

Out[6]= $\left\{\left\{ y(x) \rightarrow 2\, e^{\frac{x^2}{2}} \right\}\right\}$

Plot these two solutions jointly by typing (with the **arrow** obtained by typing – and then >)

In[7]:= `p2 = Plot[{sol1[[1, 1, 2]], sol2[[1, 1, 2]]}, {x, -3, 3},`
         `PlotRange -> {0, 3}]`

Remember that `sol1` is $\{\{y(x) \rightarrow e^{\frac{x^2}{2}}\}\}$ and `[[1, 1, 2]]` shakes off the four braces and takes the right-hand side $e^{\frac{x^2}{2}}$, so that you do obtain the function to be plotted. Similarly for `sol2`. The response (the figure) is omitted since the two curves appear in the next figure.

Finally plot the direction field and the two solution curves jointly by typing

In[8]:= `Show[p1, p2, PlotRange -> {0, 3}, AxesLabel -> {x, y}, Axes -> True]`

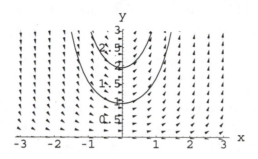

Example 1.2. Direction field and solution curves of the ODE $y' = xy$

The equation can be solved by separating variables, $y'/y = x$, and integration, $y = c \exp(x^2/2)$.

Similar Material in AEM: pp. 11, 13

EXAMPLE 1.3 | **MIXING PROBLEMS**

Mixing problems involve a tank into which brine flows, the content of the tank is stirred (this is the 'mixing'), and the mixture flows out. The model is the ODE

$$y' = \text{Salt inflow rate minus Salt outflow rate,}$$

where $y(t)$ is the amount of salt in the tank, and $y' = dy/dt$ is the time rate of change of $y(t)$. Assume the following. At $t = 0$ the tank contains 200 gal of water in which 40 lb of salt are dissolved. The inflow is 10 lb/min (5 gal of brine, each containing 2 lb of salt). 5 gal/min of mixture flow out. Hence the model is

$$y' = 10 - (5/200)y, \qquad y(0) = 40.$$

Solve this initial value problem by typing

In[1]:= `ode = y'[t] == 10 - (5/200) y[t]`

Out[1]= $y'[t] == 10 - \dfrac{y[t]}{40}$

In[2]:= `DSolve[{ode, y[0] == 40}, y[t], t]`

Out[2]= $\{\{y(t) \rightarrow e^{-t/40}(-360 + 400\, e^{t/40})\}\}$

In[3]:= `ypartic = Simplify[%]`

Out[3]= $\{\{y(t) \rightarrow 400 - 360\, e^{-t/40}\}\}$

Plot this particular solution. The command to get the function $400 - 360\,e^{-t/40}$ is `ypartic[[1, 1, 2]]`, so the plotting command is

In[4]:= `Plot[ypartic[[1, 1, 2]], {t, 0, 250}, PlotRange -> {0, 400},`
     `AxesLabel -> {t, y},   PlotLabel -> "Salt content",`
     `Ticks -> {Automatic, {0, 40, 100, 200, 300, 400}}]`

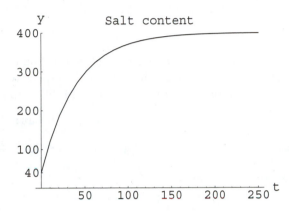

Example 1.3. Salt content $y(t)$ in the tank

The plot illustrates that $y(t)$ approaches the limit 400 lb.
   Similar Material in AEM: pp. 20, 23

| **EXAMPLE 1.4** | **INTEGRATING FACTORS** |

Integrating factors convert nonexact ODE's into exact ODE's. Let the given ODE be

$$(1) \qquad P\,dx + Q\,dy = 2\,\sin\left(y^2\right)dx + x\,y\,\cos\left(y^2\right)dy = 0.$$

Thus

In[1]:= `P = 2 Sin[y^2]`   `(* Out: 2 Sin[y²] *)`

In[2]:= `Q = x y Cos[y^2]`   `(* Out: x y Cos[y²] *)`

The exactness test fails. Indeed, the **partial derivatives** $\frac{\partial P}{\partial y}$ and $\frac{\partial Q}{\partial x}$ are typed as shown, and their difference is

In[3]:= `D[P, y] - D[Q, x]`   `(* Out: 3 y Cos [y²] *)`

In the case of exactness the response would be zero.
   Try for an integrating factor $F(x)$ depending only on x. (Ordinarily, an integrating factor (if it exists) would depend on both x and y.) The exactness condition for (1) multiplied by $F(x)$ is

In[4]:= `eq1 = D[F[x] P, y] - D[F[x] Q, x] == 0`

Out[4]= $3\,y\,\mathrm{Cos}\,[y^2]\,F[x] - x\,y\,\mathrm{Cos}\,[y^2]\,F'[x] == 0$

In[5]:= `Simplify[eq1]`

Out[5]= $y\,\mathrm{Cos}\,[y^2]\,(3\,F[x] - x\,F'[x]) == 0$

Dropping $y \cos(y^2)$, you obtain the ODE $3F(x) - xF'(x) = 0$. You can obtain this by typing `eq1[[1, 3]]`, which gives the third factor on the left side of the last response. Thus,

In[6]:= `Simplify[eq1][[1, 3]] == 0`

Out[6]= $3\,\mathtt{F[x]} - \mathtt{x}\,\mathtt{F'[x]} == 0$

In[7]:= `sol = DSolve[%, F[x], x]`

Out[7]= $\{\{\mathtt{F[x]} \rightarrow \mathtt{x}^3\,\mathtt{C[1]}\}\}$

Hence an integrating factor is x^3 (you can choose `C[1] = 1`). You can now obtain an implicit solution $u(x, y) = const$ by integrating $x^3 P$ with respect to x and $x^3 Q$ with respect to y. Do both integrations because the results will differ, in general. (Why?)

In[8]:= `Integrate[x^3 P, x]`   (* Out: $\frac{1}{2}\,\mathtt{x}^4\,\mathtt{Sin[y^2]}$ *)

In[9]:= `Integrate[x^3 Q, y]`   (* Out: $\frac{1}{2}\,\mathtt{x}^4\,\mathtt{Sin[y^2]}$ *)

Hence a solution is $x^4 \sin(y^2) = const$.

Similar Material in AEM: pp. 30, 32

EXAMPLE 1.5   BERNOULLI'S EQUATION

Bernoulli's equation includes as a special case an important population model, the **Verhulst equation** $y' = Ay - By^2$, where A and B are positive constants. Type this equation as

In[1]:= `ode = y'[x] == A y[x] - B y[x]^2`

Out[1]= $\mathtt{y'[x]} == \mathtt{A\,y[x]} - \mathtt{B\,y[x]}^2$

where x is time. Solve it by `DSolve[...]`,

In[2]:= `sol = DSolve[ode, y[x], x]`

Out[2]= $\left\{\left\{\mathtt{y[x]} \rightarrow \dfrac{\mathtt{A\,e^{A\,x}}}{\mathtt{B\,e^{A\,x}} + \mathtt{e^{C[1]}}}\right\}\right\}$

For plotting you must choose ***specific values*** of A and B. Set $A = B = 1$, for simplicity.

In[3]:= `sol2 = sol /. {A -> 1, B -> 1, Exp[C[1]] -> c}`

Out[3]= $\left\{\left\{\mathtt{y[x]} \rightarrow \dfrac{\mathtt{e^x}}{\mathtt{c} + \mathtt{e^x}}\right\}\right\}$

From this obtain and plot three typical particular solutions

In[4]:= `y1 = sol2 /. c -> 9`   (* Out: $\{\{\mathtt{y[x]} \rightarrow \frac{\mathtt{e^x}}{9 + \mathtt{e^x}}\}\}$ *)

In[5]:= `y2 = sol2 /. c -> 0`   (* Out: $\{\{\mathtt{y[x]} \rightarrow 1\}\}$ *)

In[6]:= `y3 = sol2 /. c -> -0.5`   (* Out: $\{\{\mathtt{y[x]} \rightarrow \frac{\mathtt{e^x}}{-0.5 + \mathtt{e^x}}\}\}$ *)

Now plot the three solutions on common axes by typing

In[7]:=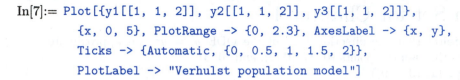

```
In[7]:= Plot[{y1[[1, 1, 2]], y2[[1, 1, 2]], y3[[1, 1, 2]]},
     {x, 0, 5}, PlotRange -> {0, 2.3}, AxesLabel -> {x, y},
     Ticks -> {Automatic, {0, 0.5, 1, 1.5, 2}},
     PlotLabel -> "Verhulst population model"]
```

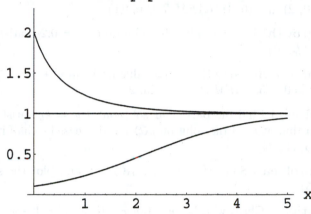

Example 1.5. Typical solution curves of the Verhulst ODE

Similar Material in AEM: p. 37, 38

EXAMPLE 1.6   *RL*-CIRCUIT

The current $i(t)$ in an *RL*-circuit with $R = 50$ ohms, $L = 1$ henry, and electromotive force $110 \cos 3t$ volts is obtained by solving the ODE

In[1]:= `ode = i'[t] + 50 i[t] == 110 Cos[3 t]`

Out[1]= $50\, i[t] + i'[t] == 110\, Cos[3\, t]$

Assume that $i(0) = 0$. The corresponding particular solution is obtained by `DSolve`,

In[2]:= `sol = DSolve[{ode, i[0] == 0}, i[t], t]`

Out[2]= $\left\{\left\{ i(t) \rightarrow \dfrac{e^{-50\,t}\,(-5500 + 5500\,e^{50\,t}\,Cos\,[3\,t] + 330\,e^{50\,t}\,Sin\,[3\,t])}{2509} \right\}\right\}$

In[3]:= `sol2 = Expand[sol[[1, 1, 2]]]`

Out[3]= $-\dfrac{5500\,e^{-50\,t}}{2509} + \dfrac{5500\,Cos\,[3\,t]}{2509} + \dfrac{330\,Sin\,[3\,t]}{2509}$

In[4]:= `N[%]`

Out[4]= $-2.19211\;\;2.71828^{-50.\,t} + 2.19211\,Cos\,[3.\,t] + 0.131527\,Sin\,[3.\,t]$

If you want, say, only 4 digits, type

In[5]:= `SetPrecision[sol2, 4]`

Out[5]= $-2.192\;\;2.718^{-50.00\,t} + 2.192\,Cos\,[3.000\,t] + 0.1315\,Sin\,[3.000\,t]$

The steady-state solution is a harmonic motion with the frequency of the electromotive force. The exponential term dies out very fast because $R/L = 50$ is large.

Similar Material in AEM: pp. 43, 47

Problem Set for Chapter 1

Pr.1.1 **(Direction field)** Plot the direction field of the ODE $y' = y^2$ and approximate solution curves through the points (0, 1), (0, 2), and (0, 3). (*AEM Ref.* pp. 11, 13 (#9))

Pr.1.2 **(Direction field)** Plot the direction field of $y' = -4x/9y$ and approximate solution curves through (0, 2) and (0, 4). (*AEM Ref.* p. 11)

Pr.1.3 **(Exponential growth)** Find and plot the solution of $y' = 0.2y$ satisfying $y(0) = 2$. (*AEM Ref.* p. 9 (#25))

Pr.1.4 **(Exponential approach)** Solve the initial value problem $y' + y = 2$, $y(0) = 0$. Plot the solution for $t = 0 \dots 5$. (*AEM Ref.* pp. 23, 24)

Pr.1.5 **(Exponential decay)** Find the particular solution of $y' = ky$ satisfying $y(0) = 4$. Determine k such that at $t = 1$ the solution $y(t)$ has decreased to half its initial value. (*AEM Ref.* pp. 6, 23, 24)

Pr.1.6 **(Initial value problem)** Solve $y' = 1 + y^2$, $y(0) = 0$ and plot the solution curve. (*AEM Ref.* p. 15)

Pr.1.7 **(Checking solutions)** Check whether $y = \tan x$ satisfies $y' = 1 + y^2$. (*AEM Ref.* p. 15)

Pr.1.8 **(Separable equation)** Solve the initial value problem $y^3 y' + x^3 = 0$, $y(0) = 1$. (*AEM Ref.* p. 18 (#14))

Pr.1.9 **(Test for exactness)** Is the following equation exact?

$$(x^3 + 3xy^2)\, dx + (3x^2 y + y^3)\, dy = 0.$$

Solve this equation. (*AEM Ref.* p. 27)

Pr.1.10 **(Integrating factor)** Show that $\cos(x + y)$ is an integrating factor of $y\, dx + (y + \tan(x + y))\, dy = 0$ and solve the exact equation. (*AEM Ref.* p. 32 (#26))

Pr.1.11 **(Linear differential equation)** Find the general solution of $y' = 2(y - 1)\tan 2x$ and from it the particular solution y_p satisfying the initial condition $y(0) = 4$. Plot y_p. (*AEM Ref.* p. 39 (#17))

Pr.1.12 **(Beats)** Find and plot the solution of the following initial value problem. (Use Csc and Cot .)

$$\csc x\, dy - (y \cot x \csc x + 20 \cos 20x)\, dx = 0, \qquad y(\pi/2) = 0.$$

Pr.1.13 **(Linear differential equation)** The general solution of $y' + p(x)y = r(x)$ is

$$y(x) = e^{-h}\left[\int e^h r\, dx + c\right], \qquad h = \int p(x)\, dx.$$

Solve $y' + ky = e^{-kx}$ by this integral formula. (*AEM Ref.* p. 34)

Pr.1.14 **(Bernoulli equation)** Solve $y' + (1/3)y = (1/3)(1 - 2x)y^4$ (a) directly by DSolve , and (b) by setting $u = 1/y^3$, simplifying the ODE in u, and then applying DSolve . (*AEM Ref.* p. 39 (#33))

Pr.1.15 (Verhulst equation) Solve the Verhulst equation $y' = 3y - 5y^2$. Find three initial conditions such that the corresponding solutions are (1) increasing, (2) constant, (3) decreasing. Plot these solutions. (*AEM Ref.* pp. 13, 37, 41)

Pr.1.16 (Orthogonal trajectories) Plot some of the hyperbolas $xy = c = const$. Find and plot some of their orthogonal trajectories, all curves and trajectories on common axes. (*AEM Ref.* p. 51 (#16))

Pr.1.17 (RC-circuit) The current $i(t)$ in an RC-circuit is governed by the ODE

$$R\, di/dt + i/C = dE/dt.$$

Solve this ODE for a general resistance R, capacitance C, and electromotive force $E(t) = E_0 \sin \omega t$. Plot $i(t)$, assuming that $R = 1$ ohm, $C = 1$ farad, $\omega = 1\,\mathrm{sec}^{-1}$, $E_0 = 220$ volts, and $i(0) = 0$ ampere. (*AEM Ref.* p. 46)

Pr.1.18 (RL-circuit) Model the current in an RL-circuit with $L = 0.1$ henry, $R = 5$ ohms, and a 12-volt battery. Determine and plot (on common axes) the current $i(t)$ when $i(0)$ equals 5, 2.5, 1, 0 amps. (*AEM Ref.* p. 43)

Pr.1.19 (Picard iteration) Integrating $y' = f(x, y)$, $y(x_0) = y_0$ gives

$$y(x) = y_0 + \int_{x_0}^{x} f(t, y(t))\, dt.$$

This suggests the Picard iteration

$$y_n(x) = y_0 + \int_{x_0}^{x} f(t, y_{n-1}(t))\, dt, \qquad n = 1, 2, \dots .$$

Solve $y' = 1 + y^2$, $y(0) = 0$ by Picard iteration, 3 steps. (*AEM Ref.* p. 57)

Pr.1.20 (Experiment on Picard iteration by a do-loop) Obtain and plot the solution of Pr.1.19 (and a few initial value problems of your choice) by the do-loop with $N = 5$ to 10 steps.

```
In[1]:= NO = 3                                    (* N is protected; write NO. *)

In[2]:= y0 = 0                                                (* for instance *)

In[3]:= pic[0] = 0

In[4]:= Do[pic[n] = y0 + Integrate[1 + (pic[n - 1] /. x -> t)^2, {t, 0, x}],
        {n, 1, NO}]

In[5]:= S = Table[pic[n], {n, 1, NO}]
```

$$\mathrm{Out[5]} = \left\{ \mathbf{x},\ \mathbf{x} + \frac{\mathbf{x}^3}{3},\ \mathbf{x} + \frac{\mathbf{x}^3}{3} + \frac{2\,\mathbf{x}^5}{15} + \frac{\mathbf{x}^7}{63} \right\}$$

```
In[6]:= Plot[Evaluate[S], {x, 0, 1}, AxesLabel -> {x, y}]
```

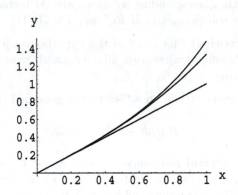

Problem 1.20. Picard approximations of the solution $y = \tan x$ of
$$y' = 1 + y^2,\ y(0) = 0$$

Chapter 2

Linear ODE's of Second and Higher Order

Content. General solutions, initial value problems (Ex. 2.1, Prs. 2.1, 2.2)
Vibrating mass on a spring (Exs. 2.2, 2.3, Prs. 2.3, 2.7, 2.8, 2.10)
Euler-Cauchy equations (Ex. 2.4, Prs. 2.12, 2.13)
Wronskian (Ex. 2.5, Prs. 2.4, 2.14)
Nonhomogeneous linear ODE's (Exs. 2.6-2.8, Prs. 2.6, 2.16, 2.18, 2.19)
Resonance, beats, electric circuits (Exs. 2.9, 2.10, Prs. 2.15, 2.17, 2.20)

Examples for Chapter 2

EXAMPLE 2.1 | GENERAL SOLUTION. INITIAL VALUE PROBLEM

Find a general solution y of the given ODE. Find and plot the particular solution y_p satisfying the given initial conditions.

$$y'' + y' - 2y = 0, \qquad y(0) = 4, \qquad y'(0) = -5.$$

Solution. Type the ODE as follows, with $''$ typed as two single primes. (Try $''$.)

In[1]:= `ode = y''[x] + y'[x] - 2 y[x] == 0`

Out[1]= $-2\,y[x] + y'[x] + y''[x] == 0$

Equivalently you may type

In[2]:= `ode2 = D[y[x], x, x] + D[y[x], x] - 2 y[x] == 0`

Out[2]= $-2\,y[x] + y'[x] + y''[x] == 0$

Obtain a general solution by `DSolve` from either one of these two equations, the results being the same:

In[3]:= `sol = DSolve[ode, y[x], x]`

Out[3]= $\{\{y[x] \to e^{-2x}\,C[1] + e^{x}\,C[2]\}\}$

In[4]:= `sol2 = DSolve[ode2, y[x], x]`

Out[4]= $\{\{y[x] \to e^{-2x}\,C[1] + e^{x}\,C[2]\}\}$

Obtain the particular solution of the **initial value problem** directly by the command

In[5]:= `ypartic = DSolve[{ode, y[0] == 4, y'[0] == -5}, y[x], x]`

Out[5]= $\{\{y[x] \to e^{-2x}(3 + e^{3x})\}\}$

In[6]:= `Simplify[%]`   (* Out: $\{\{y[x] \to 3\,e^{-2x} + e^{x}\}\}$ *)

You can check this answer by determining the arbitrary constants in the general solution as follows. For this you need the derivative of `sol`.

In[7]:= yprime = D[sol, x]

Out[7]= {{y′[x] → −2 e^{−2x} C[1] + e^x C[2]}}

and from this the values of y and y' at $x = 0$,

In[8]:= y0 = sol[[1, 1, 2]] /. x -> 0　　　　　　　(* Out: C[1] + C[2] *)

In[9]:= yprime0 = yprime[[1, 1, 2]] /. x -> 0　　　　(* Out: −2C[1] + C[2] *)

You now have to solve the linear system $y(0) = C[1]+C[2] = 4$, $y'(0) = -2C[1]+C[2] = -5$. Type, in vectoral form,

In[10]:= s = Solve[{y0, yprime0} == {4, -5}, {C[1], C[2]}]

Out[10]= {{C[1] → 3, C[2] → 1}}

Substitute this solution **s** of the linear system into the general solution **sol** of the ODE. The response will be the above particular solution.

In[11]:= sol[[1, 1, 2]] /. s　　　　　　　　　(* Out: {3 e^{−2x} + e^x} *)

If you are familiar with matrices and want to determine the two constants by using matrices, type the **coefficient matrix A** of the system and the vector **b** of the initial conditions,

In[12]:= A = {{1, 1}, {-2, 1}};

In[13]:= b = {4, -5};

and then apply the command **LinearSolve**

In[14]:= c = LinearSolve[A, b]　　　　(* Out: {3, 1}　Hence C[1] = 3, C[2] = 1. *)

You get the figure of y_p by the commands

In[15]:= Y = ypartic[[1, 1, 2]]　　　　　　　(* Out: e^{−2x} (3 + e^{3x}) *)

In[16]:= Plot[Y, {x, 0, 2}, PlotRange -> {0, 9}, AxesLabel -> {x, y}]

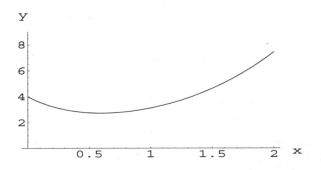

Example 2.1. Particular solution $y_p = e^x + 3 e^{−2x}$

Similar Material in AEM: p. 73

EXAMPLE 2.2 MASS-SPRING SYSTEM.
COMPLEX CHARACTERISTIC ROOTS.
DAMPED OSCILLATIONS

Solve

$$y'' + 0.2\,y' + 4.01\,y = 0, \qquad y(0) = 0, \qquad y'(0) = 2.$$

Solution. This is the model of the vertical vibrations of a weight (of mass 1) attached to the lower end of an elastic spring (of spring constant 4.01), whose upper end is fixed. This mechanical mass-spring system has damping (damping constant 0.2). It begins its motion at the displacement 0 (the position of static equilibrium) with an initial velocity 2.

Type the ODE (with " typed as two single primes '')

In[1]:= `ode = y''[t] + 0.2 y'[t] + 4.01 y[t] == 0`

Out[1]= $4.01\,\mathtt{y[t]} + 0.2\,\mathtt{y'[t]} + \mathtt{y''[t]} == 0$

where t is time. Solve the initial value problem by `DSolve`,

In[2]:= `yp = DSolve[{ode, y[0] == 0, y'[0] == 2}, y[t], t]`

Out[2]= $\{\{\mathtt{y[t]} \to e^{-0.1\,t}\,\mathtt{Sin[2.\,t]}\}\}$

In[3]:= `Plot[yp[[1, 1, 2]], {t, 0, 20}, AxesLabel -> {t, y}]`

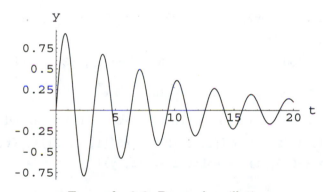

Example 2.2. Damped oscillations

This is the typical form of damped vibrations governed by a linear ODE. The oscillations lie between the exponential curves $\exp(-0.1t)$ and $-\exp(-0.1t)$. Damping takes energy from the system, so that the maximum amplitudes of the motion decrease with time and eventually go to zero. Note that the motion begins at 0 and with a positive slope, in agreement with the initial conditions.

Similar Material in AEM: p. 78

EXAMPLE 2.3 THE THREE CASES OF DAMPING

Consider the ODE $y'' + cy' + y = 0$ with $c = 1/2,\ 2,\ 3$. The term cy' is the damping term. No damping, $c = 0$, gives harmonic oscillations. For reasons of continuity you should expect (decreasing) oscillations when $c > 0$ is small ("**underdamping**"), but non-oscillatory behavior when c is large ("**critical damping**" and "**overdamping**"). Show this by solving the equation with $c = 1/2,\ 2,\ 3$ for the case that the motion starts from rest at $y = 1$.

Solution.

In[1]:= `Clear[y1, y2, y3]`

Type the equation in the three cases as

In[2]:= `ode1 = y1''[t] + (1/2) y1'[t] + y1[t] == 0`

Out[2]= $y1[t] + \dfrac{y1'[t]}{2} + y1''[t] == 0$

In[3]:= `ode2 = y2''[t] + 2 y2'[t] + y2[t] == 0`

Out[3]= $y2[t] + 2\,y2'[t] + y2''[t] == 0$

In[4]:= `ode3 = y3''[t] + 3 y3'[t] + y3[t] == 0`

Out[4]= $y3[t] + 3\,y3'[t] + y3''[t] == 0$

The initial conditions are $y(0) = 1$, $y'(0) = 0$. This gives the solutions

In[5]:= `yp1 = DSolve[{ode1, y1[0] == 1, y1'[0] == 0}, y1[t], t]`

Out[5]= $\left\{\left\{y1[t] \to e^{-t/4}\left(\mathrm{Cos}\left[\dfrac{\sqrt{15}\,t}{4}\right] + \dfrac{\mathrm{Sin}\left[\dfrac{\sqrt{15}\,t}{4}\right]}{\sqrt{15}}\right)\right\}\right\}$

In[6]:= `yp2 = DSolve[{ode2, y2[0] == 1, y2'[0] == 0}, y2[t], t]`

Out[6]= $\{\{y2[t] \to e^{-t}(1 + t)\}\}$

In[7]:= `yp3 = DSolve[{ode3, y3[0] == 1, y3'[0] == 0}, y3[t], t]`

Out[7]= $\left\{\left\{y3[t] \to \dfrac{1}{10}(5 - 3\sqrt{5})\,e^{\frac{1}{2}(-3-\sqrt{5})\,t} + \dfrac{1}{10}(5 + 3\sqrt{5})\,e^{\frac{1}{2}(-3+\sqrt{5})\,t}\right\}\right\}$

Plot these three particular solutions on common axes by the command

In[8]:= `Plot[{yp1[[1, 1, 2]], yp2[[1, 1, 2]], yp3[[1, 1, 2]]}, {t, 0, 20},`
        `PlotRange -> {-0.5, 1}, AxesLabel -> {t, y}]`

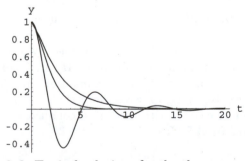

Example 2.3. Typical solutions for the three cases of damping

$c = 1/2$ (**underdamping**) gives the oscillatory solution. $c = 2$ (**critical damping**) corresponds to the **double root** of the characteristic equation $\lambda^2 + 2\lambda + 1 = 0$, and $c = 3$ (**overdamping**) gives a monotone decreasing solution (as does $c = 2$ in the present case).

If you change the initial velocity from 0 to -2, you obtain a critical solution that has a zero at $t = 1$ and is no longer monotone. Indeed, then the solution is

In[9]:= yp2b = DSolve[{ode2, y2[0] == 1, y2'[0] == -2}, y2[t], t]

Out[9]= {{y2[t] → e^{-t} (1 − t)}}

Similar Material in AEM: pp. 72, 78

EXAMPLE 2.4   **THE THREE CASES FOR AN EULER-CAUCHY EQUATION**

Solve $x^2 y'' + axy' + y = 0$.

Solution. The three cases correspond to, say, $a - 1 = 1/2,\ 2,\ 3$, thus, $a = 3/2,\ 3,\ 4$. Indeed, type the Euler-Cauchy equations as

In[1]:= ode1 = x^2 y1''[x] + (3/2) x y1'[x] + y1[x] == 0

Out[1]= $y1[x] + \dfrac{3}{2}x\,y1'[x] + x^2 y1''[x] == 0$

In[2]:= ode2 = x^2 y2''[x] + 3 x y2'[x] + y2[x] == 0

Out[2]= $y2[x] + 3x\,y2'[x] + x^2 y2''[x] == 0$

In[3]:= ode3 = x^2 y3''[x] + 4 x y3'[x] + y3[x] == 0

Out[3]= $y3[x] + 4x\,y3'[x] + x^2 y3''[x] == 0$

Choose initial conditions $y(1) = 1$, $y'(1) = 0$. You cannot choose conditions at $x = 0$, where the coefficients of the equation divided by x^2, that is, a/x and $1/x^2$, are singular. Apply DSolve to obtain

In[4]:= yp1 = DSolve[{ode1, y1[1] == 1, y1'[1] == 0}, y1[x], x]

Out[4]= $\left\{\left\{y1[x] \to \dfrac{\mathrm{Cos}\left[\frac{1}{4}\sqrt{15}\,\mathrm{Log}\,[x]\right]}{x^{1/4}} + \dfrac{\mathrm{Sin}\left[\frac{1}{4}\sqrt{15}\,\mathrm{Log}\,[x]\right]}{\sqrt{15}\,x^{1/4}}\right\}\right\}$

In[5]:= yp2 = DSolve[{ode2, y2[1] == 1, y2'[1] == 0}, y2[x], x]

Out[5]= $\left\{\left\{y2[x] \to \dfrac{1}{x} + \dfrac{\mathrm{Log}\,[x]}{x}\right\}\right\}$

In[6]:= yp3 = DSolve[{ode3, y3[1] == 1, y3'[1] == 0}, y3[x], x]

Out[6]= $\left\{\left\{y3[x] \to \dfrac{1}{10}\left(5 - 3\sqrt{5}\right)x^{\frac{1}{2}\left(-3-\sqrt{5}\right)} + \dfrac{\left(3+\sqrt{5}\right)x^{\frac{1}{2}\left(-3+\sqrt{5}\right)}}{2\sqrt{5}}\right\}\right\}$

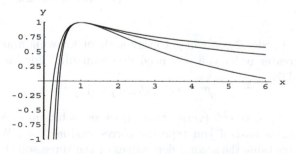

Example 2.4. The three cases of solutions for an Euler-Cauchy equation
(On the right, yp1 is the lowest curve.)

The three solutions were plotted simultaneously by typing the following

In[7]:= Plot[{yp1[[1, 1, 2]], yp2[[1, 1, 2]], yp3[[1, 1, 2]]}, {x, 0, 6},
       PlotRange -> {-1, 1}, AxesLabel -> {x, y}]

Similar Material in AEM: pp. 94, 95

EXAMPLE 2.5   **WRONSKIAN**

For $y'' + p_1(x)y' + p_0(x)y = 0$ the **Wronskian** is the determinant

$$W(y_1, y_2) = \begin{vmatrix} y_1 & y_2 \\ y_1' & y_2' \end{vmatrix} = y_1\, y_2' - y_2\, y_1'.$$

$\{y_1, y_2\}$ is a basis of solutions of the equation (with continuous p_1 and p_0) if and only if $W(y_1, y_2)$ is different from zero. Show this for the basis $y_1 = \cos kx$, $y_2 = \sin kx$ of the ODE $y'' + k^2 y = 0$ with $k \neq 0$.

Solution. The basis is

In[1]:= y1 = Cos[k x];   y2 = Sin[k x];

From this you obtain the Wronskian by typing

In[2]:= W = y1 D[y2, x] - y2 D[y1, x]   (* Out: k Cos[k x]^2 + k Sin[k x]^2 *)

In[3]:= Simplify[%]   (* Out: k *)

If $k = 0$, then $y_1 = 1$, $y_2 = 0$, which is not a basis. If k is not 0, then y_1, y_2 form a basis.

     If you know about determinants and matrices, you can use them in connection with Mathematica. The Wronskian is the determinant of the 2×2 matrix **A** which you can type as

In[4]:= A = {{y1, y2}, {D[y1, x], D[y2, x]}}

Out[4]= {{Cos[k x], Sin[k x]}, {-k Sin[k x], k Cos[k x]}}

In[5]:= %//MatrixForm   (* Out: $\begin{pmatrix} \text{Cos}[k\,x] & \text{Sin}[k\,x] \\ -k\,\text{Sin}[k\,x] & k\,\text{Cos}[k\,x] \end{pmatrix}$ *)

and obtain $W = \det \mathbf{A}$ (suggesting "determinant" of **A**) simply by typing

In[6]:= W = Det[A]   (* Out: k Cos[k x]^2 + k Sin[k x]^2 *)

In[7]:= Simplify[%]   (* Out: k *)

     For an ODE of order 2 you can get away without knowing about determinants. For an ODE of greater order you will need determinants. For instance, a basis of solutions of the ODE

$$y''' - 2y'' - y' + 2y = 0$$

is $y_1 = e^{-x}$, $y_2 = e^{x}$, $y_3 = e^{2x}$ (verify that these are solutions). You can see that these solutions form a basis if you type the corresponding 3×3 Wronskian matrix (whose third row contains the second derivatives of the three solutions). Hence type the solutions

In[8]:= y1 = Exp[-x];   y2 = Exp[x];   y3 = Exp[2 x];

then

In[9]:= y = {y1, y2, y3}　　　　　　　　　　　(* Out: $\{e^{-x}, e^x, e^{2x}\}$ *)

In[10]:= A = {y, D[y, x], D[y, x, x]}

Out[10]= $\{\{e^{-x}, e^x, e^{2x}\}, \{-e^{-x}, e^x, 2e^{2x}\}, \{e^{-x}, e^x, 4e^{2x}\}\}$

In[11]:= %//MatrixForm　　　　　　　　(* Out: $\begin{pmatrix} e^{-x} & e^x & e^{2x} \\ -e^{-x} & e^x & 2e^{2x} \\ e^{-x} & e^x & 4e^{2x} \end{pmatrix}$ *)

In[12]:= Det[A]　　　　　　　　　　　　　　　　(* Out: $6e^{2x}$ *)

Since this is not zero, the three solutions form a basis of solutions for the given ODE on any interval.

      Similar Material in AEM: pp. 97, 127

EXAMPLE 2.6　　**NONHOMOGENEOUS LINEAR ODE's**

Nonhomogeneous linear ODE's can be solved by DSolve practically in the same way as homogeneous ODE's. For instance, to solve

$$y'' + 2y' + 101y = 10.4\,e^x, \qquad y(0) = 1.1, \qquad y'(0) = -0.9,$$

type the equation as before, say,

In[1]:= Clear[y]　　　　　　　(* Unassign y (used in the previous example). *)

In[2]:= ode = y''[x] + 2 y'[x] + 101 y[x] == 10.4 Exp[x]

Out[2]= $101\,y[x] + 2\,y'[x] + y''[x] == 10.4\,e^x$

Then DSolve, applied to the initial value problem, gives the particular solution

In[3]:= yp = DSolve[{ode, y[0] == 1.1, y'[0] == -0.9}, y[x], x]

Out[3]= $\left\{\left\{y[x] \to 0.1\,e^{-1.\,x}(e^{2.\,x} + 10.\,\text{Cos}\,[10.\,x])\right\}\right\}$

To plot this solution, type

In[4]:= Plot[yp[[1, 1, 2]], {x, 0, 3.5}, Ticks -> {{1, 2, 3}, Automatic},
　　　　AxesLabel -> {x, y}]

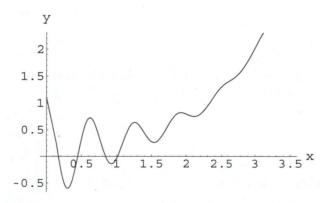

Example 2.6. Particular solution $y = 0.1e^x + e^{-x}\cos 10\,x$

You see that the oscillatory effect of the second term decreases and the solution curve approaches the curve of the exponential function $0.1\,e^x$.

Similar Material in AEM: pp. 103, 127

EXAMPLE 2.7 **SOLUTION BY UNDETERMINED COEFFICIENTS**

Write a given nonhomogeneous ODE, say,

$$y'' - 3y' + 2y = e^x,$$

in the form $Ly = r$, where Ly (L suggesting 'linear operator') is the left-hand side of the given ODE,

In[1]:= `Clear[y]`

In[2]:= `Ly = y''[x] - 3 y'[x] + 2 y[x]`

Out[2]= $2\,y[x] - 3\,y'[x] + y''[x]$

and r is the right-hand side,

In[3]:= `r = Exp[x]`   (* Out: e^x *)

To see whether the modification rule applies, you must first solve the homogeneous equation $Ly = 0$; thus,

In[4]:= `yh = DSolve[Ly == 0, y[x], x]`   (* Out: $\{\{y[x] \to e^x C[1] + e^{2x} C[2]\}\}$ *)

You see that r is a solution of the homogeneous ODE, so that the modification rule (for a simple root) does apply. That is, instead of $C\,e^x$ you have to use

In[5]:= `yp = C x Exp[x]`   (* Out: $C\,e^x\,x$ *)

(In the case of a double root you would have to multiply e^x by x^2 instead of x.) The constant C is unknown. You find its value from the given nonhomogeneous equation with $y = $ `yp` ; that is,

In[6]:= `Ly == r`   (* Out: $2\,y[x] - 3\,y'[x] + y''[x] == e^x$ *)

In[7]:= `Ly == r /. {y[x] -> yp, y'[x] -> D[yp, x], y''[x] -> D[yp, x, x]}`

Out[7]= $2\,C\,e^x + 3\,C\,e^x\,x - 3\,(C\,e^x + C\,e^x\,x) == e^x$

In[8]:= `sol = Simplify[%]`   (* Out: $-C\,e^x == e^x$ *)

You see that $C = -1$. The command for obtaining this (in a more involved case) would be

In[9]:= `C0 = Solve[sol, C]`   (* Out: $\{\{C \to -1\}\}$ *)

Substituting this into `yp` and adding `yh` , you obtain a general solution of the given ODE. Perhaps you do this step by step.

In[10]:= `yh[[1, 1, 2]]`   (* Out: $e^x C[1] + e^{2x} C[2]$ *)

In[11]:= `yp`   (* Out: $C\,e^x\,x$ *)

In[12]:= `yp2 = yp /. C -> C0[[1, 1, 2]]`   (* Out: $-e^x\,x$ *)

In[13]:= `ygen = yh[[1, 1, 2]] + yp2`   (* Out: $-e^x\,x + e^x C[1] + e^{2x} C[2]$ *)

This example served to explain the method. Clearly, you can obtain the answer directly by typing

In[14]:= `DSolve[Ly == r, y[x], x]`

Out[14]= $\{\{\texttt{y[x]} \rightarrow -e^x - e^x\,\texttt{x} + e^x\,\texttt{C[1]} + e^{2x}\,\texttt{C[2]}\}\}$

This agrees with your previous result, except for the notation.

Similar Material in AEM: p. 105

| EXAMPLE 2.8 | **SOLUTION BY VARIATION OF PARAMETERS**

To solve

$$y'' + y = \sec x$$

by variation of parameters, write it as $Ly = r$, typing

In[1]:= `Clear[y]`

In[2]:= `Ly = y''[x] + y[x]`   `(* Type two single primes! *)`

Out[2]= $\texttt{y[x]} + \texttt{y}''\texttt{[x]}$

In[3]:= `r = Sec[x]`   `(* Out: Sec[x] *)`

Obtain a general solution of the homogeneous ODE by `DSolve` ,

In[4]:= `yh = DSolve[Ly == 0, y[x], x]`   `(* General solution of` $Ly = 0$ `*)`

Out[4]= $\{\{\texttt{y[x]} \rightarrow \texttt{C[2] Cos[x]} - \texttt{C[1] Sin[x]}\}\}$

Hence a basis of solutions of the homogeneous ODE, as needed in the present method, is

In[5]:= `y1 = Cos[x];`   `y2 = Sin[x];`

For this basis the Wronskian is (see Example 2.5 in this Guide, if necessary)

In[6]:= `W = y1 D[y2, x] - y2 D[y1, x]`

Out[6]= $\texttt{Cos[x]}^2 + \texttt{Sin[x]}^2$

In[7]:= `Simplify[%]`   `(* Out: 1 *)`

The formula for a particular solution is

$$y_p = -y_1 \int \frac{y_2 r}{W}\,dx + y_2 \int \frac{y_1 r}{W}\,dx.$$

For the present ODE you obtain from it

In[8]:= `yp = -y1 Integrate[y2 r/W, x] + y2 Integrate[y1 r/W, x]`

Out[8]= $\texttt{Cos[x] Log[Cos[x]]} + \texttt{x Sin[x]}$

Hence a general solution of the given nonhomogeneous equation is

In[9]:= `ygen = yh[[1, 1, 2]] + yp`

Out[9]= $\texttt{C[2] Cos[x]} + \texttt{Cos[x] Log[Cos[x]]} + \texttt{x Sin[x]} - \texttt{C[1] Sin[x]}$

This example served to explain the method. Clearly, you can obtain the solution directly by typing

In[10]:= `DSolve[Ly == r, y[x], x]`

Out[10]= $\{\{\texttt{y[x]} \rightarrow \texttt{C[2] Cos[x]} + \texttt{Cos[x] Log[Cos[x]]} + \texttt{x Sin[x]} - \texttt{C[1] Sin[x]}\}\}$

Similar Material in AEM: p. 109

| EXAMPLE 2.9 | FORCED VIBRATIONS. RESONANCE. BEATS

Resonance occurs in an **undamped vibrating system** if the frequency of the driving force equals the natural frequency of the free vibrations of the system. For instance, let

$$y'' + y = \cos t.$$

Type this ODE and solve it.

Solution.

In[1]:= Clear[y]

In[2]:= ode = y''[t] + y[t] == Cos[t]

Out[2]= y[t] + y''[t] == Cos[t]

In[3]:= sol = DSolve[ode, y[t], t]

Out[3]= $\left\{ \left\{ y[t] \to C[2] \, Cos\,[t] + \dfrac{Cos\,[t]^3}{2} - C[1]\,Sin\,[t] - Sin\,[t] \left(-\dfrac{t}{2} - \dfrac{1}{4} Sin\,[2\,t] \right) \right\} \right\}$

In[4]:= sol2 = Simplify[%]

Out[4]= $\left\{ \left\{ y[t] \to \dfrac{1}{2} \, (Cos\,[t] + 2\,C[2]\,Cos\,[t] + t\,Sin[t] - 2\,C[1]\,Sin\,[t]) \right\} \right\}$

The third term on the right will be growing without bound as time t tends to infinity. You can single it out by choosing C[1] $= 0$ and C[2] $= -1/2$, or by the command

In[5]:= y1 = sol2[[1, 1, 2, 2, 3]]   (* Out: t Sin[t] *)

In[6]:= Plot[y1, {t, 0, 50}, AxesLabel -> {t, "y1"}]

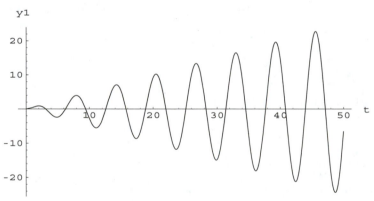

Example 2.9. Resonance

Beats occur if the frequency of the driving force of the undamped system is close to the natural frequency of the free vibrations. For example, consider the equation $y'' + y = 0.19 \cos 0.9t$. (The factor 0.19 has been included to have a simpler final answer; it is of no other importance.) Type and solve this ODE.

Solution.

In[7]:= ode2 = y2''[t] + y2[t] == 0.19 Cos[0.9 t]

Out[7]= y2[t] + y2''[t] == 0.19 Cos [0.9 t]

In[8]:= sol2 = DSolve[{ode2, y2[0] == 0, y2'[0] == 0}, y2[t], t]

Out[8]= {{y2[t] → Cos[0.9 t] − 1. Cos [t]}}

To comprehend this form of the solution ("**beats**"), note that the difference of two cosine functions can be written as the product of two sine functions, one with a high frequency (this gives the rapid oscillations shown in the figure) and one with a low frequency ($0.05/(2\pi)$, giving a period of about 126). The formula is obtained by

In[9]:= 2 TrigReduce[Sin[0.95 t] Sin[0.05 t]]   (* Out: Cos[0.9 t] − Cos[1. t] *)

In[10]:= Plot[sol2[[1, 1, 2]], {t, 0, 200}, AxesLabel -> {t, y2}]

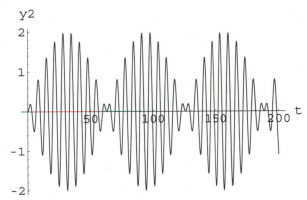

Example 2.9. Beats given by $y = 2\sin 0.95t \sin 0.05t$

Similar Material in AEM: pp. 114, 115

| **EXAMPLE 2.10** | ***RLC*-CIRCUIT** |

The current $i(t)$ in an *RLC*-circuit with sinusoidal electromotive force is obtained by solving

$$Li' + Ri + \frac{1}{C} \int i\, dt = E_0 \sin \omega t$$

where $Q = \int i\, dt$ is the charge on the capacitor. Differentiation gives the ODE

$$Li'' + Ri' + \frac{1}{C}i = \omega E_0 \cos \omega t.$$

(Write i since Mathematica uses I for $\sqrt{-1}$.) For instance, let $L = 0.1$ henry, $R = 100$ ohms, $C = 0.001$ farad, $E_0 = 155$, and $\omega = 377$ (thus 60 hertz). Solve the ODE for these data.

Solution. Write the ODE as $Mi = r$, where Mi denotes the left-hand side. (Note that L here is used for the inductance.)

In[1]:= Mi = 0.1 i''[t] + 100 i'[t] + 1000 i[t]

Out[1]= 1000 i[t] + 100 i'[t] + 0.1 i''[t]

In[2]:= r = 377 155 Cos[377 t]   (* Out: 58435 Cos[377 t] *)

A general solution of the homogeneous ODE is

In[3]:= ihom = DSolve[Mi == 0, i[t], t]

Out[3]= $\{\{i[t] \to e^{-989.898\,t}\,C[1] + e^{-10.1021\,t}\,C[2]\}\}$

The particular solution of the nonhomogeneous ODE satisfying $i(0) = 0$, $i'(0) = 0$ is (see the figure)

In[4]:= `ipartic = DSolve[{Mi == r, i[0] == 0, i'[0]== 0}, i[t], t]`

Out[4]= $\{\{i[t] \to (6.26617 \times 10^{-12} + 0.\,i)\,(8.39696 \times 10^{10}\,e^{-989.898\,t}$

$-6.76006 \times 10^9\,e^{-10.1021\,t} - 7.72096 \times 10^{10}\,\mathrm{Cos}\,[377.\,t] +$

$2.203 \times 10^{11}\,\mathrm{Sin}\,[377.\,t])\}\}$

In[5]:= `Plot[ipartic[[1, 1, 2]], {t, 0, 0.13}, AxesLabel -> {t, i}]`

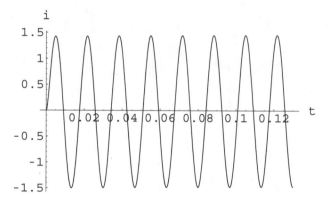

Example 2.10. Current in the RLC-circuit

The oscillation is very rapid. The **transient solution** becomes practically zero after a very short time, so that a transient period is practically not visible in the figure. $i(0) = 0$ means no current at $t = 0$. Assume that the capacitor is uncharged at $t = 0$; thus $Q(0) = 0$ in the first formula in this example. Solve that formula algebraically for i', obtaining

$$i' = \frac{1}{L}(E_0 \sin \omega t - R\,i - \frac{Q}{C}).$$

For $t = 0$ all three terms on the right are 0. This shows that the initial condition $i'(0) = 0$ physically means that the capacitor is initially uncharged.

    Similar Material in AEM: pp. 118-121

Problem Set for Chapter 2

Pr.2.1 (General solution, initial value problem, harmonic oscillations) Find a general solution of $y'' + 3y' + 2y = 0$. Find and plot the particular solution satisfying the initial conditions $y(0) = 4$, $y'(0) = -2$. (*AEM Ref.* pp. 72, 73)

Pr.2.2 (Maximum of solution) Find and plot the solution y_p of the initial value problem $y'' - 2y' + y = 0$, $y(0) = 4$, $y'(0) = 3$. Find the location and the value of the maximum of y_p, first from the graph and then by calculation. (*AEM Ref.* p. 74)

Pr.2.3 **(Experiment on damped oscillations, dependence on initial conditions)** Experiment with Example 2.2 in this Guide by systematically changing the initial conditions (e.g., by keeping $y'(0)$ constant and varying $y(0)$, etc.) in order to find out whether and how solutions depend on initial conditions. (*AEM Ref.* pp. 87-90)

Pr.2.4 **(Linear independence, Wronskian)** Show that the solution in Pr.2.1 is a general solution. (*AEM Ref.* pp. 73, 97)

Pr.2.5 **(Verification of solution)** Show that a linear combination of e^{-x}, $x e^{-x}$, $x^2 e^{-x}$ is a solution of $y''' + 3y'' + 3y' + y = 0$. (*AEM Ref.* p. 135)

Pr.2.6 **(Nonhomogeneous equation, complex roots)** Find a general solution of

$$y'' + 2y' + 226y = e^{-0.001\,t}.$$

Find and plot the particular solution which starts at $y = 0$ with initial velocity 0. (*AEM Ref.* pp. 79, 101)

Pr.2.7 **(Experiment on overdamping)** Find a general solution of $y'' + 10y' + 16y = 0$. Find and plot (on common axes) the three particular solutions starting from $y = 1$ with different initial velocities, namely, $-1, 0, 1$. Can you find an initial velocity such that the particular solution becomes 0 for some positive t? For two different positive t? What is the effect of changing the initial displacement? (*AEM Ref.* p. 87)

Pr.2.8 **(Critical damping)** Find the value of the damping constant c such that the mass-spring system
$$y'' + cy' + 37.21y = 0$$
is critically damped. Characterize the other two cases (underdamping, overdamping) in terms of the values of c. (*AEM Ref.* p. 88)

Pr.2.9 **(Pendulum)** Find the motion of a pendulum of mass m and length 1 meter for small angle $|T|$ (so that $\sin T$ can be approximated by T rather accurately). Assume that the pendulum starts from the equilibrium position $(T = 0)$ with velocity 1. (*AEM Ref.* p. 91 (#10))

Pr.2.10 **(Logarithmic decrement)** Prove on the computer that the ratio of two consecutive maximum amplitudes of a damped oscillation $y = C \exp(-at) \cos(wt - d)$ is constant, the natural logarithm of this ratio being called the *logarithmic decrement* and being given by $LD = 2\pi a/w$. (C, a, w, d are constant.) (*AEM Ref.* p. 92 (#18))

Pr.2.11 **(Boundary value problem)** Solve $y'' + y = 0$, $y(0) = 3$, $y(\pi) = -3$. (*AEM Ref.* p. 80)

Pr.2.12 **(Euler-Cauchy equation)** Determine a and b in the Euler-Cauchy equation $x^2 y'' + axy' + by = 0$ so that the auxiliary equation has a double root and determine a general solution for this "critical case". (*AEM Ref.* p. 93)

Pr.2.13 **(Euler-Cauchy equation)** Find the Euler-Cauchy equation with $x \cos(\ln x)$ and $x \sin(\ln x)$ as a basis of solutions. (*AEM Ref.* p. 95)

Pr.2.14 **(Wronskian)** Show that $\{e^{-2x}, e^{2x}, e^{4x}\}$ is a basis of solutions of the ODE $y''' - 4y'' - 4y' + 16y = 0$. (*AEM Ref.* p. 127)

Pr.2.15 **(Resonance)** Find and plot the solution of $y'' + 4y = -12\sin 2t$, $y(0) = 1$, $y'(0) = 3$. (*AEM Ref.* p. 114)

Pr.2.16 **(Nonhomogeneous ODE)** Show that $-e^{-2x} \ln x$ is a particular solution of

$$y'' + 4y' + 4y = e^{-2x}/x^2.$$

Find a general solution of this ODE and determine its arbitrary constants so that you obtain a particular solution satisfying the initial conditions $y(1) = 1/e^2$, $y'(1) = -2/e^2$. (*AEM Ref.* p. 104 (#15))

Pr.2.17 **(Beats)** Find and plot the solution of the initial value problem $y'' + 100y = 36 \cos 8t$, $y(0) = 0$, $y'(0) = 0$. Can you see by looking at the ODE that you will obtain beats, because the frequency of the driving force is close to that of the free harmonic oscillations? (*AEM Ref.* p. 115)

Pr.2.18 **(Undetermined coefficients)** Find a particular solution of $y'' + 3y' - 18y = 9 \sinh 3x$ by the method of undetermined coefficients. Check your result by DSolve. (*AEM Ref.* pp. 104, 107 (#12))

Pr.2.19 **(Variation of parameters)** Find a particular solution of the ODE

$$y'' - 2y' + y = 3x^{3/2} e^x$$

by the method of variation of parameters. (*AEM Ref.* pp. 108, 111 (#7))

Pr.2.20 **(*RLC*-circuit)** Find the current in the *RLC*-circuit with $R = 16$ ohms, $L = 8$ henrys, $C = 1/8$ farad, and $E = 100 \cos 2t$ volts. Also find and plot the particular solution satisfying $i(0) = 0$, $i'(0) = 0$. (*AEM Ref.* pp. 118-122)

Chapter 3

Systems of Differential Equations.
Phase Plane, Qualitative Methods

Content. Solution of systems, use of matrices (Exs. 3.1, 3.2)
Critical points, pendulum ODE (Exs. 3.3-3.5, Prs. 3.1-3.4, 3.6)
Nonhomogeneous systems (Exs. 3.6, 3.7)
Mixing problems, networks (Prs. 3.8-3.10)

Examples for Chapter 3

EXAMPLE 3.1 **SOLVING A SYSTEM OF ODE's BY DSolve.**
INITIAL VALUE PROBLEM

The application of the command DSolve to a system of ODE's or to an initial value problem involving such a system is quite similar to that in the case of a single ODE. A linear system of ODE's

$$\begin{aligned} y_1' &= -3y_1 + y_2 \\ y_2' &= y_1 - 3y_2 \end{aligned}$$

can be typed as

In[1]:= sys = {y1'[t] == -3 y1[t] + y2[t],
            y2'[t] == y1[t] - 3 y2[t]}

Out[1]= {y1'[t] == -3 y1[t] + y2[t], y2'[t] == y1[t] - 3 y2[t]}

If for some reason you need the two equations separately, you can access them by typing

In[2]:= sys[[1]]   (* Out: y1'[t] == -3 y1[t] + y2[t] *)

and

In[3]:= sys[[2]]   (* Out: y2'[t] == y1[t] - 3 y2[t] *)

For obtaining a general solution of the system, type

In[4]:= sol = DSolve[sys, {y1[t], y2[t]}, t]

Out[4]= $\{\{y1[t] \to \frac{1}{2} e^{-4t} (C[1] + e^{2t} C[1] - C[2] + e^{2t} C[2]),$

$y2[t] \to \frac{1}{2} e^{-4t} (-C[1] + e^{2t} C[1] + C[2] + e^{2t} C[2])\}\}$

Inspection shows that this can be written

$$y_1 = c_1 e^{-2t} + c_2 e^{-4t}$$

$$y_2 = c_1 e^{-2t} - c_2 e^{-4t}$$

where $c_1 = (\mathtt{C[1]} + \mathtt{C[2]})/2$, $c_2 = (\mathtt{C[1]} - \mathtt{C[2]})/2$. The command `FullSimplify` is not very useful here; it gives

In[5]:= `sol2 = FullSimplify[sol]`

Out[5]= {{y1[t] $\rightarrow$ e^{-3t} (C[1] Cosh[t] + C[2] Sinh[t]),

       y2[t] $\rightarrow$ e^{-3t} (C[2] Cosh[t] + C[1] Sinh[t])}}

Combining it with a substitution (`C[1]` and `C[2]` expressed in terms of c_1 and c_2) gives

In[6]:= `FullSimplify[sol /. {C[1] ->  c1 + c2, C[2] -> c1 - c2}]`

Out[6]= {{y1[t] $\rightarrow$ e^{-4t} (c2 + c1 e^{2t}), y2[t] $\rightarrow$ e^{-4t} (- c2 + c1 e^{2t})}}

From this you can more easily see the above result obtained by inspection.

Initial value problems. The above system, together with initial conditions, say, $y_1(0) = 1$, $y_2(0) = 5$, can be solved by typing

In[7]:= `yp = DSolve[{sys[[1]], sys[[2]], y1[0] == 1, y2[0] == 5},`

       `{y1[t], y2[t]}, t]`

Out[7]= {{y1[t] $\rightarrow$ e^{-4t} (- 2 + 3 e^{2t}), y2[t] $\rightarrow$ e^{-4t} (2 + 3 e^{2t})}}

For plotting, access the two functions in this solution by typing

In[8]:= `f1 = yp[[1]][[1, 2]]`   (* Out: e^{-4t} ($-2 + 3e^{2t}$) *)

and

In[9]:= `f2 = yp[[1]][[2, 2]]`   (* Out: e^{-4t} ($2 + 3e^{2t}$) *)

Then give the plotting command

In[10]:= `Plot[{f1, f2}, {t, 0, 5}, PlotRange -> {0, 5}]`

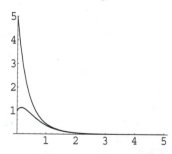

Example 3.1. Solutions y_1 (lower curve) and y_2 of the initial value problem

To plot the solution of the initial value problem as a trajectory in the $y_1 y_2$-plane (the phase plane), type

In[11]:= `ParametricPlot[{f1, f2}, {t, 0, 2}]`

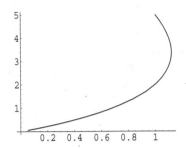

Example 3.1. Solution of the initial value problem
as a trajectory in the phase plane

You see that the trajectory starts at $(1, 5)$, in agreement with the intial conditions given at $t = 0$, and rapidly approaches the origin $(0, 0)$, which corresponds to $t \to \infty$. Try $t = 5$, $t = 10$ as an upper limit in the command.
Similar Material in AEM: p. 163

EXAMPLE 3.2   USE OF MATRICES IN SOLVING SYSTEMS OF ODE's

A general solution of a homogeneous linear system of ODE's is completely determined by the eigenvalues and eigenvectors of the coefficient matrix of the system and can readily be obtained from these eigenvalues and eigenvectors, without resorting to trials of achieving simplifications in terms of various commands of your software. We explain this for the system considered in the previous example, that is,

$$y_1' = -3y_1 + y_2$$
$$y_2' = y_1 - 3y_2$$

In[1]:= `Clear[y1, y2]`

In[2]:= `sys = {y1'[t] == -3 y1[t] + y2[t],`
    `y2'[t] == y1[t] - 3 y2[t]}`

Out[2]= {y1′[t] == −3 y1[t] + y2[t], y2′[t] == y1[t] − 3 y2[t]}

Type the coefficient matrix, call it **A**,

In[3]:= `A = {{-3, 1}, {1, -3}}; MatrixForm[A]` $\left(* \text{ Out: } \begin{pmatrix} -3 & 1 \\ 1 & -3 \end{pmatrix} *\right)$

An **eigenvalue** of **A** is a number λ such that the vector equation $\mathbf{Ax} = \lambda\mathbf{x}$ has a vector solution **x** not the zero vector. (The zero vector is a solution of this equation for any value of λ, and is called the '***trivial solution***'.) This nonzero **x** is called an **eigenvector** of **A** corresponding to that eigenvalue λ. The eigenvalues are the solutions of the **characteristic equation** $\det(\mathbf{A} - \lambda\mathbf{I}) = 0$. Here, **I** is the 2×2 **unit matrix**. For the eigenvalues you may type

In[4]:= `Eigenvalues[A]`   (* Out: {−4, −2} *)

For the eigenvectors you may type

In[5]:= `Eigenvectors[A]`   (* Out: {{−1, 1}, {1, 1}} *)

You can obtain eigenvalues and eigenvectors jointly by the command

In[6]:= `eig = Eigensystem[A]`   (* Out: {{−4, −2}, {{−1, 1}, {1, 1}}} *)

Hence $\lambda = -4$ has as an eigenvector $\mathbf{x}_1 = [1 \ -1]^T$, where T indicates the transpose; so this is a column vector. The other vector, $\mathbf{x}_2 = [1 \ 1]^T$, is an eigenvector of $\mathbf{A}$ corresponding to $\lambda = -2$.

In[7]:= `eig[[1]]`   (* This gives the eigenvalues. *)

Out[7]= {−4, −2}

In[8]:= `L1 = eig[[1, 1]]`   (* First eigenvalue -4 *)

In[9]:= `x1 = eig[[2, 1]]`   (* Eigenvector of A for -4 *)

Out[9]= {−1, 1}

In[10]:= `L2 = eig[[1, 2]]`   (* Second eigenvalue -2 *)

In[11]:= `x2 = eig[[2, 2]]`   (* Eigenvector of A for -2 *)

Out[11]= {1, 1}

This is all you need for making up a general solution by typing

In[12]:= `y = c1 x1 Exp[L1 t] + c2 x2 Exp[L2 t]`

Out[12]= $\{-$c1$\,e^{-4\,t} +$ c2$\,e^{-2\,t},\,$ c1$\,e^{-4\,t} +$ c2$\,e^{-2\,t}\}$

This vector solution automatically appears in terms of components. It agrees with the solution in the previous example, except for the notation of the arbitrary constants. (To reach complete agreement with the previous example, set $c_1 = -c_2$, $c_2 = c_1$ in the present example.)

Initial value problem. Substituting the initial values $y_1(0) = 1$, $y_2(0) = 5$ (as in the previous example), you obtain a linear system of equations for the arbitrary constants

In[13]:= `const = (y /. t -> 0) == {1, 5}`

Out[13]= {−c1 + c2, c1 + c2} == {1, 5}

To solve this system, type

In[14]:= `Solve[const, {c1, c2}]`   (* Out: {{c1 → 2, c2 → 3}} *)

By inserting these values into the general solution you obtain the particular solution also obtained in the previous example, namely,

In[15]:= `y /. {c1 -> 2, c2 -> 3}`   (* Out: $\{-2\,e^{-4\,t} + 3\,e^{-2\,t},\, 2\,e^{-4\,t} + 3\,e^{-2\,t}\}$ *)

Similar Material in AEM: pp. 151, 163

EXAMPLE 3.3   **CRITICAL POINTS. NODE**

From a given linear system of ODE's

$$\mathbf{y}' = \mathbf{A}\mathbf{y}$$

where the prime denotes the derivative with respect to t, in components

$$y_1' = a_{11}\,y_1 + a_{12}\,y_2$$
$$y_2' = a_{21}\,y_1 + a_{22}\,y_2$$

you obtain

$$\frac{dy_2}{dy_1} = \frac{dy_2}{dt} \bigg/ \frac{dy_1}{dt} = \frac{a_{21}\,y_1 + a_{22}\,y_2}{a_{11}\,y_1 + a_{12}\,y_2}.$$

At each point (y_1, y_2) in the $y_1 y_2$-plane (the **phase plane**) this determines a tangent direction of the solution drawn as a curve in the phase plane. Such a curve is called a **trajectory** (or **path** or **orbit**). An exception is the point $(y_1, y_2) = (0, 0)$, at which the numerator and the denominator on the right are both zero. This is called a **critical point** of the system of ODE's. Critical points can be classified in terms of three quantities related to the coefficient matrix $\mathbf{A} = [a_{jk}]$ of the system. These quantities are

$$p = a_{11} + a_{22} \qquad \text{the \textbf{trace} of the matrix}$$
$$q = \det \mathbf{A} = a_{11}\,a_{22} - a_{12}\,a_{21} \quad \text{the \textbf{determinant} of the matrix}$$
$$\delta = p^2 - 4\,q.$$

The critical point is a:

Node	if $q > 0$ and $\delta \geq 0$,
Saddle point	if $q < 0$,
Center	if $p = 0$ and $q > 0$,
Spiral point	if $p \neq 0$ and $\delta < 0$.

For example, let the matrix of the system be (see the previous example)

In[1]:= `A = {{-3, 1}, {1, -3}}; MatrixForm[A]`

Out[1]//MatrixForm=

$$\begin{pmatrix} -3 & 1 \\ 1 & -3 \end{pmatrix}$$

Then the quantities needed for classifying the critical point are

In[2]:= `p = Tr[A]`   (* Out: −6 *)

In[3]:= `q = Det[A]`   (* Out: 8 *)

In[4]:= `delta = p^2 - 4 q`   (* Out: 4 *)

Since q and δ are positive, the critical point at the origin is a node. More precisely, it is an **improper node,** characterized by the property that at that point all trajectories, except for two of them, have the same tangent direction, as the figure shows. (A **proper node** is one at which for each direction there is a trajectory having it as the tangent direction at the node.) Each trajectory shown is obtained by choosing a pair of constants c_1, c_2 in the general solution in the preceding example. For $c_1 = 0$ you get one of the two straight lines and for $c_2 = 0$ the other. For any other choice of c_1, c_2 you get a quadratic parabola because in that general solution, e^{-4t} is the square of e^{-2t}. In the plot commands, use $\tau = e^{-t}$ as a simpler parameter and write again t for τ^2. That is, for instance, in p3 below, with $c_1 = -1/10$, $c_2 = 1$, we have
$$y = \{-c_1\,e^{-4t} + c_2\,e^{-2t},\ c_1\,e^{-4t} + c_2\,e^{-2t}\} = \{-c_1\,\tau^4 + c_2\,\tau^2,\ c_1\,\tau^4 + c_2\,\tau^2\}$$
$$= \{t^2/10 + t,\ -t^2/10 + t\}.$$

In[5]:= `p1 = Plot[{t, t}, {t, -10, 10}]`

In[6]:= `p2 = Plot[{t, -t}, {t, -10, 10}]`

In[7]:= `p3 = Plot[{t + t^2/10, t - t^2/10}, {t, -10, 10}]`

In[8]:= `p4 = Plot[{t - t^2/10, t + t^2/10}, {t, -10, 10}]`

In[9]:= `p5 = Plot[{t + t^2/20, t - t^2/20}, {t, -10, 10}]`

In[10]:= `p6 = Plot[{t - t^2/20, t + t^2/20}, {t, -10, 10}]`

In[11]:= `Show[{p1, p2, p3, p4, p5, p6}, PlotLabel -> "Trajectories near a node",`
     `AspectRatio -> Automatic, PlotRange -> {-10,10}]`

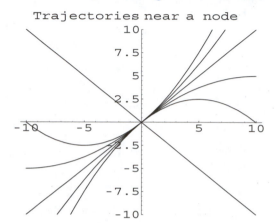

Example 3.3. Trajectories near an improper node

Similar Material in AEM: pp. 163, 165 (Fig. 78), 172

| EXAMPLE 3.4 | **PROPER NODE, SADDLE POINT, CENTER, SPIRAL POINT** |

Improper nodes were discussed and shown in the previous example. Phase plane plots ("**phase portraits**") of the other types of critical points are given on pp. 165 and 166 of AEM. We present here the corresponding calculations for typical examples.

A ***proper node*** with trajectories $y_2 = k y_1$ and $y_1 = 0$ (straight lines) is obtained for the system $y_1' = y_1$, $y_2' = y_2$, whose matrix is the unit matrix.

A ***saddle point*** with trajectories $y_1 y_2 = const$ (hyperbolas) is obtained for $y_1' = y_1$, $y_2' = -y_2$ with matrix

In[1]:= `A = {{1, 0}, {0, -1}}; MatrixForm[A]`

Out[1]//MatrixForm=

$$\begin{pmatrix} 1 & 0 \\ 0 & -1 \end{pmatrix}$$

for which

In[2]:= `q = Det[A]`   (* Out: −1 *)

so that we do have a saddle point. The eigenvalues of **A** have opposite sign, as is typical,

In[3]:= `Eigenvalues[A]`   (* Out: {−1, 1} *)

A **center** with trajectories $y_1{}^2 + y_2{}^2/4 = const$ (ellipses) is obtained for $y_1' = y_2$, $y_2' = -4y_1$ with matrix

In[4]:= `A ={{0, 1}, {-4, 0}}; MatrixForm[A]`

Out[4]//MatrixForm=

$$\begin{pmatrix} 0 & 1 \\ -4 & 0 \end{pmatrix}$$

for which $p = 0$ and

In[5]:= `q = Det[A]`   (* Out: 4 *)

which gives a center. The eigenvalues of **A** are pure imaginary, which is typical,

In[6]:= `Eigenvalues[A]`   (* Out: {-2 i, 2 i} *)

A **spiral point** with trajectories $r = c \exp(-\theta)$ (spirals, in polar coordinates) is obtained for the system with matrix

In[7]:= `A ={{-1, 1}, {-1, -1}}; MatrixForm[A]`

Out[7]//MatrixForm=

$$\begin{pmatrix} -1 & 1 \\ -1 & -1 \end{pmatrix}$$

for which $p = -2 \neq 0$ and

In[8]:= `delta = (-2)^2 - 4 Det[A]`   (* Out: -4 *)

which gives a spiral point. In this case, **A** has complex eigenvalues,

In[9]:= `Eigenvalues[A]`   (* Out: {-1 - i, -1 + i} *)

Similar Material in AEM: pp. 164-167

| EXAMPLE 3.5 | **PENDULUM EQUATION**

The ODE governing the motion of a pendulum (e.g., in a pendulum clock) is $y'' + k\sin y = 0$ (here we have disregarded damping) and is a classical case of an ODE not exactly solvable in terms of finitely many elementary functions. Physically, $y(t)$ is the angle of displacement from rest (the position when the pendulum hangs vertically downward), and t is time. k is a positive constant, and we choose $k = 1$ for simplicity. We set $y = y_1$ and $y' = y_1' = y_2$. Then $y'' = y_2' = -k\sin y = -\sin y = -\sin y_1$. Hence you can write the given equation (with $k = 1$) as a system

$$y_1' = y_2$$
$$y_2' = -\sin y_1.$$

This illustrates the standard method of ***transforming an ODE of second order into a system of two first-order ODE's.*** To plot a direction field of this system, load the package

In[1]:= `<<Graphics'PlotField'`

and then give the plotting command

In[2]:= `PlotVectorField[{y2, -Sin[y1]}, {y1, -7, 7}, {y2, -4, 4}, Axes -> True,`
`Ticks -> {{-2 Pi, -Pi, Pi, 2 Pi}, Automatic}, AxesLabel -> {y1, y2}]`

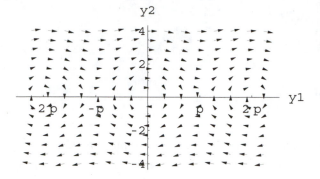

Example 3.5. Direction field for the pendulum equation (where $p = \pi$)

From the system you see that -2π , 0, 2π are critical points, and the plot suggests that these are centers. You also see that $-\pi$ and π are further critical points and the plot suggests (although less distinctly) that these are saddle points. (See also the figure on p. 176 of AEM.)

Similar Material in AEM: p. 176

EXAMPLE 3.6 | NONHOMOGENEOUS SYSTEM

Find a general solution of the nonhomogeneous linear system

$$\mathbf{y}' = \mathbf{A}\mathbf{y} + \mathbf{g} = \begin{bmatrix} 2 & -4 \\ 1 & -3 \end{bmatrix}\mathbf{y} + \begin{bmatrix} 2t^2 + 10t \\ t^2 + 9t + 3 \end{bmatrix}.$$

Solution. Type the system as shown and obtain a general solution by the command `DSolve`.

In[1]:= `Clear[y1, y2]`

In[2]:= `sys = {y1'[t] == 2 y1[t] - 4 y2[t] + 2 t^2 + 10 t,`
`y2'[t] == y1[t] - 3 y2[t] + t^2 + 9 t + 3}`

Out[2]= $\{y1'[t] == 10\,t + 2\,t^2 + 2\,y1[t] - 4\,y2[t],$

$y2'[t] == 3 + 9\,t + t^2 + y1[t] - 3\,y2[t]\}$

In[3]:= `sol = DSolve[sys, {y1[t], y2[t]}, t]`

Out[3]= $\left\{\left\{y1[t] \rightarrow \frac{1}{3}\,e^{-2\,t}\,(-3\,e^{2\,t}\,t^2 - C[1] + 4\,e^{3\,t}\,C[1] + 4\,C[2] - 4\,e^{3\,t}\,C[2]),\right.\right.$

$\left.\left.y2[t] \rightarrow \frac{1}{3}\,e^{-2\,t}\,(9\,e^{2\,t}\,t - C[1] + e^{3\,t}\,C[1] + 4\,C[2] - e^{3\,t}\,C[2])\right\}\right\}$

In[4]:= `FullSimplify[%]`

Out[4]= $\left\{\left\{y1[t] \rightarrow -t^2 - \frac{1}{3}\,e^{-2\,t}\,(C[1] - 4\,C[2]) + \frac{4}{3}\,e^{t}\,(C[1] - C[2]),\right.\right.$

$\left.\left.y2[t] \rightarrow 3\,t + \frac{1}{3}\,e^{-2\,t}\,(-C[1] + e^{3\,t}\,(C[1] - C[2]) + 4\,C[2])\right\}\right\}$

From this you see that a particular solution of the *nonhomogeneous* system is
$y_{1p} = -t^2$, $y_{2p} = 3t$. You can obtain a simpler form of the general solution of the
homogeneous system by typing the coefficient matrix

In[5]:= A = {{2, -4}, {1, -3}}   (* Out: {{2, −4}, {1, −3}} *)

and its eigensystem,

In[6]:= e = Eigensystem[A]   (* Out: {{−2, 1}, {{1, 1}, {4, 1}}} *)

and then proceeding as in Example 3.2. This gives

In[7]:= yhom = c1 e[[2, 1]] Exp[e[[1, 1]] t] + c2 e[[2, 2]] Exp[e[[1, 2]] t]

Out[7]= $\{c1\,e^{-2t} + 4\,c2\,e^{t},\ c1\,e^{-2t} + c2\,e^{t}\}$

Verify that this agrees with the above solution, except for the notation of the arbitrary
constants.

Similar Material in AEM: pp. 184, 185

| EXAMPLE 3.7 | **METHOD OF UNDETERMINED COEFFICIENTS**

Find a particular solution of the nonhomogeneous system in the previous example by
the method of undetermined coefficients.

Solution. The system is $\mathbf{y}' = \mathbf{Ay} + \mathbf{g}$, where

In[1]:= A = {{2, -4}, {1, -3}}   (* Out: {{2, −4}, {1, −3}} *)

and

In[2]:= g = {2 t^2 + 10 t, t^2 + 9 t + 3}   (* Out: $\{10\,t + 2\,t^2, 3 + 9\,t + t^2\}$ *)

The nonhomogeneous part g suggests choosing

In[3]:= u = {u1, u2}; v = {v1, v2}; w = {w1, w2};

In[4]:= yp = u + v t + w t^2   (* Out: $\{u1 + t\,v1 + t^2\,w1, u2 + t\,v2 + t^2\,w2\}$ *)

with the components of u, v, w to be determined. Substitute this into the system,
obtaining

In[5]:= sys = D[yp, t] == A.yp + g

Out[5]= $\{v1 + 2\,t\,w1, v2 + 2\,t\,w2\}$ ==

$$\{10\,t + 2\,t^2 + 2\,(u1 + t\,v1 + t^2\,w1) - 4\,(u2 + t\,v2 + t^2\,w2),$$

$$3 + 9\,t + t^2 + u1 + t\,v1 + t^2\,w1 - 3\,(u2 + t\,v2 + t^2\,w2)\}$$

Now there are six components to be determined. Since sys is a vector equation with
two components, you can obtain from it six scalar equations by choosing three values
of t, most conveniently, $t = 0, 1, -1$. This gives

In[6]:= eq1 = sys /. t -> 0   (* Out: {v1, v2} == {2 u1−4 u2, 3 + u1−3 u2} *)

In[7]:= eq2 = sys /. t -> 1

Out[7]= $\{v1 + 2\,w1, v2 + 2\,w2\}$ ==

$$\{12 + 2\,(u1 + v1 + w1) - 4\,(u2 + v2 + w2), 13 + u1 + v1 + w1 - 3\,(u2 + v2 + w2)\}$$

In[8]:= `eq3 = sys /. t -> -1`

Out[8]= $\{v1 - 2\,w1,\ v2 - 2\,w2\} == \{-8 + 2\,(u1 - v1 + w1) - 4\,(u2 - v2 + w2),$

$-5 + u1 - v1 + w1 - 3\,(u2 - v2 + w2)\}$

Solve this system of three vector equations (six equations for the components) by typing

In[9]:= `sol = Solve[{eq1, eq2, eq3}, {u1, u2, v1, v2, w1, w2}]`

Out[9]= $\{\{u1 \to 0,\ u2 \to 0,\ v1 \to 0,\ v2 \to 3,\ w1 \to -1,\ w2 \to 0\}\}$

It is somewhat strange that you cannot use `u`, `v`, `w` on the right. Try it. Substituting `sol` into `yp`, you obtain the answer which agrees with the result in the previous example, namely,

In[10]:= `answer = yp /. sol`   (* Out: $\{\{-t^2, 3\,t\}\}$ *)

Similar Material in AEM: pp. 184, 185

Problem Set for Chapter 3

Pr.3.1 (Node) Using the criteria, show that $y_1' = y_1$, $y_2' = 2y_2$ has a node. Find a general solution. Plot a phase portrait. (*AEM Ref.* pp. 164, 165, 172)

Pr.3.2 (Saddle point) Find and plot the solution of the initial value problem $y_1' = 2\,y_1 - 4\,y_2$, $y_2' = y_1 - 3\,y_2$, $y_1(0) = 3$, $y_2(0) = 0$ as a trajectory. (*AEM Ref.* p. 165)

Pr.3.3 (Center) Solve $y'' + 9y = 0$ by first converting it to a system of equations. Plot a phase portrait. (*AEM Ref.* pp. 165, 166)

Pr.3.4 (Spiral point) Find a general solution of the system $y_1' = -y_1 + 4\,y_2$, $y_2' = -y_1 - y_2$. From it find the particular solution satisfying the initial conditions $y_1(0) = 1$, $y_2(0) = 1$. Plot this solution as a trajectory in a phase portrait of the system.
(*AEM Ref.* p. 166)

Pr.3.5 (No basis of eigenvectors available) Determine the eigenvalues and eigenvectors of the coefficient matrix of the system

$$y_1' = 4\,y_1 + y_2, \qquad y_2' = -y_1 + 2\,y_2.$$

Show that this gives no basis of eigenvectors, so that you get at first only one solution. It is shown in AEM, pp. 167, 168, that a second independent solution is of the form

$$\mathbf{y} = (\mathbf{x}t + \mathbf{u})e^{\lambda t}.$$

Show that `DSolve` does indeed give a general solution of the corresponding form. (*AEM Ref.* pp. 167-169)

Pr.3.6 (Damped pendulum) Add a damping term cy' to the pendulum equation in Example 3.5 in this Guide, choosing $c = 1/4$. Plot a phase portrait. Compare with Example 3.5. (*AEM Ref.* pp. 177, 178)

Pr.3.7 (Harmonic oscillations) Indicate the kind of trajectories of $y'' + (1/9)y = 0$ in the phase plane. (*AEM Ref.* p. 174 (#12))

Pr.3.8 **(Mixing problem involving two tanks)** Tank T_1 initially contains 100 gal of pure water. Tank T_2 initially contains 100 gal of water in which 150 lb of fertilizer are dissolved. Liquid circulates through the tanks (by means of two tubes connecting the tanks) at a constant rate of 2 gal/min and the mixture is kept uniform by stirring. Find the amounts of fertilizer $y_1(t)$ and $y_2(t)$ in T_1 and T_2, respectively, where t is time. (*AEM Ref.* pp. 152-154)

Pr.3.9 **(Electrical network)** Find and plot the currents $i_1(t)$, $i_2(t)$ in a network governed by the equations

$$i_1' + 4(i_1 - i_2) = 12$$

$$6i_2 + 4(i_2 - i_1) + 4\int i_2\, dt = 0,$$

assuming $i_1(0) = 0$, $i_2(0) = 0$. The equations result from the two loops of the network. Loop 1 contains an inductor of $L = 1$ henry (giving i_1'), a resistor of $R = 4$ ohms (common to both loops), and a battery of 12 volts. Loop 2 contains another resistor of 6 ohms and a capacitor of 1/4 farad. (Differentiate the second equation to get rid of the integral.) (*AEM Ref.* pp. 154-156)

Pr.3.10 **(Electrical network)** Find a general solution in Pr.3.9 by using matrices. (*AEM Ref.* p. 155)

Chapter 4

Series Solutions of Differential Equations

Content. Solving in series, plotting (Ex. 4.1, Prs. 4.1-4.3)
  Legendre's ODE, orthogonality, Fourier-Legendre series
    (Exs. 4.2-4.4, Prs. 4.4-4.8, 4.20)
  Frobenius method, hypergeometric ODE (Ex. 4.5, Prs. 4.9-4.13)
  Bessel's ODE (Ex. 4.6, Prs. 4.14-4.19)

A power series of a function $f(x)$ is obtained by the command `Series[f, {x, a, M}]`. Here `a` is the center and `M` the exponent of the highest power of `x` shown. For instance,

In[1]:= `f = Exp[x]`   (* Out: e^x *)

In[2]:= `Series[f, {x, 0, 9}]`

Out[2]= $1 + x + \dfrac{x^2}{2} + \dfrac{x^3}{6} + \dfrac{x^4}{24} + \dfrac{x^5}{120} + \dfrac{x^6}{720} + \dfrac{x^7}{5040} + \dfrac{x^8}{40320} + \dfrac{x^9}{362880} + O[x]^{10}$

Or you may use `Table` for a sequence and `Sum` for a sum (a polynomial). For example,

In[3]:= `Table[x^n/n!, {n, 0, 9}]`

Out[3]= $\left\{1, x, \dfrac{x^2}{2}, \dfrac{x^3}{6}, \dfrac{x^4}{24}, \dfrac{x^5}{120}, \dfrac{x^6}{720}, \dfrac{x^7}{5040}, \dfrac{x^8}{40320}, \dfrac{x^9}{362880}\right\}$

In[4]:= `Sum[x^n/n!, {n, 0, 9}]`

Out[4]= $1 + x + \dfrac{x^2}{2} + \dfrac{x^3}{6} + \dfrac{x^4}{24} + \dfrac{x^5}{120} + \dfrac{x^6}{720} + \dfrac{x^7}{5040} + \dfrac{x^8}{40320} + \dfrac{x^9}{362880}$

In[5]:= `N[%]`

Out[5]= $1. + x + 0.5 x^2 + 0.166667 x^3 + 0.0416667 x^4 + 0.00833333 x^5 + 0.00138889 x^6 +$

$0.000198413 x^7 + 0.0000248016 x^8 + 2.75573 \times 10^{-6} x^9$

Special functions, elementary and higher, are known to Mathematica. For instance, to obtain the Maclaurin series of the Bessel function $J_0(x)$, use the command

In[6]:= `Series[BesselJ[0, x], {x, 0, 12}]`

Out[6]= $1 - \dfrac{x^2}{4} + \dfrac{x^4}{64} - \dfrac{x^6}{2304} + \dfrac{x^8}{147456} - \dfrac{x^{10}}{14745600} + \dfrac{x^{12}}{2123366400} + O[x]^{13}$

Plotting and calculating values from a series will be explained in the examples.

Examples for Chapter 4

| EXAMPLE 4.1 | **POWER SERIES SOLUTIONS. PLOTS FROM THEM. NUMERICAL VALUES**

Find a **power series solution** of the initial value problem $(1 - x^2)\, y'' - 2x\, y' + 72y = 0$, $y(0) = 0$, $y'(0) = 1$. Plot the solution and find its value at $x = 0.8$.

Solution. Type the ODE.

$\text{In}[1]:=$ `Clear[y]`

$\text{In}[2]:=$ `ode = (1 - x^2) y''[x] - 2 x y'[x] + 72 y[x] == 0`

$\text{Out}[2]=$ $72\,\texttt{y[x]} - 2\,\texttt{x}\,\texttt{y}'\texttt{[x]} + (1 - \texttt{x}^2)\,\texttt{y[x]} == 0$

Solve the initial value problem by `DSolve`.

$\text{In}[3]:=$ `sol = DSolve[{ode, y[0] == 0, y'[0] == 1}, y[x], x]`

$\text{Out}[3]=$ $\left\{\left\{\texttt{y[x]} \rightarrow \texttt{x}\,\texttt{Hypergeometric2F1}\!\left[-\dfrac{7}{2},\, 5,\, \dfrac{3}{2},\, \texttt{x}^2\right]\right\}\right\}$

You see that the solution is given in terms of a hypergeometric function. This should not disturb you because you can obtain its Maclaurin series by typing

$\text{In}[4]:=$ `ser = Series[sol[[1, 1, 2]], {x, 0, 11}]`

$\text{Out}[4]=$ $\texttt{x} - \dfrac{35\,\texttt{x}^3}{3} + 35\,\texttt{x}^5 - 35\,\texttt{x}^7 + \dfrac{70\,\texttt{x}^9}{9} + \dfrac{14\,\texttt{x}^{11}}{11} + \texttt{O[x]}^{12}$

and use this series for plotting (and for calculating values, see below).

Plotting. Convert the last expression to a polynomial by dropping the remainder (the error term).

$\text{In}[5]:=$ `p = Normal[ser]`   (* Out: $\texttt{x} - \dfrac{35\,\texttt{x}^3}{3} + 35\,\texttt{x}^5 - 35\,\texttt{x}^7 + \dfrac{70\,\texttt{x}^9}{9} + \dfrac{14\,\texttt{x}^{11}}{11}$ *)

$\text{In}[6]:=$ `Plot[p, {x, -1, 1}, AxesLabel -> {x, y}]`

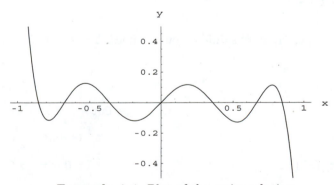

Example 4.1. Plot of the series solution

Numerical values can be obtained by substitution. For instance,

$\text{In}[7]:=$ `p /. x -> 0.8`   (* Out: 0.108677 *)

The zeros of the polynomial `p` are

In[8]:= `NSolve[p == 0, x]`

Out[8]= $\{\{x \to -1.39522\}, \{x \to -0.852542\}, \{x \to -0.675059\}, \{x \to -0.360622\},$

$\{x \to -3.0611\,i\}, \{x \to 3.0611\,i\}, \{x \to 0.\}, \{x \to 0.360622\},$

$\{x \to 0.675059\}, \{x \to 0.852542\}, \{x \to 1.39522\}\}$

If you want only four digits, type

In[9]:= `SetPrecision[Solve[p == 0, x], 4]`

Out[9]= $\{\{x \to 0\}, \{x \to 0. \times 10^{-4} - 3.061\,i\}, \{x \to 0. \times 10^{-4} + 3.061\,i\},$

$\{x \to -0.3606\}, \{x \to 0.3606\}, \{x \to -0.6751\}, \{x \to 0.6751\},$

$\{x \to -0.8525\}, \{x \to 0.8525\}, \{x \to -1.395\}, \{x \to 1.395\}\}$

We mention that the ODE is Legendre's equation with parameter $n = 8$, but the series solution obtained is not a Legendre polynomial (to be discussed in the next example).

Similar Material in AEM: p. 205

EXAMPLE 4.2   LEGENDRE POLYNOMIALS

The Legendre polynomials $P_0(x)$, $P_1(x)$, $P_2(x)$, ... are, after the Bessel functions, probably the most important special functions in applications. You can obtain the Legendre polynomials by typing, for instance,

In[1]:= `LegendreP[3, x]`   (* Out: $-\frac{3\,x}{2} + \frac{5\,x^3}{2}$ *)

A formula for defining any of these polynomials is ***Rodrigues's formula***

$$P_n(x) = \frac{1}{2^n\,n!}\frac{d^n}{d\,x^n}[(x^2 - 1)^n].$$

This can be used for obtaining individual Legendre polynomials as follows.

In[2]:= `Leg[n_] = 1/(2^n n!) D[(x^2 - 1)^n, {x, n}]`

Out[2]= $\dfrac{2^{-n}\,\partial_{\{x,n\}}\,(-1 + x^2)^n}{n!}$

In this command, `{x, n}` means differentiate n times. You can now simply type

In[3]:= `Leg[0]`   (* Out: 1 *)

In[4]:= `Leg[1]`   (* Out: x *)

In[5]:= `Leg[2]`   (* Out: $\frac{1}{8}\,(8\,x^2 + 4\,(-1 + x^2))$ *)

In[6]:= `Simplify[%]`   (* Out: $\frac{1}{2}\,(-1 + 3\,x^2)$ *)

In[7]:= `Simplify[Leg[3]]`   (* Out: $\frac{1}{2}\,x\,(-3 + 5\,x^2)$ *)

In[8]:= `Apart[Leg[3]]`   (* Out: $-\frac{3\,x}{2} + \frac{5\,x^3}{2}$ *)

Note that for $n = 0$ Mathematica interpreted the zeroth derivative as the function $(x^2 - 1)^0 = 1$, thus $P_0(x) = 1/(2^0 0!)(x^2 - 1)^0 = 1$. You may now use the command `Table` and obtain the first few Legendre polynomials at once, and plot them on common axes. That is,

In[9]:= `S = Table[Simplify[Leg[n]], {n, 0, 4}]`

Out[9]= $\left\{1, x, \frac{1}{2}(-1+3x^2), \frac{1}{2}x(-3+5x^2), \frac{1}{8}(3-30x^2+35x^4)\right\}$

In[10]:= `Plot[Evaluate[S], {x, -1.2, 1.2}, PlotRange -> {-1.1, 1.1},`
        `PlotLabel -> "Legendre polynomials",`
        `Ticks -> {{-1, 0, 1}, {-1, -0.5, 0, 0.5, 1}}]`

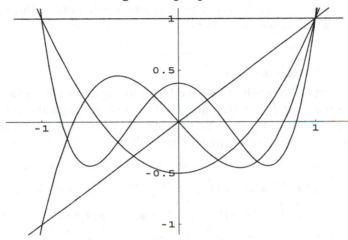

Legendre polynomials

Example 4.2. Legendre polynomials $P_0, \ldots, P_4$

Similar Material in AEM: pp. 207, 208

EXAMPLE 4.3   LEGENDRE'S DIFFERENTIAL EQUATION

The Legendre polynomials $P_n(x), n = 0, 1, 2, \ldots$, just considered in Example 4.2 are solutions of **Legendre's differential equation**

$$(1 - x^2)y'' - 2xy' + n(n + 1)y = 0$$

satisfying suitable initial conditions (depending on n). Type

In[1]:= `Clear[y, n, m, s, a]`

In[2]:= `LegEq = (1 - x^2) y''[x] - 2 x y'[x]  + n (n + 1) y[x] == 0`

Out[2]= $n(1+n)y[x] - 2xy'[x] + (1-x^2)y''[x] == 0$

For instance, consider $n = 4$,

In[3]:= `LegEq4 = LegEq /. n -> 4`

Out[3]= $20y[x] - 2xy'[x] + (1-x^2)y''[x] == 0$

To obtain the particular solution $P_4(x)$ of this ODE, note that $P_4(0) = 3/8$ and its derivative at $x = 0$ is 0. Thus, type

In[4]:= `DSolve[{LegEq4, y[0] == 3/8, y'[0] == 0}, y[x], x]`

Out[4]= $\left\{\left\{y[x] \to \frac{3}{8}\left(1 - 10x^2 + \frac{35x^4}{3}\right)\right\}\right\}$

Legendre polynomials are very convenient in connection with **boundary value problems** because at -1 and 1 they equal 1 when n is even, or -1 and 1 when n is odd. Thus you also obtain P_4 from

In[5]:= DSolve[{LegEq4, y[-1] == 1, y[1] == 1}, y[x], x]

Out[5]= $\left\{\left\{\mathtt{y[x]} \rightarrow \dfrac{3}{8}\left(1-10\,\mathtt{x}^2 + \dfrac{35\,\mathtt{x}^4}{3}\right)\right\}\right\}$

Coefficient recursion. For a power series solution of the Legendre equation with arbitrary positive integer n the coefficient recursion will involve three subsequent powers x^s, x^{s+1}, x^{s+2}. Accordingly, type

In[6]:= ser = Sum[a[m] x^m, {m, s, s + 2}]

Out[6]= $\mathtt{x}^s\,\mathtt{a[s]} + \mathtt{x}^{1+s}\,\mathtt{a[1+s]} + \mathtt{x}^{2+s}\,\mathtt{a[2+s]}$

Substitute this and its derivatives into the Legendre equation with general n, where Simplify multiplies out $1 - x^2$ (the first coefficient of the Legendre equation times the second derivative of ser) and $-2x$ (the other coefficient of the equation) times the first derivative of ser, and simplifies the result.

In[7]:= (1 - x^2) D[ser, x, x] - 2 x D[ser, x] + n (n + 1) ser == 0

Out[7]= $(1-\mathtt{x}^2)\,((-1+s)\,s\,\mathtt{x}^{-2+s}\,\mathtt{a[s]} + s\,(1+s)\,\mathtt{x}^{-1+s}\,\mathtt{a[1+s]} + (1+s)\,(2+s)\,\mathtt{x}^s\,\mathtt{a[2+s]}) -$

$2\,\mathtt{x}\,(s\,\mathtt{x}^{-1+s}\,\mathtt{a[s]} + (1+s)\,\mathtt{x}^s\,\mathtt{a[1+s]} + (2+s)\,\mathtt{x}^{1+s}\,\mathtt{a[2+s]}) +$

$\mathtt{n}\,(1+\mathtt{n})\,(\mathtt{x}^s\,\mathtt{a[s]} + \mathtt{x}^{1+s}\,\mathtt{a[1+s]} + \mathtt{x}^{2+s}\,\mathtt{a[2+s]}) == 0$

In[8]:= temp = Simplify[%]

Out[8]= $\mathtt{x}^{-2+s}\,((\mathtt{n}\,(1+\mathtt{n})\,\mathtt{x}^2 - s^2\,(-1+\mathtt{x}^2) - s\,(1+\mathtt{x}^2))\,\mathtt{a[s]} +$

$\mathtt{x}\,((s + (-2+\mathtt{n}+\mathtt{n}^2)\,\mathtt{x}^2 - 3\,s\,\mathtt{x}^2 - s^2\,(-1+\mathtt{x}^2))\,\mathtt{a[1+s]} +$

$\mathtt{x}\,(2 + (-6+\mathtt{n}+\mathtt{n}^2)\,\mathtt{x}^2 + s\,(3-5\,\mathtt{x}^2) - s^2\,(-1+\mathtt{x}^2))\,\mathtt{a[2+s]})) == 0$

The coefficient of x^s (a sum of expressions) is now obtained by the command

In[9]:= co = Coefficient[Expand[temp[[1]]], x^s]

Out[9]= $\mathtt{n}\,\mathtt{a[s]} + \mathtt{n}^2\,\mathtt{a[s]} - s\,\mathtt{a[s]} - s^2\,\mathtt{a[s]} + 2\,\mathtt{a[2+s]} + 3\,s\,\mathtt{a[2+s]} + s^2\,\mathtt{a[2+s]}$

This gives the coefficient recursion

In[10]:= Solve[co == 0, a[s + 2]]

Out[10]= $\left\{\left\{\mathtt{a[2+s]} \rightarrow -\dfrac{\mathtt{n}\,\mathtt{a[s]} + \mathtt{n}^2\,\mathtt{a[s]} - s\,\mathtt{a[s]} - s^2\,\mathtt{a[s]}}{2 + 3\,s + s^2}\right\}\right\}$

In[11]:= a[s+2] == Factor[%[[1, 1, 2]]]   (* Factors the right-hand side. *)

Out[11]= $\mathtt{a[2+s]} == \dfrac{(-\mathtt{n}+s)\,(1+\mathtt{n}+s)\,\mathtt{a[s]}}{(1+s)\,(2+s)}$

This is the basic recursion relation for the coefficients of the power series solution of the Legendre equation with positive integer parameter n that leads to the Legendre polynomials.

Similar Material in AEM: pp. 206, 207

EXAMPLE 4.4   ORTHOGONALITY. FOURIER-LEGENDRE SERIES

The importance of the Legendre polynomials results in part from their orthogonality. These polynomials are orthogonal on the interval from -1 to 1. By definition this means that the integral of the product of two Legendre polynomials of different orders, integrated from -1 to 1, is zero,

$$\int_{-1}^{1} P(m,\,x)\,P(n,\,x)\,dx = 0 \qquad (m \neq n).$$

It can be shown that for $n = m$ the integral equals $2/(2m+1)$, as is stated on p. 242 of AEM (without proof, which is tricky). For specific cases, e.g. for all m and n up to 100 (or more), you can verify these statements, using the following command (which does it for 0, 1, ..., 10).

In[1]:= `Table[Table[Integrate[LegendreP[m, x] LegendreP[n, x], {x, -1, 1}],`
    `{m, 0, n}], {n, 0, 10}]`

Out[1]= $\left\{\{2\},\ \left\{0, \dfrac{2}{3}\right\},\ \left\{0, 0, \dfrac{2}{5}\right\},\ \left\{0, 0, 0, \dfrac{2}{7}\right\},\ \left\{0, 0, 0, 0, \dfrac{2}{9}\right\},\ \left\{0, 0, 0, 0, 0, \dfrac{2}{11}\right\},\right.$

$\left\{0, 0, 0, 0, 0, 0, \dfrac{2}{13}\right\},\ \left\{0, 0, 0, 0, 0, 0, 0, \dfrac{2}{15}\right\},\ \left\{0, 0, 0, 0, 0, 0, 0, 0, \dfrac{2}{17}\right\},$

$\left.\left\{0, 0, 0, 0, 0, 0, 0, 0, 0, \dfrac{2}{19}\right\},\ \left\{0, 0, 0, 0, 0, 0, 0, 0, 0, 0, \dfrac{2}{21}\right\}\right\}$

You see the values $2/(2m+1) = 2, 2/3, 2/5, \ldots$ for $m = n = 0, 1, 2, \ldots$ as well as the zeros for the index pairs (0,1), (0,2), (1,2), (0,3), (1,3), (2,3), and so on.

Orthogonal expansions are series for a given function f in terms of orthogonal functions, in which orthogonality helps determining the coefficients in a simple fashion – this is a main point of orthogonality. For the Legendre polynomials this gives **Fourier-Legendre series**, as follows. Take, for instance $f(x) = \sin \pi x$,

In[2]:= `f = Sin[Pi x]`   `(* Out: Sin[π x] *)`

The series will be $f = a_0 P_0(x) + a_1 P_1(x) + \ldots$ with coefficients a_m equal $(2m+1)/2$ times the integral of $f P_m$ from $x = -1$ to 1. The double primes in the next command will have the effect that the Legendre polynomials will not be shown in terms of powers of x. (Try without.)

In[3]:= `S = Sum[(2 m + 1)/2 Integrate[f LegendreP[m, x], {x, -1, 1}]`
    `"LegendreP"[m, x], {m, 1, 5}]`

Out[3]= $\dfrac{3\,\text{LegendreP}[1,\,x]}{\pi} + \dfrac{7\,(-15 + \pi^2)\,\text{LegendreP}[3,\,x]}{\pi^3} +$

$\dfrac{11\,(945 - 105\,\pi^2 + \pi^4)\,\text{LegendreP}[5,\,x]}{\pi^5}$

You obtain this series with coefficients as decimal fractions by typing

In[4]:= `S1 = N[%]`

Out[4]= $0.95493\,\text{LegendreP}[1.,\,x] - 1.15824\,\text{LegendreP}[3.,\,x] + 0.21929\,\text{LegendreP}[5.,\,x]$

To obtain the coefficients of `S1`, type

In[5]:= `coeffs = Table[S1[[n, 1]], {n, 1, 3}]`

Out[5]= {0.95493, −1.15824, 0.21929}

The terms of S1 with the Legendre polynomials in terms of powers of x can be obtained by the following command, in which you have to observe that only the odd-numbered Legendre polynomials occur because the sine function f under discussion is an odd function, so that p is the number of the term and $2p - 1$ is the parameter of the corresponding Legendre polynomial.

In[6]:= terms = Table[coeffs[[p]] LegendreP[2 p - 1, x], {p, 1, 3}]

Out[6]= $\left\{0.95493\,x, -1.15824\left(-\dfrac{3\,x}{2} + \dfrac{5\,x^3}{2}\right), 0.21929\left(\dfrac{15\,x}{8} - \dfrac{35\,x^3}{4} + \dfrac{63\,x^5}{8}\right)\right\}$

An even smaller number of our series will give you a good approximation, as the corresponding plot will show. Take just the sum of the first two terms

In[7]:= S2 = terms[[1]] + terms[[2]]

Out[7]= $0.95493\,x - 1.15824\left(-\dfrac{3\,x}{2} + \dfrac{5\,x^3}{2}\right)$

The accuracy is surprising. The areas under the curves seem almost equal, and so are the locations and size of the extrema. The zeros of $\sin \pi x$ are not too well approximated. In terms of powers of x this approximation of $\sin \pi x$ is

In[8]:= Expand[S2]   (* Out: $2.69229\,x - 2.8956\,x^3$ *)

In[9]:= Plot[{f, S2}, {x, -1, 1}, AxesLabel -> {x, y}]

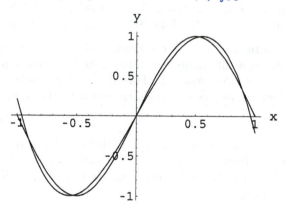

Example 4.4. Sine function and approximation by $a_1\,P_1(x) + a_3\,P_3(x)$

Similar Material in AEM: p. 242

| EXAMPLE 4.5 | **FROBENIUS METHOD**

In this example we solve some ODE's by DSolve for which the Frobenius method is needed, because the coefficients of the equations are singular when the equations are written in standard form with 1 as the coefficient of y''. You will see in what form the computer will give a general solution.

In[1]:= ode1 = x^2 y''[x] + b x y'[x] + c y[x] == 0

Out[1]= $c\,y[x] + b\,x\,y'[x] + x^2\,y''[x] == 0$

For this Euler-Cauchy equation $x^2 y'' + bxy' + cy = 0$ with arbitrary constant b and c you obtain solutions x^μ, where μ is a solution of the auxiliary equation as given by the usual formula for a quadratic equation; that is,

In[2]:= `DSolve[ode1, y[x], x]`

Out[2]= $\left\{\left\{\text{y[x]} \rightarrow x^{\frac{1}{2}\left(1-b-\sqrt{(-1+b)^2-4c}\right)}\text{C[1]} + x^{\frac{1}{2}\left(1-b+\sqrt{(-1+b)^2-4c}\right)}\text{C[2]}\right\}\right\}$

For the next equation, $x(x-1)y'' + (3x-1)y' + y = 0$, type

In[3]:= `ode2 = x (x - 1) y''[x] + (3 x - 1) y'[x] + y[x] == 0`

Out[3]= $\text{y[x]} + (-1 + 3x)\,\text{y}'\text{[x]} + (-1+x)\,x\,\text{y}''\text{[x]} == 0$

and obtain by `DSolve` a general solution

In[4]:= `DSolve[ode2, y[x], x]`

Out[4]= $\left\{\left\{\text{y[x]} \rightarrow \dfrac{\text{C[1]}}{1-x} + \text{C[2]}\,\text{MeijerG[\{\{0,0\},\{\}\},\{\{0,0\},\{\}\},x]}\right\}\right\}$

It is not necessary that in the present case you know something about the somewhat obscure Meijer G function because you can type the following and see what is going on. (First try `Simplify`.)

In[5]:= `FullSimplify[%]`

Out[5]= $\left\{\left\{\text{y[x]} \rightarrow \dfrac{-\text{C[1]} + \text{C[2]}\,\text{Log[x]}}{-1+x}\right\}\right\}$

This response involving a logarithm shows that the **indicial equation** (see AEM, p. 213) has a double root, which has the value zero. You may try a power series solution, expecting a geometric series with sum $1/(1-x)$. To accomplish this, proceed as in the previous example. Type

In[6]:= `ser = Sum[a[m] x^m, {m, s, s + 2}]`

Out[6]= $x^s\,\text{a[s]} + x^{1+s}\,\text{a[1 + s]} + x^{2+s}\,\text{a[2 + s]}$

In[7]:= `x (x - 1) D[ser, x, x] + (3 x - 1) D[ser, x] + ser`

`== 0` (* Long response *)

In[8]:= `FullSimplify[%]`

Out[8]= $x^{-1+s}\,((s^2\,(-1+x) + x + 2sx)\,\text{a[s]} +$

$x\,((-(1+s)^2 + (2+s)^2\,x)\,\text{a[1+s]} + x\,(-(2+s)^2 + (3+s)^2\,x)\,\text{a[2+s]})) == 0$

In[9]:= `co = Coefficient[Expand[%[[1]]], x^s]`

Out[9]= $\text{a[s]} + 2s\,\text{a[s]} + s^2\,\text{a[s]} - \text{a[1+s]} - 2s\,\text{a[1+s]} - s^2\,\text{a[1+s]}$

In[10]:= `Solve[% == 0, a[s + 1]]` (* Out: {{a[1 + s] → a[s]}} *)

This is the recursion for the coefficients of a geometric series, as expected.

As a third ODE, consider $x(1-x)y'' + 2(1-2x)y' - 2y = 0$. Type and solve it.

In[11]:= `ode3 = x (1 - x) y''[x] + 2 (1 - 2 x) y'[x] - 2 y[x] == 0`

Out[11]= $-2\,y[x] + 2\,(1 - 2\,x)\,y'[x] + (1 - x)\,x\,y''[x] == 0$

In[12]:= `DSolve[ode3, y[x], x]`

Out[12]= $\left\{\left\{y[x] \to \dfrac{C[1]}{1 - x} + \dfrac{C[2]}{(1 - x)\,x}\right\}\right\}$

This indicates that the indicial equation has roots differing by an integer (these are 0 and -1). A simple basis of solutions is $1/(1 - x)$, $1/x$. Can you see this?

Similar Material in AEM: p. 212-216

EXAMPLE 4.6 BESSEL'S EQUATION. BESSEL FUNCTIONS

Bessel's differential equation

$$x^2 y'' + xy' + (x^2 - \nu^2)y = 0$$

appears in various practical problems, in particular in those that exhibit cylindrical symmetry. ν (Greek nu) is a given real number (a real parameter suggested by the kind of problem). Type the equation as

In[1]:= `BES = x^2 y''[x] + x y'[x] + (x^2 - ν^2) y[x] == 0`

Out[1]= $(x^2 - \nu^2)\,y[x] + x\,y'[x] + x^2\,y''[x] == 0$

Here the **Greek** ν is obtained by typing `\[Nu]`. Now obtain a general solution by the command

In[2]:= `sol = DSolve[BES, y[x], x]`

Out[2]= $\{\{y[x] \to \text{BesselJ}[-\nu, x]\,C[1] + \text{BesselJ}[\nu, x]\,C[2]\}\}$

The first term involves the **Bessel function of the first kind** $J_{-\nu}(x)$ and the second term the **Bessel function of the first kind** $J_\nu(x)$. When ν is an integer, one writes n instead of ν. (For integer n the solution **sol** is no longer a general solution; see AEM, p. 222.)

The functions J_0 and J_1 are particularly important. J_0 satisfies Bessel's equation with parameter 0, whose left-hand side is

In[3]:= `BES[[1]]/x /. ν -> 0`   (* Out: $\dfrac{x^2 y[x] + x\,y'[x] + x^2\,y''[x]}{x}$ *)

In[4]:= `Simplify[%]`   (* Out: $x\,y[x] + y'[x] + x\,y''[x]$ *)

Mathematica did not drop the factor x automatically, hence our trick. From **sol** (or equally well from the Bessel equation with $n = 0$) you can obtain J_0, its Maclaurin series, and the conversion of the response to a polynomial by the command **Normal**.

In[5]:= `sol /. {C[1] -> 0, C[2] -> 1, ν -> 0}`

Out[5]= $\{\{y[x] \to \text{BesselJ}[0, x]\}\}$

In[6]:= `ser0 = Series[%[[1, 1, 2]], {x, 0, 12}]`

Out[6]= $1 - \dfrac{x^2}{4} + \dfrac{x^4}{64} - \dfrac{x^6}{2304} + \dfrac{x^8}{147456} - \dfrac{x^{10}}{14745600} + \dfrac{x^{12}}{2123366400} + O[x]^{13}$

In[7]:= `p0 = Normal[%]`

Out[7]= $1 - \dfrac{x^2}{4} + \dfrac{x^4}{64} - \dfrac{x^6}{2304} + \dfrac{x^8}{147456} - \dfrac{x^{10}}{14745600} + \dfrac{x^{12}}{2123366400}$

Similarly for the Bessel function $J_1(x)$

In[8]:= `sol /. {C[1] -> 0, C[2] -> 1, ν -> 1}`

Out[8]= $\{\{y[x] \rightarrow \text{BesselJ}[1, x]\}\}$

In[9]:= `ser1 = Series[%[[1, 1, 2]], {x, 0, 12}]`

Out[9]= $\dfrac{x}{2} - \dfrac{x^3}{16} + \dfrac{x^5}{384} - \dfrac{x^7}{18432} + \dfrac{x^9}{1474560} - \dfrac{x^{11}}{176947200} + \text{O}[x]^{13}$

In[10]:= `p1 = Normal[%]`

Out[10]= $\dfrac{x}{2} - \dfrac{x^3}{16} + \dfrac{x^5}{384} - \dfrac{x^7}{18432} + \dfrac{x^9}{1474560} - \dfrac{x^{11}}{176947200}$

Plot the two polynomials and the two functions jointly, so that you see that the approximation seems good for x to about 3 or 4. For larger x you would need more terms. The other figure shows that the curves of the two functions look similar to those of cosine and sine, due to the similarity of the respective series.

In[11]:= `Plot[{p0, p1, BesselJ[0, x], BesselJ[1, x]}, {x, 0, 6}]`

In[12]:= `Plot[{BesselJ[0, x], BesselJ[1, x]}, {x, 0, 15},`

      `Ticks -> {{0, 5, 10, 15}, Automatic}]`

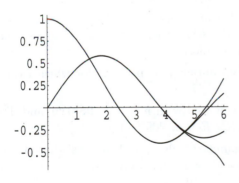

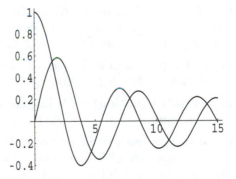

Example 4.6. Partial sum approximations of the Bessel functions J_0 and J_1

Example 4.6. Bessel functions J_0 (starting from 1) and J_1 (starting from 0)

Zeros of Bessel functions are often needed in applications (vibration of membranes, celestial mechanics, etc.) and have been extensively tabulated, as have been the functions themselves, see Ref. [1] in Appendix 1 of this Guide. You obtain the zeros by the command `NSolve`. For instance, for the first positive zero of J_0 you obtain as an approximation 2.40483 (exact to 5 digits) from the command

In[13]:= `x01appr = NSolve[p0 == 0, x]`

Out[13]= {{x → −6.78143 − 5.6004 i}, {x → −6.78143 + 5.6004 i},

{x → −6.54585 − 1.72484 i}, {x → −6.54585 + 1.72484 i},

{x → −5.40592}, {x → −2.40483}, {x → 2.40483}, {x → 5.40592},

{x → 6.54585 − 1.72484 i}, {x → 6.54585 + 1.72484 i},

{x → 6.78143 − 5.6004 i}, {x → 6.78143 + 5.6004 i}}

and an exact 10-digit value from the command

In[14]:= `x01 = SetPrecision[FindRoot[BesselJ[0, x] == 0, {x, 1}], 10]`

Out[14]= {x → 2.404825543}

For the second zero (near 5.5) you see from the first figure that you cannot expect great accuracy. Try it.

Similar Material in AEM: pp. 218-220

Problem Set for Chapter 4

Pr.4.1 (Power series) Find the Maclaurin series of $\tan x$ (5 nonzero terms). Find the coefficient of x^7 (as a decimal fraction with 6 digits) by applying a suitable command to that series. (*AEM Ref.* p. 195)

Pr.4.2 (Power series) Find a partial sum of the Maclaurin series of $f = \cos \pi x$ on the computer (powers up to x^{18}, inclusively). Plot f and the sum jointly to see for what x the sum gives useful approximations. (*AEM Ref.* p. 195)

Pr.4.3 (Coefficient recursion.) Obtain the coefficient recursion for the power series solution of $y' = 2xy$ with $a_0 = 1$ and write the series thus obtained. (For another coefficient recursion, see Example 4.3 in this Guide.)

Pr.4.4 (Legendre polynomials) Find and solve an initial value problem for which $P_4(x)$ is a solution. (*AEM Ref.* p. 208)

Pr.4.5 (Legendre's equation) Solve Legendre's equation with $n = 0$ by `DSolve` and develop the solution in a Maclaurin series. (*AEM Ref.* p. 209 (#1))

Pr.4.6 (Boundary value problem) Solve the boundary value problem consisting of the Legendre equation with parameter $n = 5$ and boundary values $y(-1) = -1$ and $y(1) = 1$. Plot the solution. (*AEM Ref.* pp. 205-210)

Pr.4.7 (Experiment on Fourier-Legendre series) Represent e^x on the interval from -3 to 3 by partial sums S_n of a Fourier-Legendre series. Plot e^x and the approximating partial sums on common axes, beginning with 2 terms and taking more and more terms stepwise until the two curves practically coincide in the plot. Describe what you can see regarding the increase of accuracy (a) on the subinterval from -1 to 1, (b) on the entire interval from -3 to 3. Calculate e approximately from S_n with increasing n. Find an empirical formula for the (approximate) size of the error as a function of n. (*AEM Ref.* pp. 242, 246 (#6))

Pr.4.8 (Fourier-Legendre series) Develop x^6 in a Fourier-Legendre series. Why will this series reduce to a polynomial? Plot x^6 and the corresponding Fourier-Legendre series of powers up to x^4 and comment on the quality of approximation. Show graphically that the error (as a function of x) is oscillating. (*AEM Ref.* pp. 242, 246)

Pr.4.9 **(Frobenius method)** Using `DSolve`, solve the initial value problem

$$xy'' + (1 - 2x)y' + (x - 1)y = 0, \qquad y(1) = 1, \qquad y'(1) = 1,$$

involving an ODE for which the Frobenius method is needed. Also obtain a general solution and find out whether you could prescribe initial values at $x = 0$.
(*AEM Ref.* p. 216 (#7))

Pr.4.10 **(Frobenius method)** Find a general solution of $4xy'' + 2y' + y = 0$. Set up two initial value problems that give the two solutions of the basis separately, and find series in powers of x for these solutions. You cannot prescribe initial conditions at $x = 0$. Why? Prescribe them at $x = \pi^2$. (*AEM Ref.* p. 216 (#4))

Pr.4.11 **(Hypergeometric equation)** Apply the command `DSolve` to the hypergeometric equation

$$x(1 - x)y'' + [c - (a + b + 1)x]y' - aby = 0.$$

In many cases, the solutions of this equation become elementary functions. Show that this is the case when $a = -n$, $c = b$, $x = -t$. (*AEM Ref.* pp. 216, 217)

Pr.4.12 **(Frobenius method)** The equation $xy'' + 2y' + xy = 0$ requires the Frobenius method (why?). Solve it by `DSolve` in terms of sin and cos. (*AEM Ref.* p. 216 (#5))

Pr.4.13 **(Hypergeometric equation)** Solve the initial value problem

$$x(1 - x)y'' + (2 - 4x)y' - 2y = 0, \qquad y(0) = 1, \qquad y'(0) = 1.$$

Give the solution also in the form of a series. To what a, b, c in the general hypergeometric equation does the present equation correspond? (*AEM Ref.* p. 216)

Pr.4.14 **(Bessel functions)** The Bessel functions J_ν satisfy the basic relations

$$[x^\nu J_\nu(x)]' = x^\nu J_{\nu-1}(x), \qquad [x^{-\nu} J_\nu(x)]' = -x^{-\nu} J_{\nu+1}(x),$$

where $' = d/dx$. Try to obtain these relations on the computer, either directly or, if this will not work, in integrated form on both sides. (*AEM Ref.* p. 223)

Pr.4.15 **(Zeros of Bessel functions)** Find approximations of the first two positive zeros of $J_1(x)$ from the Maclaurin series, using powers up to x^{19}, and compare with the values obtained directly from $J_1(x) = 0$. (*AEM Ref.* p. 220)

Pr.4.16 **(Experiment on asymptotics of Bessel functions)** For large x the Bessel function $J_n(x)$ is approximately equal to

$$\sqrt{\frac{2}{\pi x}} \cos\left(x - \frac{n\pi}{2} - \frac{\pi}{4}\right).$$

Experiment with this formula for various n; find out empirically from the graphs how accurate the formula is for small integers n and for larger ones.
(*AEM Ref.* pp. 220, 228 (#30))

Pr.4.17 **(Bessel functions Y_0 and Y_1)** Verify that Y_0 and Y_1 satisfy Bessel's ODE with parameters 0 and 1, respectively, and plot these functions on common axes.
(*AEM Ref.* p. 231)

Pr.4.18 **(Bessel's equation)** A large number of ODE's can be reduced to Bessel's equation. Show that $x^2 y'' + x y' + (4x^4 - 1/4) y = 0$ is of this kind, by solving it by DSolve, which will give a general solution in terms of Bessel functions. (*AEM Ref.* p. 226 (#3))

Pr.4.19 **(Reduction to Bessel's equation)** Find out from the solution by DSolve what transformation will reduce $y'' + x^2 y = 0$ to Bessel's equation.
(*AEM Ref.* p. 232 (#5))

Pr.4.20 **(Orthogonality)** The functions $\sin nx$ with positive integer are orthogonal on any interval of length 2π and have the norm $\sqrt{\pi}$. Verify this for $n = 1, ..., 10$ by integration.
(*AEM Ref.* pp. 234-236)

Chapter 5

Laplace Transform Method for Solving ODE's

Content. Transforms, inverse transforms (Ex. 5.1, Prs. 5.1-5.5)
Solving ODE's (Exs. 5.2-5.4, Prs. 5.6, 5.7, 5.11-5.14)
Forced oscillations, resonance (Ex. 5.3)
Unit step, Dirac delta (Ex. 5.4, Prs. 5.9-5.15)
Solving systems of ODE's (Ex. 5.5)
Differentiation, s-shifting, t-shifting, convolution, etc. (Ex. 5.6, Pr. 5.8)
Electric circuits, rectifiers (Prs. 5.11, 5.12, 5.14, 5.15)

Laplace transforms. Type `LaplaceTransform[f, t, s]`, f the given function, t its variable, s the variable of the transform F. For instance,

In[1]:= `LaplaceTransform[t^2 + t + 1 + Exp[a t] + Sin[Pi t] + Cos[`ω` t], t, s]`

Out[1]= $\dfrac{2}{s^3} + \dfrac{1}{s^2} + \dfrac{1}{s} + \dfrac{1}{-a+s} + \dfrac{\pi}{\pi^2+s^2} + \dfrac{s}{s^2+\omega^2}$

Here the **Greek** ω is obtained by typing `\[Omega]`.

Inverse transforms. Type `InverseLaplaceTransform[F, s, t]`. For instance,

In[2]:= `InverseLaplaceTransform[1/s^4 + s/(s^2 + 25), s, t]`

Out[2]= $\dfrac{t^3}{6} + \text{Cos}\,[5\,t]$

Transforms of the derivatives $\mathcal{L}(f') = s\mathcal{L}(f) - f(0)$ and $\mathcal{L}(f'') = s^2\mathcal{L}(f) - sf(0) - f'(0)$ (where $f' = df/dt$, etc.) are typed as

In[3]:= `LaplaceTransform[f'[t], t, s]`

Out[3]= $-\mathtt{f[0]} + \mathtt{s}\,\text{LaplaceTransform}[\mathtt{f[t]}, \mathtt{t}, \mathtt{s}]$

In[4]:= `LaplaceTransform[f''[t], t, s]`

Out[4]= $-\mathtt{s}\,\mathtt{f[0]} + \mathtt{s}^2\,\text{LaplaceTransform}[\mathtt{f[t]}, \mathtt{t}, \mathtt{s}] - \mathtt{f}'[0]$

Examples for Chapter 5

EXAMPLE 5.1 | TRANSFORMS AND INVERSE TRANSFORMS

For e^{at}, $\cos\omega t$, and $\sin\omega t$ see before. Further frequently used functions $\cosh at$, $\sinh at$, $e^{at}\cos\omega t$, $e^{at}\sin\omega t$ have the transforms

In[1]:= `LaplaceTransform[Cosh[a t], t, s]` (* Out: $\dfrac{s}{-a^2+s^2}$ *)

In[2]:= `LaplaceTransform[Sinh[a t], t, s]` (* Out: $\dfrac{a}{-a^2+s^2}$ *)

55

In[3]:= LaplaceTransform[Exp[a t] Cos[ω t], t, s] (* Out: $\dfrac{-a+s}{a^2-2as+s^2+\omega^2}$ *)

In[4]:= FullSimplify[%]   (* Simplify is not enough. Try it. *)

Out[4]= $\dfrac{-a+s}{(a-s)^2+\omega^2}$

In[5]:= LaplaceTransform[Exp[a t] Sin[ω t], t, s]

Out[5]= $\dfrac{\omega}{a^2-2as+s^2+\omega^2}$

Apply FullSimplify.

Powers of t are transformed as follows.

In[6]:= LaplaceTransform[1, t, s]   (* Out: $\frac{1}{s}$ *)

In[7]:= LaplaceTransform[t, t, s]   (* Out: $\frac{1}{s^2}$ *)

In[8]:= LaplaceTransform[t^n, t, s]   (* Out: s^{-1-n} Gamma[1+n] *)

Hence t^n with $n = 0, 1, 2, \ldots$ has the transform $n!/s^{(n+1)}$ because $\Gamma(n+1) = n!$ when n is a positive integer. Here, Γ is the Gamma function (see AEM, p. A54).

Inverse transforms are more difficult to obtain than transforms. Corresponding tables are available (e.g. in AEM, pp. 297-299), and you can use them just as you would use a table of integrals in integration. The Mathematica command InverseLaplaceTransform(F, s, t) gives help. (Note that now, s comes first in the command!) For example,

In[9]:= InverseLaplaceTransform[(s + 2)/(s^2 + 4 s + 5), s, t]

Out[9]= $\dfrac{1}{2}\,e^{(-2-i)\,t}\left(1+e^{2\,i\,t}\right)$

In[10]:= FullSimplify[%]   (* Try Simplify. *)

Out[10]= Cos[t] (Cosh[2t] − Sinh[2t])

In[11]:= InverseLaplaceTransform[1/s^4, s, t]   (* Out: $\frac{t^3}{6}$ *)

In[12]:= InverseLaplaceTransform[(s − 4)/(s^2 − 4), s, t]

Out[12]= $-\dfrac{1}{2}\,e^{-2\,t}\left(-3+e^{4\,t}\right)$

In[13]:= InverseLaplaceTransform[LaplaceTransform[f, t, s], s, t]

Out[13]= f

In[14]:= LaplaceTransform[InverseLaplaceTransform[F, s, t], t, s]

Out[14]= F

Thus the two commands are inverses of each other, as had to be expected.

Similar Material in AEM: pp. 251-255

EXAMPLE 5.2 DIFFERENTIAL EQUATIONS

Consider the differential equation and initial conditions

$$y'' + 2y' + 2y = 0, \qquad y(0) = 1, \quad y'(0) = -1.$$

I. Find a general solution $y(t)$ in two ways, first by DSolve , and then by using the subsidiary equation and transforming its solution $Y(s)$ back.

II. Apply the same two methods to the initial value problem.

Solution. I. Type the equation in the following form, and then apply DSolve.

In[1]:= `ode = y''[t] + 2 y'[t] + 2 y[t] == 0`

Out[1]= $2\,y[t] + 2\,y'[t] + y''[t] == 0$

In[2]:= `DSolve[ode, y[t], t]`

Out[2]= $\{\{y[t] \to e^{-t}C[2]\,Cos\,[t] - e^{-t}\,C[1]\,Sin\,[t]\}\}$

In[3]:= `Simplify[%]`   (* Out: $\{\{y[t] \to e^{-t}\,(C[2]\,Cos\,[t] - C[1]\,Sin\,[t])\}\}$ *)

This is a general solution in the form expected.

In the second method you get the **subsidiary equation** by applying the command LaplaceTransform to the differential equation,

In[4]:= `Clear[t, s]`

In[5]:= `subsid = LaplaceTransform[ode, t, s]`

Out[5]= $2\,LaplaceTransform[y[t], t, s] + s^2\,LaplaceTransform[y[t], t, s] +$

$2\,(s\,LaplaceTransform[y[t], t, s] - y[0]) - s\,y[0] - y'[0] == 0$

As the next step, solve the subsidiary equation *algebraically* for the transform, call it **Y**, of the unknown solution; thus,

In[6]:= `Y = Solve[subsid, LaplaceTransform[y[t], t, s]]`

Out[6]= $\left\{\left\{LaplaceTransform[y[t], t, s] \to -\dfrac{-2\,y[0] - s\,y[0] - y'[0]}{2 + 2\,s + s^2}\right\}\right\}$

Finally, obtain $y(t)$ itself by taking the inverse of the expression for Y. This gives the general solution

In[7]:= `sol = InverseLaplaceTransform[Y, s, t]`

Out[7]= $\left\{\left\{y[t] \to -\dfrac{1}{2}\,i\,e^{(-1-i)\,t}\,((1+i)\,(i+e^{2\,i\,t})\,y[0] + (-1+e^{2\,i\,t})\,y'[0])\right\}\right\}$

In[8]:= `FullSimplify[%]`   (* Try Simplify *)

Out[8]= $\{\{y[t] \to e^{-t}(Cos[t]\,y[0] + Sin[t]\,(y[0] + y'[0]))\}\}$

This form of the solution should not surprise you. It is needed to make the solution equal to $y(0)$ when $t = 0$ (so that the sine terms are zero) and its derivative equal to $y'(0)$.

II. Now apply the same two methods to the initial value problem, remembering from Chap. 2 what will change when initial conditions are given. Basically, not much; put braces {} around as shown.

In[9]:= `DSolve[{ode, y[0] == 1, y'[0] == -1}, y[t], t]`

Out[9]= $\{\{y[t] \to e^{-t}\,Cos\,[t]\}\}$

In the second method use the previous subsid and substitute the initial condition into it.

In[10]:= `subsid2 = subsid /. {y[0] -> 1, y'[0] -> -1}`

Out[10]= $1 - s + 2\,\text{LaplaceTransform}[y[t], t, s] + s^2\,\text{LaplaceTransform}[y[t], t, s] +$

$\qquad 2\,(-1 + s\,\text{LaplaceTransform}[y[t], t, s]) == 0$

In[11]:= Y2 = Solve[subsid2, LaplaceTransform[y[t], t, s]]

Out[11]= $\left\{\left\{\text{LaplaceTransform}[y[t], t, s] \to -\dfrac{-1 - s}{2 + 2\,s + s^2}\right\}\right\}$

In[12]:= y2 = InverseLaplaceTransform[Y2, s, t]

Out[12]= $\left\{\left\{y[t] \to \dfrac{1}{2}\,e^{(-1 - i)\,t}\,\left(1 + e^{2\,i\,t}\right)\right\}\right\}$

In[13]:= FullSimplify[%]   (* Out: $\left\{\left\{y[t] \to e^{-t}\,\text{Cos}\,[t]\right\}\right\}$ *)

Similar Material in AEM: pp. 260-262

EXAMPLE 5.3 | FORCED VIBRATIONS. RESONANCE

Resonance occurs in undamped mechanical or electrical systems if the driving force (the electromotive force, respectively) has the frequency equal to the natural frequency of the free vibrations of the system. This is the case for the differential equation

$$y'' + \omega_0{}^2\,y = K\sin\omega_0\,t.$$

Using the Laplace transform method, find the solution of this equation satisfying the initial conditions $y(0) = 0$, $y'(0) = 0$. Specifying K and ω_0 to be 1, plot that solution.

Solution. Type the equation; then obtain the subsidiary equation. (Here the **Greek** ω is obtained by typing \[Omega] .)

In[1]:= ode = y''[t] + ω[0]^2 y[t] == K Sin[ω[0] t]

Out[1]= $y[t]\,\omega[0]^2 + y''[t] == K\,\text{Sin}[t\,\omega[0]]$

In[2]:= subsid = LaplaceTransform[ode, t, s]

Out[2]= $s^2\,\text{LaplaceTransform}[y[t], t, s] - s\,y[0] +$

$\qquad \text{LaplaceTransform}[y[t], t, s]\,\omega[0]^2 - y'[0] == \dfrac{K\,\omega[0]}{s^2 + \omega[0]^2}$

Now include the initial conditions in the subsidiary equation,

In[3]:= subsid2 = subsid /. {y[0] -> 0, y'[0] -> 0}

Out[3]= $s^2\,\text{LaplaceTransform}[y[t], t, s] +$

$\qquad \text{LaplaceTransform}[y[t], t, s]\,\omega[0]^2 == \dfrac{K\,\omega[0]}{s^2 + \omega[0]^2}$

Next solve the subsidiary equation *algebraically* for the Laplace transform, call it Y, of the unknown solution y:

In[4]:= Y = Solve[subsid2, LaplaceTransform[y[t], t, s]]

Out[4]= $\left\{\left\{\text{LaplaceTransform}[y[t], t, s] \to \dfrac{K\,\omega[0]}{(s^2 + \omega[0]^2)^2}\right\}\right\}$

The inverse Laplace transform then gives the solution

In[5]:= yp = InverseLaplaceTransform[Y, s, t]

Out[5]= $\left\{\left\{ \text{y[t]} \to \dfrac{\text{K}\,(\text{Sin}\,[\text{t}\,\omega\,[0]] - \text{t}\,\text{Cos}\,[\text{t}\,\omega\,[0]]\,\omega\,[0])}{2\,\omega\,[0]^2}\right\}\right\}$

For plotting choose $K = 1$ and $\omega_0 = 1$ by the command

In[6]:= `yp0 = yp /. {K -> 1, `ω`[0] -> 1}`

Out[6]= $\left\{\left\{ \text{y[t]} \to \dfrac{1}{2}\,(-\,\text{t}\,\text{Cos}\,[\text{t}] + \text{Sin}\,[\text{t}])\right\}\right\}$

The maximum amplitude of the solution grows beyond bound as t increases indefinitely.

In[7]:= `Plot[yp0[[1, 1, 2]], {t, 0, 100}, AxesLabel -> {t, y}]`

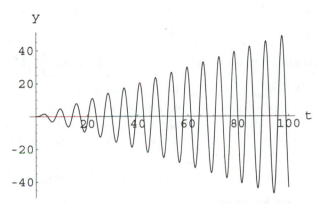

Example 5.3. Particular solution exhibiting resonance

Similar Material in AEM: p. 288

$\boxed{\textbf{EXAMPLE 5.4}}$ **UNIT STEP FUNCTION (HEAVISIDE FUNCTION), DIRAC'S DELTA**

The **unit step function** $u(t - a)$ (or **Heaviside function**) is 0 for $t < a$, has a jump of size 1 at $t = a$ (where we need not assign a value to it) and is 1 for $t > a$. Here, $a \geq 0$. The Mathematica notation is `UnitStep[t - a]`. The transform of $u(t - a)$ is obtained by typing

In[1]:= `LaplaceTransform[UnitStep[t - a], t, s]`   (* Out: $\dfrac{e^{-a\,s}}{s}$ *)

The unit step function is the basic building block for representing piecewise continuous functions that are given by different expressions over different intervals. For instance, type (see the figure)

In[2]:= `f = Which[t  <  Pi, 2, t < 2 Pi, 0, t > 2 Pi, Sin[t]]`

Out[2]= `Which[t`$< \pi$`, 2, t`$< 2\pi$`, 0, t`$> 2\pi$`, Sin[t]]`

To explain: for $t < \pi$ the function equals 2. For π to 2π it equals zero. For $t > 2\pi$ it equals $\sin t$. See the figure. To represent **f** in terms of unit step functions, use the command

In[3]:= `f = 2 - 2 UnitStep[t - Pi] + UnitStep[t - 2 Pi] Sin[t]`

Out[3]= $2 + \text{Sin[t]}\,\text{UnitStep}[-2\pi + \text{t}] - 2\,\text{UnitStep}[-\pi + \text{t}]$

To plot f, type the following, where the (optional) command `AspectRatio -> Automatic` gives equal scales on both axes.

In[4]:= `Plot[f, {t, 0, 6 Pi}, AxesLabel -> {t, y}, AspectRatio -> Automatic,`
            `Ticks -> {{Pi, 2 Pi, 4 Pi}, {0, 1, 2}}]`

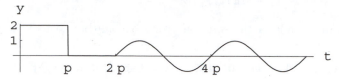

Example 5.4. "Piecewise function" (where $p = \pi$)

The transform of f is

In[5]:= `F = LaplaceTransform[f, t, s]`   (* Out: $\dfrac{2}{s} - \dfrac{2\,e^{-\pi s}}{s} + \dfrac{e^{-2\pi s}}{1+s^2}$ *)

Functions such as f may occur as driving forces in mechanics or as electromotive forces in circuits. For instance, the response of an undamped mass-spring system of mass 1 and spring constant 1 (initially at rest at equilibrium $y = 0$) to a single rectangular wave, say, $g(t) = 4$ if $2 < t < 10$ and 0 otherwise, is the solution of the ODE

In[6]:= `ode = y''[t] + y[t] == 4 UnitStep[t - 2]- 4 UnitStep[t - 10]`

Out[6]= $y[t] + y''[t] == -4\,\text{UnitStep}[-10+t] + 4\,\text{UnitStep}[-2+t]$

In[7]:= `yp = DSolve[{ode, y[0] == 0, y'[0] == 0}, y[t], t]`

Out[7]= $\{\{y[t] \rightarrow -4\,\text{UnitStep}[-10+t] + 4\,\text{Cos}[10-t]\,\text{UnitStep}[-10+t] +$

       $4\,\text{UnitStep}[-2+t] - 4\,\text{Cos}[2-t]\,\text{UnitStep}[-2+t]\}\}$

In[8]:= `Plot[yp[[1, 1, 2]], {t, 0, 20}, Ticks -> {{2, 5, 10, 15, 20}, Automatic}]`

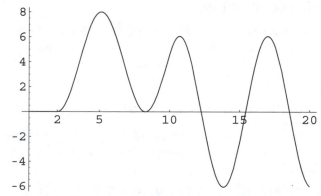

Example 5.4. Response of an undamped mass-spring system
to a single rectangular wave

You see that the motion begins at $t = 2$. Until $t = 10$ there is a transition period (sinusoidal), followed by the steady-state harmonic oscillation that begins at $t = 10$ when the driving force shuts off.

The **Dirac delta function** $\delta(t - a)$ (also known as the **unit impulse function**)

models an impulse at $t = a$ (practically, for instance, a hammer blow). The command is `DiracDelta[t - a]`.

In[9]:= `DiracDelta[t - a]`   (* Out: `DiracDelta[a−t]` *)

In[10]:= `LaplaceTransform[DiracDelta[t - a], t, s]`

Out[10]= e^{-as} `UnitStep [a]`

     For instance, if at $t = 1$ a hammer blow is imposed on an undamped mass-spring system of mass 1 and spring constant 1, the model and the solution (the displacement $y(t)$) are obtained as follows.

In[11]:= `ode = y''[t] + y[t] == DiracDelta[t - 1]`

Out[11]= `y[t] + y″[t] == DiracDelta[ − 1 + t]`

In[12]:= `sol = DSolve[{ode, y[0] == 0, y'[0] == 0}, y[t], t]`

Out[12]= `{{y[t] → −Sin [1 − t] UnitStep [ − 1 + t]}}`

In[13]:= `Plot[sol[[1, 1, 2]], {t, 0, 10}, Ticks -> {{0, 1, 5, 10}, Automatic},`
       `AxesLabel -> {t, y}]`

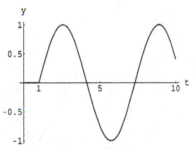

Example 5.4. Response of an undamped mass-spring system
to a hammer blow at $t = 1$

You see that the physical system remains at rest until the blow happens, and then immediately begins its steady-state harmonic oscillation.

     Similar Material in AEM: pp. 266-273

EXAMPLE 5.5 | SOLUTION OF SYSTEMS BY LAPLACE TRANSFORM

The process of solving systems of ODE's by Laplace transform is quite similar to that for single differential equations. Consider the following system, with unit step functions on the right, representing constant driving forces acting from $t = 0$ to $t = 1$ only. For instance,

In[1]:= `Clear[y1, y2]`

In[2]:= `sys = {y1'[t] == -y2[t] + 1 - UnitStep[t - 1],`
     `y2'[t] == y1[t] + 1 - UnitStep[t - 1]}`

Out[2]= `{y1′[t] == 1 − UnitStep [ − 1 + t] − y2[t],`

     `y2′[t] == 1 − UnitStep [ − 1 + t] + y1[t]}`

Applying `DSolve` and prescribing initial conditions, say, $y_1(0) = 0$, $y_2(0) = 0$, you obtain particular solutions, involving unit step functions,

In[3]:= `sol = DSolve[{sys[[1]], sys[[2]], y1[0] == 0, y2[0] == 0},`
    `{y1[t], y2[t]}, t]`

Out[3]= $\{\{$y1$[t] \rightarrow -1 + Cos[t] + Sin[t] + $UnitStep$[-1+t] - $

    Cos$[1-t]$ UnitStep$[-1+t] + Sin[1-t]$ UnitStep$[-1+t]$,

    y2$[t] \rightarrow 1 - Cos[t] + Sin[t] - $UnitStep$[-1+t] + $

    Cos$[1-t]$ UnitStep$[-1+t] + Sin[1-t]$ UnitStep$[-1+t]\}\}$

In[4]:= `Simplify[%]`

Out[4]= $\{\{$y1$[t] \rightarrow -1 + Cos[t] + Sin[t] + $

    $(1 - Cos[1-t] + Sin[1-t])$ UnitStep$[-1+t]$,

    y2$[t] \rightarrow 1 - Cos[t] + Sin[t] + $

    $(-1 + Cos[1-t] + Sin[1-t])$ UnitStep$[-1+t]\}\}$

In[5]:= `Plot[{sol[[1, 1, 2]], sol[[1, 2, 2]]}, {t, 0, 5}]`

Example 5.5. Solutions $y_1(t)$ (lower curve for t from 0 to 3) and
$y_2(t)$ of the second system `sys2`

At $t = 1$, where the driving force jumps from 1 down to 0, the curves have a cusp, where the derivative (the slope) suddenly decreases, as you can see.

    Similar Material in AEM: pp. 291-294

EXAMPLE 5.6	**FORMULAS ON GENERAL PROPERTIES OF THE LAPLACE TRANSFORM**

The Laplace transform has various general properties that are essential for its practical usefulness. In this example we collect some of the most important corresponding formulas.

In[1]:= `Clear[f, g, a]`

In[2]:= `LaplaceTransform[a f[t] + b g[t], t, s]`   `(* Linearity *)`

Out[2]= `a LaplaceTransform[f[t], t, s] + b LaplaceTransform[g[t], t, s]`

In[3]:= `LaplaceTransform[f'[t], t, s]`   `(* Differentiation *)`

Out[3]= $-$`f[0]` $+$ `s LaplaceTransform[f[t],t,s]`

In[4]:= `LaplaceTransform[D[f[t], t], t, s]`

Out[4]= $-$`f[0]` $+$ `s LaplaceTransform[f[t],t,s]`

In[5]:= `LaplaceTransform[D[f[t], t, t], t, s]`   `(* Second derivative *)`

Out[5]= $-$`s f[0]` $+$ `s`^{2} `LaplaceTransform[f[t],t,s]` $-$ `f`$'$`[0]`

Thus, roughly speaking, differentiation of a function $f(t)$ corresponds to the multiplication of its transform by s. And integration of $f(t)$ from 0 to t (with the variable of integration denoted by τ) corresponds to division of the transform by s, namely,

In[6]:= `LaplaceTransform[Integrate[f[tau], {tau, 0, t}], t, s]`

Out[6]= $\dfrac{\texttt{LaplaceTransform[f[t],t,s]}}{\texttt{s}}$

Multiplying a function by e^{at} corresponds to replacing s by $s - a$ in the transform. This is called **s-shifting**. For instance,

In[7]:= `LaplaceTransform[Exp[a t] Cos[t], t, s]`

Out[7]= $\dfrac{1}{\left(1 + \dfrac{1}{(\texttt{a}-\texttt{s})^2}\right)(-\texttt{a}+\texttt{s})}$

In[8]:= `FullSimplify[%]`

Out[8]= $\dfrac{-\texttt{a}+\texttt{s}}{1+(\texttt{a}-\texttt{s})^2}$

Multiplying a transform by e^{-as} corresponds to replacing t by $t - a$ in the function and to equating it to 0 for $t < a$. This is called **t-shifting**. For instance,

In[9]:= `InverseLaplaceTransform[Exp[-a s] LaplaceTransform[Cos[t], t, s], s, t]`

Out[9]= `Cos [a` $-$ `t] UnitStep [` $-$ `a` $+$ `t]`

In[10]:= `D[LaplaceTransform[f[t], t, s], s]`

Out[10]= $-$`LaplaceTransform[t f[t],t,s]`

Hence differentiating the transform of a function $f(t)$ corresponds to taking the transform of $t f(t)$ (and multiplying it by -1), that is, $\mathcal{L}(t f(t)) = -F'(s)$. Taking the inverse transform on both sides (and interchanging the two sides), you thus have $\mathcal{L}^{-1}(F'(s)) = -t f(t)$; indeed,

In[11]:= `InverseLaplaceTransform[D[LaplaceTransform[f[t], t, s], s], s, t]`

Out[11]= $-$`t f[t]`

Finally, taking the product of two transforms corresponds to taking the transform of the **convolution** (the integral shown) of the two corresponding functions,

In[12]:= `LaplaceTransform[Integrate[f[tau] g[t - tau], {tau, 0, t}], t, s]`

Out[12]= `LaplaceTransform[f[t],t,s] LaplaceTransform[g[t],t,s]`

For instance, if $f(t) = \cos t$ and $g(t) = e^t$, their transforms are $s/(s^2+1)$ and $1/(s-1)$, and you should get the product of these two transforms on the right; indeed,

In[13]:= `LaplaceTransform[Integrate[Cos[tau] Exp[t - tau], {tau, 0, t}], t, s]`

Out[13]= $\dfrac{1}{2\,(-1+s)} + \dfrac{1}{2\,(1+s^2)} - \dfrac{s}{2\,(1+s^2)}$

In[14]:= Simplify[%]   (* Out: $\dfrac{s}{(-1+s)\,(1+s^2)}$ *)

Similar Material in AEM: p. 296

Problem Set for Chapter 5

Pr.5.1 **(Transform)** Find $\mathcal{L}(\sin \pi t)$ by evaluating the defining integral.
(*AEM Ref.* p. 257 (#3))

Pr.5.2 **(Transform by integration)** Find $\mathcal{L}(\cos^2 \omega t)$ by evaluating the defining integral of the transform. (*AEM Ref.* p. 257 (#4))

Pr.5.3 **(Transform by integration)** Find the transform of $f(t) = k$ if $0 < t < c$, $f(t) = 0$ otherwise, by integration. (*AEM Ref.* p. 257 (#11))

Pr.5.4 **(Inverse transform)** Using InverseLaplaceTransform, find the function whose transform is $(s - 4)/(s^2 - 4)$. (*AEM Ref.* p. 257 (#20))

Pr.5.5 **(Inverse transform)** Find the inverse transform of $\dfrac{s}{L^2 s^2 + n^2 \pi^2}$.
(*AEM Ref.* p. 257 (#23))

Pr.5.6 **(Initial value problem, subsidary equation)** Find the solution of the initial value problem $y' - 5y = 1.5\,e^{-4t}$, $y(0) = 1$ by obtaining the subsidiary equation, solving it, and transforming the solution back. (*AEM Ref.* p. 264 (#2))

Pr.5.7 **(Initial value problem, subsidiary equation)** Solve the initial value problem $y'' + 2y' - 3y = 6e^{-2t}$, $y(0) = 2$, $y'(0) = -14$ by using the subsidiary equation.
(*AEM Ref.* p. 264 (#9))

Pr.5.8 **(*t*-shifting)** Plot the function $4u(t - \pi)\cos t$ and find its transform.
(*AEM Ref.* p. 273 (#7))

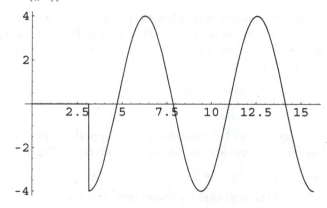

Problem 5.8. Given function $f(t)$

Pr.5.9 **(Unit step function)** Find and plot the inverse of $3(1 - e^{-\pi s})/(s^2 + 9)$.
(*AEM Ref.* p. 273 (#17))

Pr.5.10 (**Unit step function**) Find and plot the inverse transform of $e^{-2\pi s}/(s^2 + 2s + 2)$. (*AEM Ref.* p. 273 (#18))

Pr.5.11 (***RC*-circuit, unit step**) Find the subsidiary equation of the equation

$$R\,i(t) + \frac{1}{C}\int_0^t i(\tau)\,d\tau = v(t), \qquad i(0) = 0, \qquad i'(0) = 0$$

for the current $i(t)$ in an *RC*-circuit, assuming that $v(t) = K = const$ for t from 1 to 3 and $v(t) = 0$ otherwise. Find $i(t)$. Plot $i(t)$ when $R = 1$ ohm, $C = 1$ farad, and $K = 110$ volts. (*AEM Ref.* p. 274)

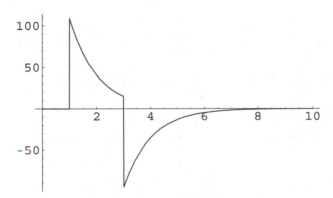

Problem 5.11. Current in an *RC*-circuit with a single rectangular wave as electromotive force

Pr.5.12 (***RC*-circuit, Dirac's delta**) Solve Pr.5.11 when $v(t) = K\,\delta(t-1)$, the other data being as before. (*AEM Ref.* p. 271)

Pr.5.13 (**Experiment on repeated Dirac's delta**) Solve $y'' + y = \delta(t-\pi) - \delta(t - 2\pi)$, $y(0) = 0$, $y'(0) = 1$. First guess what the solution may look like. Then solve and plot. What will happen if you add further terms $\delta(t - 3\pi) - \delta(t - 4\pi) \pm \dots$? (*AEM Ref.* p. 274 (#27)).

Pr.5.14 (***RL*-circuit**)   Using `LaplaceTransform`, solve $L\,i' + R\,i = v(t)$, $i(0) = 0$, where $v(t) = \sin t$ if $0 < t < 2\pi$ and 0 otherwise. Plot the solution with $R = 1$, $L = 1$ and comment. (*AEM Ref.* p. 274 (#32))

Pr.5.15 (**Full-wave rectifier**) Plot the function obtained by the full-wave rectification of $\sin t$, that is, by multiplying the negative half-waves by -1. Let t go from 0 to 5π. (*AEM Ref.* p. 290 (#16))

PART B. LINEAR ALGEBRA, VECTOR CALCULUS

Content. Matrices, vectors, determinants, linear systems of equations (Chap. 6)
Matrix eigenvalue problems (Chap. 7)
Vectors in R^2 and R^3, dot and cross products, grad, div, curl (Chap. 8)
Vector integral calculus, integral theorems (Chap. 9)

Chapter 6

Matrices, Vectors, Determinants.
Linear Systems of Equations

Content. Addition, scalar multiplication, matrix multiplication, determinants
(Ex. 6.1, Prs. 6.1-6.6)
Special matrices, composing matrices (Exs. 6.2, 6.3, Prs. 6.7, 6.9, 6.17)
Transpose, inverse, linear transformations (Prs. 6.8, 6.10-6.12)
Orthogonality, norm (Prs. 6.13-6.15)
Solution of linear systems of equations (Exs. 6.4-6.6, Pr. 6.14)
Rank of a matrix, row space, linear independence (Prs. 6.16, 6.18-6.20)

Numerical methods for matrices see Chap 18.

Examples for Chapter 6

EXAMPLE 6.1 | MATRIX ADDITION, SCALAR MULTIPLICATION, MATRIX MULTIPLICATION. VECTORS

In[1]:= `A = {{5, -3, 0}, {6, 1, -4}}`

Out[1]= {{5, −3, 0}, {6, 1, −4}}

This shows how to type a matrix. You may equally well type the six entries in two rows, the response remaining the same.

In[2]:= `A = {{5, -3,  0},`
         `{6,  1, -4}}`

To display the response as an array in the usual matrix form, type

In[3]:= `%//MatrixForm`   (* Out: $\begin{pmatrix} 5 & -3 & 0 \\ 6 & 1 & -4 \end{pmatrix}$ *)

66

More compactly, you can type

In[4]:= `A = {{5, -3,  0}, {6,  1, -4}}; MatrixForm[A]`

Out[4]//MatrixForm=

$$\begin{pmatrix} 5 & -3 & 0 \\ 6 & 1 & -4 \end{pmatrix}$$

Addition of matrices. Let **A** be as before, and let

In[5]:= `B = {{-2, 4, -1}, {1, 1, 5}}`   `(* Out:  {{−2, 4, −1}, {1, 1, 5}} *)`

The sum is typed with a + sign,

In[6]:= `C = A + B`

        `Set::wrsym : Symbol C is Protected`

Out[6]= {{3, 1, −1}, {7, 2, 1}}

The error message indicates that you cannot use `C` as a notation for the sum because `C` is protected. Use some other symbol, for instance, `C1`.

In[7]:= `C1 = A + B`   `(* Out:  {{3, 1, −1}, {7, 2, 1}} *)`

In[8]:= `%//MatrixForm`   `(* Out:` $\begin{pmatrix} 3 & 1 & -1 \\ 7 & 2 & 1 \end{pmatrix}$ `*)`

Scalar multiplication. Transposition. Multiplication of matrices. The **product** of a matrix by a scalar is typed with or without an asterisk. For instance,

In[9]:= `4 C1`   `(* Out:  {{12, 4, −4}, {28, 8, 4}} *)`

In[10]:= `4*C1`   `(* Out:  {{12, 4, −4}, {28, 8, 4}} *)`

The product of two matrices is typed with a dot. For instance, for the above matrices **A** and **B**, type

In[11]:= `A.B`

        `Dot::dotsh : Tensors {{5,-3,0},{6,1,-4}} and {{-2,4,-1},{1,1,5}} have`
            `incompatible shapes`

Out[11]= {{5, −3, 0}, {6, 1, −4}} . {{−2, 4, −1}, {1, 1, 5}}

The error message reminds you that the product of a 2×3 matrix by another 2×3 matrix is not defined. But you can multiply **A** by the transpose of **B**, which is a 3×2 matrix.

In[12]:= `F = Transpose[B]`   `(* Transpose of a matrix *)`

Out[12]= {{−2, 1}, {4, 1}, {−1, 5}}

In[13]:= `%//MatrixForm`   `(* Out:` $\begin{pmatrix} -2 & 1 \\ 4 & 1 \\ -1 & 5 \end{pmatrix}$ `*)`

In[14]:= `G = A.F`   `(* Product of two matrices *)`

Out[14]= {{−22, 2}, {−4, −13}}

In[15]:= `%//MatrixForm`

$$\text{Out[15]}= \begin{pmatrix} -22 & 2 \\ -4 & -13 \end{pmatrix}$$

Another command for the product of two matrices is `Dot[ , ]`, but do not call **AF** a dot product; call it a *matrix product* and use the term *dot product* only for vectors (see below).

In[16]:= `Dot[A, F]`   `(* Out: {{-22, 2}, {-4, -13}} *)`

To obtain $\mathbf{G}^2$, try

In[17]:= `G^2`   `(* Wrong *)`

Out[17]= {{484, 4}, {16, 169}}

You see that this is not the square of **G** but the matrix whose entries are the squares of the entries of **G**. To obtain $\mathbf{G}^2$, type

In[18]:= `G.G`   `(* Square of a matrix *)`

Out[18]= {{476, -70}, {140, 161}}

Inverse of a matrix. Determinant. Trace

In[19]:= `GI = Inverse[G]; MatrixForm[GI]`   `(* Inverse of a matrix *)`

Out[19]//MatrixForm=

$$\begin{pmatrix} -\dfrac{13}{294} & -\dfrac{1}{147} \\ \dfrac{2}{147} & -\dfrac{11}{147} \end{pmatrix}$$

To check this response, type

In[20]:= `GI.G//MatrixForm`   `(* Out:` $\begin{pmatrix} 1 & 0 \\ 0 & 1 \end{pmatrix}$ `*)`

The **determinant** of **G** and the trace of **G** are obtained by typing

In[21]:= `Det[G]`   `(* Out: 294 *)`

In[22]:= `Tr[G]`   `(* trace -22 - 13 = -35 *)`

Out[22]= -35

Vectors occur together with matrices, particularly in connection with linear systems of equations. Let **A** be as before and $\mathbf{v} = [3\ 5]$. Then the 1×3 matrix (the row vector) `vA` is obtained by typing

In[23]:= `v = {3, 5}`   `(* Out: {3, 5} *)`

In[24]:= `v.A`   `(* Out: {45, -4, -20} *)`

or equallly well by the command `Dot[v, A]` (which is not a dot product of two vectors in the usual sense, as has been noted before),

In[25]:= `Dot[v, A]`   `(* Out: {45, -4, -20} *)`

The product **A.v** is undefined. Indeed,

In[26]:= `A.v`

`Dot::dotsh : Tensors {{5,-3,0},{6,1,-4}} and {3,5} have incompatible shapes`

Out[26]= {{5, −3, 0}, {6, 1, −4}}.{3, 5}

The **inner product (dot product)** of the vector **v** and another vector, say,

In[27]:= u = {4, -2}   (* Out: {4, −2} *)

is obtained by either of the following two commands.

In[28]:= u.v   (* Out: 2 *)

In[29]:= Dot[u, v]   (* Out: 2 *)

Similarly, you obtain the matrix **vG** (a row vector with 2 components) by either of the following two commands.

In[30]:= v.G   (* Out: {−86, −59} *)

In[31]:= Dot[v, G]   (* Out: {−86, −59} *)

Similar Material in AEM: pp. 307-309, 311-313

EXAMPLE 6.2   SPECIAL MATRICES

Matrices of the subsequent form will be needed quite frequently.

In[1]:= <<LinearAlgebra'MatrixManipulation'   (* Load the package. *)

In[2]:= ZeroMatrix[3, 5]//MatrixForm   (* Out: $\begin{pmatrix} 0 & 0 & 0 & 0 & 0 \\ 0 & 0 & 0 & 0 & 0 \\ 0 & 0 & 0 & 0 & 0 \end{pmatrix}$ *)

In[3]:= Table[0, {3}, {5}]//MatrixForm   (* Out: $\begin{pmatrix} 0 & 0 & 0 & 0 & 0 \\ 0 & 0 & 0 & 0 & 0 \\ 0 & 0 & 0 & 0 & 0 \end{pmatrix}$ *)

In[4]:= DiagonalMatrix[{a, b, c}]   (* A **diagonal matrix** *)

Out[4]= {{a, 0, 0}, {0, b, 0}, {0, 0, c}}

In[5]:= DiagonalMatrix[{1, 1, 1}]   (* The 3×3 **unit matrix** *)

Out[5]= {{1, 0, 0}, {0, 1, 0}, {0, 0, 1}}

Matrices whose entries are given by a formula of the subscripts of the entries can be obtained as illustrated by a famous example (the 3×3 **Hilbert matrix**):

In[6]:= Table[1/(j + k - 1), {j, 1, 3}, {k, 1, 3}]

Out[6]= $\left\{\left\{1, \frac{1}{2}, \frac{1}{3}\right\}, \left\{\frac{1}{2}, \frac{1}{3}, \frac{1}{4}\right\}, \left\{\frac{1}{3}, \frac{1}{4}, \frac{1}{5}\right\}\right\}$

In[7]:= %//MatrixForm

Out[7]//MatrixForm=

$$\begin{pmatrix} 1 & \frac{1}{2} & \frac{1}{3} \\ \frac{1}{2} & \frac{1}{3} & \frac{1}{4} \\ \frac{1}{3} & \frac{1}{4} & \frac{1}{5} \end{pmatrix}$$

Similar Material in AEM: p. 314

EXAMPLE 6.3   CHANGING AND COMPOSING MATRICES, ACCESSING ENTRIES. SUBMATRICES

Having discussed the basic operations with matrices and vectors in Example 6.1, we now turn to operations of accessing entries and of accessing, interchanging or changing rows or columns of a matrix.

In[1]:= <<LinearAlgebra'MatrixManipulation'   (* Load the package. *)

In[2]:= A = {{0, 1, -1, 2}, {1, -2, 4, -3}, {3, -4, 0, 0}}

Out[2]= {{0, 1, −1, 2}, {1, −2, 4, −3}, {3, −4, 0, 0}}

In[3]:= %//MatrixForm

Out[3]//MatrixForm=

$$\begin{pmatrix} 0 & 1 & -1 & 2 \\ 1 & -2 & 4 & -3 \\ 3 & -4 & 0 & 0 \end{pmatrix}$$

To extract (access) an **entry** a_{jk} of $\mathbf{A} = [a_{jk}]$, for instance, a_{23}, type

In[4]:= A[[2,3]]   (* Out: 4 *)

To extract a whole **row**, for instance, the third row of A, type

In[5]:= A[[3]]   (* Out: {3, −4, 0, 0} *)

To extract (access) a submatrix, for instance, the **submatrix** whose entries belong to Rows 2 and 3 and simultaneously to Columns 2, 3, 4 of A, type

In[6]:= A[[{2, 3}, {2, 3, 4}]] // MatrixForm

Out[6]//MatrixForm=

$$\begin{pmatrix} -2 & 4 & -3 \\ -4 & 0 & 0 \end{pmatrix}$$

This command can also be written in terms of Range, namely

In[7]:= A[[Range[2, 3], Range[2, 4]]]//MatrixForm

Out[7]//MatrixForm=

$$\begin{pmatrix} -2 & 4 & -3 \\ -4 & 0 & 0 \end{pmatrix}$$

Here, Range[2, 4] stands for {2, 3, 4}. And if you type Range[3], you obtain {1, 2, 3}. Try it.

Furthermore, single **columns** of **A** can be accessed as follows.

In[8]:= A[[{1, 2, 3}, 1]]   (* First column {0, 1, 3} *)

In[9]:= A[[{1, 2, 3}, 2]]   (* Second column {1, -2, -4}. Etc. *)

Changing rows and columns

You can insert a row r into a given matrix A by the command Insert. For instance (using A from before),

In[10]:= `A//MatrixForm`   (* Out: $\begin{pmatrix} 0 & 1 & -1 & 2 \\ 1 & -2 & 4 & -3 \\ 3 & -4 & 0 & 0 \end{pmatrix}$ *)

In[11]:= `r = {5, 0, -6, 1}`   (* Out: {5, 0, −6, 1} *)

`r` becomes the second row of the new matrix if you type

In[12]:= `Insert[A, r, 2] //MatrixForm`   (* Out: $\begin{pmatrix} 0 & 1 & -1 & 2 \\ 5 & 0 & -6 & 1 \\ 1 & -2 & 4 & -3 \\ 3 & -4 & 0 & 0 \end{pmatrix}$ *)

The command for **deleting** a row is `Delete`. For instance, if you want to delete the third row of the new matrix, type

In[13]:= `Delete[%, 3]//MatrixForm`   (* Out: $\begin{pmatrix} 0 & 1 & -1 & 2 \\ 5 & 0 & -6 & 1 \\ 3 & -4 & 0 & 0 \end{pmatrix}$ *)

You can use the commands `Insert` and `Delete` for **interchanging (swapping)** rows. This process is important, for instance, in solving linear systems of equations. Suppose you want to interchange Rows 1 and 3 of **A**. Then do the following. Insert Row 3 of **A**, that is, `A[[3]]`, as new Row 1, and Row 1 of **A**, that is, `A[[1]]`, as Row 3 of the new 5×4 matrix **F** thus obtained.

In[14]:= `B = Insert[A, A[[3]], 1];MatrixForm[B]` (* Out: $\begin{pmatrix} 3 & -4 & 0 & 0 \\ 0 & 1 & -1 & 2 \\ 1 & -2 & 4 & -3 \\ 3 & -4 & 0 & 0 \end{pmatrix}$ *)

In[15]:= `F = Insert[B, A[[1]], 4];MatrixForm[F]` (* Out: $\begin{pmatrix} 3 & -4 & 0 & 0 \\ 0 & 1 & -1 & 2 \\ 1 & -2 & 4 & -3 \\ 0 & 1 & -1 & 2 \\ 3 & -4 & 0 & 0 \end{pmatrix}$ *)

The insertion is in Row 4, not 3, because of the first insertion. You see that Rows 2 and 5 should be deleted. Note that Row 5 becomes Row 4 after the first deletion. Accordingly, type

In[16]:= `Delete[F, 2]`

Out[16]= {{3, −4, 0, 0}, {1, −2, 4, −3}, {0, 1, −1, 2}, {3, −4, 0, 0}}

In[17]:= `Delete[%, 4]//MatrixForm`   (* Out: $\begin{pmatrix} 3 & -4 & 0 & 0 \\ 1 & -2 & 4 & -3 \\ 0 & 1 & -1 & 2 \end{pmatrix}$ *)

You see that Rows 1 and 3 of `A` have been interchanged (swapped). You can also swap more simply by using the command `ReplacePart`, as follows.

In[18]:= `ReplacePart[A, A[[1]], 3]`

In[19]:= `ReplacePart[%, A[[3]], 1]//MatrixForm`

Out[19]//MatrixForm=

$$\begin{pmatrix} 3 & -4 & 0 & 0 \\ 1 & -2 & 4 & -3 \\ 0 & 1 & -1 & 2 \end{pmatrix}$$

You may combine the two commands, typing

In[20]:= `ReplacePart[ReplacePart[A, A[[1]], 3], A[[3]], 1]`

Out[20]= {{3, −4, 0, 0}, {1, −2, 4, −3}, {0, 1, −1, 2}}

This suggests defining a **module** for swapping any two rows r1 and r2 , as follows.

In[21]:= `SwapRow[A_, r1_, r2_] := Module[{t1, t2},`
    `t1 = A[[r1]]; t2 = A[[r2]];`
    `ReplacePart[ReplacePart[A, t1, r2], t2, r1]]`

The {t1, t2} indicates that t1 and t2 are **local variables**. That is, they are known only within the scope of the module. Any values for t1 and t2 assigned during the current session of Mathematica are not available within the module, and the values assigned within the module are not available outside the module.

In[22]:= `SwapRow[A, 1, 3]//MatrixForm`   (* Out: $\begin{pmatrix} 3 & -4 & 0 & 0 \\ 1 & -2 & 4 & -3 \\ 0 & 1 & -1 & 2 \end{pmatrix}$ *)

In[23]:= `SwapRow[A, 2, 1]//MatrixForm`   (* Out: $\begin{pmatrix} 1 & -2 & 4 & -3 \\ 0 & 1 & -1 & 2 \\ 3 & -4 & 0 & 0 \end{pmatrix}$ *)

You will need this in the Gauss elimination. From this module you obtain a module for swapping *columns* by noting that the columns of **A** are the rows of its transpose.

In[24]:= `SwapCol[A_, c1_, c2_] := Module[{},`
    `Transpose[SwapRow[Transpose[A], c1, c2]]]`

The {} indicates that there are no local variables.

In[25]:= `SwapCol[A, 2, 3]//MatrixForm`   (* Out: $\begin{pmatrix} 0 & -1 & 1 & 2 \\ 1 & 4 & -2 & -3 \\ 3 & 0 & -4 & 0 \end{pmatrix}$ *)

Composition of matrices from vectors

A matrix with given vectors as row vectors can be obtained by the command `Join`. For instance,

In[26]:= `a = {{2, 1}};   b = {{3, 8}};   c = {{-1, 2}};`

In[27]:= `Join[a, b, c]//MatrixForm`   (* Out: $\begin{pmatrix} 2 & 1 \\ 3 & 8 \\ -1 & 2 \end{pmatrix}$ *)

In[28]:= `Transpose[%]//MatrixForm`   (* Out: $\begin{pmatrix} 2 & 3 & -1 \\ 1 & 8 & 2 \end{pmatrix}$ *)

More generally, by the command Join you can compose matrices and vectors. For instance, let

In[29]:= B = {{-2, 4, -1}, {1, 1, 5}};

In[30]:= Join[B, {{1, 3, -3}}]

Out[30]= {{−2, 4, −1}, {1, 1, 5}, {1, 3, −3}}

As a particularly important case, the **augmented matrix** of a matrix **A** (see above) and a column vector **b**, say, $\mathbf{b} = [4\ 5\ 2]^T$ can be obtained as follows.

In[31]:= A = {{0, 1, -1, 2}, {1, -2, 4, -3}, {3, -4, 0, 0}};

In[32]:= b = {4, 5, 2}   (* Out: {4, 5, 2} *)

In[33]:= %//MatrixForm $\left(\ast\ \text{Out:}\ \begin{pmatrix} 4 \\ 5 \\ 2 \end{pmatrix}\ \ast\right)$

In[34]:= Transpose[Join[Transpose[A], {b}]]//MatrixForm

Out[34]//MatrixForm=

$$\begin{pmatrix} 0 & 1 & -1 & 2 & 4 \\ 1 & -2 & 4 & -3 & 5 \\ 3 & -4 & 0 & 0 & 2 \end{pmatrix}$$

Similar Material in AEM: pp. 322-326

EXAMPLE 6.4   **SOLUTION OF A LINEAR SYSTEM**

Solve the linear system

$$\begin{aligned} -x_1 &+ x_2 + 2x_3 = 2 \\ 3x_1 &- x_2 + x_3 = 6 \\ -x_1 &+ 3x_2 + 4x_3 = 4 \end{aligned}$$

Solution. First method. Write the system in matrix form $\mathbf{Ax} = \mathbf{b}$ and type the **coefficient matrix A** and the vector **b** as shown. Then apply the command LinearSolve.

In[1]:= A = {{-1, 1, 2}, {3, -1, 1}, {-1, 3, 4}}; MatrixForm[A]

Out[1]//MatrixForm=

$$\begin{pmatrix} -1 & 1 & 2 \\ 3 & -1 & 1 \\ -1 & 3 & 4 \end{pmatrix}$$

In[2]:= b = {2, 6, 4}   (* Out: {2, 6, 4} *)

In[3]:= x = LinearSolve[A, b]   (* Out: {1, −1, 2} *)

Second method. In this method, the augmented matrix [**A b**] is row-reduced, so that the solution can then be read immediately. The command Join, applied to the transpose, produces the augmented matrix.

In[4]:= A1 = Transpose[Join[Transpose[A], {b}]]

Out[4]= {{−1, 1, 2, 2}, {3, −1, 1, 6}, {−1, 3, 4, 4}}

In[5]:= %//MatrixForm   (* Out: $\begin{pmatrix} -1 & 1 & 2 & 2 \\ 3 & -1 & 1 & 6 \\ -1 & 3 & 4 & 4 \end{pmatrix}$ *)

In[6]:= B = LinearSolve[A, b]   (* Out: {1, −1, 2} *)

In[7]:= RowReduce[A1]//MatrixForm   (* Out: $\begin{pmatrix} 1 & 0 & 0 & 1 \\ 0 & 1 & 0 & -1 \\ 0 & 0 & 1 & 2 \end{pmatrix}$ *)

From Row 1 you see that $x_1 = 1$, from Row 2 that $x_2 = −1$, and from Row 3 that $x_3 = 2$.

Similar Material in AEM: p. 327

EXAMPLE 6.5   **LINEAR SYSTEMS: A FURTHER CASE**

Infinitely many solutions. The system is

$$1\,x_1 + 1\,x_2 = 3$$
$$0.3\,x_1 + 0.3\,x_2 = 0.9$$

Obviously, the solutions are $x_1 = 3 − x_2$, x_2 arbitrary. Now try to obtain these solutions by LinearSolve, as in the previous example. Type

In[1]:= A = {{1, 1}, {0.3, 0.3}}; MatrixForm[A]

Out[1]//MatrixForm=

$$\begin{pmatrix} 1 & 1 \\ 0.3 & 0.3 \end{pmatrix}$$

In[2]:= b = {3, 0.9}   (* Out: {3, 0.9} *)

The following two results have been observed. Correct (but not complete) is

In[3]:= LinearSolve[A, b]   (* Out: {3., −1.85037 × 10^{-16}} *)

that is, $x_1 = 3$, $x_2 = 0$ (except for round-off error); this is the solution $x_1 = 3 − x_2$ with $x_2 = 0$.

The error message that the system has no solutions at all is not correct. In the present case the discovery of the disaster was easy. In real-life problems it may remain unnoticed. Hence ***checking of computer results is important***. If you suspect a failure, you may be able to avoid it by using ordinary fractions in **A** and **b**. In the present case, try (using other symbols)

In[4]:= A2 = {{1, 1}, {3/10, 3/10}}

Out[4]= {{1, 1}, {$\frac{3}{10}$, $\frac{3}{10}$}}

In[5]:= b2 = {3, 9/10}   (* Out: {3, $\frac{9}{10}$} *)

In[6]:= LinearSolve[A2, b2]   (* Out: {3, 0} *)

This is one of the infinitely many solutions, namely, the solution obtained by taking the arbitrary x_2 equal to zero. This is typical of other cases, too.

Similar Material in AEM: p. 327

EXAMPLE 6.6   GAUSS ELIMINATION; BACK SUBSTITUTION

Consider the linear system $\mathbf{Ax} = \mathbf{b}$, where

In[1]:= A = {{3, 2, 2, -5}, {6/10, 15/10, 15/10, -54/10},
           {12/10, -3/10, -3/10, 24/10}}; MatrixForm[A]

Out[1]//MatrixForm=

$$\begin{pmatrix} 3 & 2 & 2 & -5 \\ \dfrac{3}{5} & \dfrac{3}{2} & \dfrac{3}{2} & -\dfrac{27}{5} \\ \dfrac{6}{5} & -\dfrac{3}{10} & -\dfrac{3}{10} & \dfrac{12}{5} \end{pmatrix}$$

and

In[2]:= b = {8, 27/10, 21/10}   (* Out: $\left\{8, \frac{27}{10}, \frac{21}{10}\right\}$ *)

In[3]:= x = LinearSolve[A, b]   (* Out: {2, 1, 0, 0} *)

You will see that the system has infinitely many solutions, with two of the four unknowns arbitrary, and you obtain the solution [2, 1, 0, 0] if you equate the two arbitrary constants to zero. Solve the system by Gauss elimination (without pivoting) and back substitution.

Solution. First we need the augmented matrix

In[4]:= B = Transpose[Join[Transpose[A],{b}]]

Out[4]= $\left\{\{3, 2, 2, -5, 8\}, \left\{\dfrac{3}{5}, \dfrac{3}{2}, \dfrac{3}{2}, -\dfrac{27}{5}, \dfrac{27}{10}\right\}, \left\{\dfrac{6}{5}, -\dfrac{3}{10}, -\dfrac{3}{10}, \dfrac{12}{5}, \dfrac{21}{10}\right\}\right\}$

Now comes the Gauss elimination program, with some explanations afterward.

```
In[5]:= m = Dimensions[B][[1]]
    Do [
      Do [
        Print[B = ReplacePart[B,
              B[[j]] - B[[k]] B[[j,k]]/B[[k,k]], j]],
            {j, k + 1, m}
        ],
      {k, 1, m - 1}
      ]
```

Out[5]= 3

$$\left\{\{3, 2, 2, -5, 8\}, \left\{0, \frac{11}{10}, \frac{11}{10}, -\frac{22}{5}, \frac{11}{10}\right\}, \left\{\frac{6}{5}, -\frac{3}{10}, -\frac{3}{10}, -\frac{12}{5}, -\frac{21}{10}\right\}\right\}$$

$$\left\{\{3, 2, 2, -5, 8\}, \left\{0, \frac{11}{10}, \frac{11}{10}, -\frac{22}{5}, \frac{11}{10}\right\}, \left\{0, \frac{11}{10}, \frac{11}{10}, -\frac{22}{5}, \frac{11}{10}\right\}\right\}$$

$$\left\{\{3, 2, 2, -5, 8\}, \left\{0, \frac{11}{10}, \frac{11}{10}, -\frac{22}{5}, \frac{11}{10}\right\}, \{0, 0, 0, 0, 0\}\right\}$$

Explanations. Dimensions[B] gives $\{m, n+1\}$ the size $m \times (n+1)$ of the augmented matrix B. Hence Dimensions[B][[1]] gives m, the first of the two elements of the list.

The program has two do-loops. Each begins with a bracket [and ends with a closing bracket] . The two closing brackets stand on lines by themselves. The first do-loop is called the *outer do-loop* and the second the *inner do-loop* because it stands within the outer.

Print causes all the intermediate steps to be shown. You could drop the command Print[...] if this were not wanted.

ReplacePart has been used in Example 6.3. ReplacePart[expr, new, j] replaces the jth part of expr by new; that is the jth row of B by the new row obtained according to the Gauss elimination. B[[j]] is the jth row of the current matrix.

$\{j, \ k+1, \ m\}$ means that j runs from $k+1$ to m in steps of size 1. It runs in the inner do-loop through rows. Similarly, $\{k, \ 1, \ m-1\}$ shows that k runs from 1 to $m-1$. It runs in the outer do-loop through columns. Indeed, this is how the Gauss elimination proceeds.

Namely, in each step you stick to a certain row (the pivot row), and you eliminate a certain unknown from all the equations below the pivot equation. When this is done, you go to the next row (the next value of k) and eliminate the next unknown from the equations below the next pivot equation, and so on. Through all these changes the notation for the matrix remains the same, namely, B .

Back substitution. The program for back substitution involves a single do-loop. t1, ..., t4 are notations for the unknowns, some of which may remain arbitrary, as for your present system. $\{j, m, 1, -1\}$ in the program means that j **decreases** from m to 1 in steps of size 1.

```
In[6]:= x = {t1, t2, t3, t4}
    m = Dimensions[A][[1]]              (* Number of rows of A *)
    n = Dimensions[A][[2]]              (* Number of columns of A *)
    Do [
        If[B[[j,j]] ≠ 0,               (* Type the ≠ as ! followed by =. *)
          x[[j]] = (B[[j,n + 1]] -
                  Sum[B[[j,k]] x[[k]], {k, j + 1, n}])/B[[j,j]]],
              {j, m, 1, - 1}
          ];
    x
```

Out[6]= {t1, t2, t3, t4}

Out[6]= 3

Out[6]= 4

Out[6]= $\left\{ \frac{1}{3}\left(8 - 2\,t3 + 5\,t4 - \frac{20}{11}\left(\frac{11}{10} - \frac{11\,t3}{10} + \frac{22\,t4}{5} \right) \right), \ \frac{10}{11}\left(\frac{11}{10} - \frac{11\,t3}{10} + \frac{22\,t4}{5} \right), \ t3, \ t4 \right\}$

In[7]:= Simplify[x]

Out[7]= $\{2 - t4, \ 1 - t3 + 4\,t4, \ t3, \ t4\}$

You see that $x_3 = t_3$ and $x_4 = t_4$ remain arbitrary.

Similar Material in AEM: pp. 327-329

Problem Set for Chapter 6

Pr.6.1 **(Addition, scalar multiplication)** In Example 6.1 of this Guide, compute $3\mathbf{A}$, $\mathbf{A} - \mathbf{B}$, and $(2\mathbf{A} - (1/2)\mathbf{B})^T$. (*AEM Ref.* p. 308)

Pr.6.2 **(Matrix multiplication)** In Example 6.1 compute $\mathbf{A}\mathbf{A}^T$, $\mathbf{A}^T\mathbf{A}$, $(\mathbf{A}^T\mathbf{A})^2$, $(\mathbf{A} + \mathbf{B})(\mathbf{A} - \mathbf{B})^T$. (*AEM Ref.* pp. 311-313)

Pr.6.3 **(Matrix multiplication)** Let

$$\mathbf{A} = \begin{bmatrix} 1 & 3 & 2 \\ 3 & 5 & 0 \\ 2 & 0 & 4 \end{bmatrix}, \qquad \mathbf{B} = \begin{bmatrix} 0 & 2 & 1 \\ -2 & 0 & -3 \\ -1 & 3 & 0 \end{bmatrix}, \qquad \mathbf{c} = \begin{bmatrix} 1 \\ 0 \\ -2 \end{bmatrix}, \qquad \mathbf{d} = \begin{bmatrix} 3 \\ 1 \\ 2 \end{bmatrix}.$$

Compute $\mathbf{A}^2$, $\mathbf{A}^4$, $\mathbf{B}^2$, $\mathbf{B}^4$, $\mathbf{A}\mathbf{B} - \mathbf{B}\mathbf{A}$, $\mathbf{A} - \mathbf{A}^T$, $\mathbf{B} + \mathbf{B}^T$, $\det \mathbf{A}$, $\det \mathbf{B}$, $\mathbf{A}\mathbf{c}$, $\mathbf{B}(\mathbf{c} - 3\mathbf{d})$, $\mathbf{c} \bullet \mathbf{d}$. (*AEM Ref.* pp. 311-316)

Pr.6.4 **(Matrices and vectors)** Using the data given in Pr.6.3, compute $\mathbf{B}\mathbf{c}$, $\mathbf{c}^T\mathbf{A}\mathbf{c}$, $\mathbf{A}\mathbf{B}\mathbf{c}$, $(\mathbf{A}\mathbf{c}) \bullet (\mathbf{B}\mathbf{d})$, $\mathbf{d}^T\mathbf{B}\mathbf{d}$. Why is the last result 0?

Pr.6.5 **(Experiments on multiplication)** By experimenting with 3×3 or 4×4 *symmetric* and *skew-symmetric* matrices whose entries are numbers or general letters, try to answer the following questions. Are sums, products, and powers of symmetric matrices symmetric? Study the same questions for skew-symmetric matrices. What can you say about products of a symmetric matrix times a skew-symmetric one? Is $\det \mathbf{A} = 0$ for every skew-symmetric matrix? For some symmetric matrices? (*AEM Ref.* p. 307)

Pr.6.6 **(Associativity, distributivity)** Verify the associativity of matrix multiplication and the distributivity by 3×3 matrices of your own choice. (*AEM Ref.* p. 313)

Pr.6.7 **(Rotation)** If

$$\mathbf{A} = \begin{bmatrix} \cos\theta & -\sin\theta \\ \sin\theta & \cos\theta \end{bmatrix}, \quad \text{verify that} \quad \mathbf{A}^n = \begin{bmatrix} \cos n\theta & -\sin n\theta \\ \sin n\theta & \cos n\theta \end{bmatrix}$$

for $n = 2, 3, 4$. What does this mean in terms of rotations through an angle θ? (*AEM Ref.* p. 320)

Pr.6.8 **(Transposition rule for products)** Prove $(\mathbf{A}\mathbf{B})^T = \mathbf{B}^T\mathbf{A}^T$ for general 2×2 matrices on the computer. (*AEM Ref.* p. 315)

Pr.6.9 **(Experiment on Hankel matrices)** Find empirically a law for the smallest n as a function of m (> 0, integer) such that $\det \mathbf{A} = 0$, where the $n \times n$ matrix $\mathbf{A} = [a_{jk}]$ has the entries $a_{jk} = (j + k)^m$. (Enjoy these special Hankel matrices, whose determinants have very fast growing values, but all of sudden become 0 from some n on. This is of practical interest in connection with the so-called Padé approximation.)

Pr.6.10 **(Inverse)** Using the computer, find the formula for the inverse of a 2×2 matrix $\mathbf{A} = [a_{jk}]$ in terms of a_{jk} and $\det \mathbf{A}$. (*AEM Ref.* p. 353)

Pr.6.11 **(Inverse of a product)** Verify the basic relation $(\mathbf{A}\mathbf{B})^{-1} = \mathbf{B}^{-1}\mathbf{A}^{-1}$ for the matrices

$$\mathbf{A} = \begin{bmatrix} 0 & -2 & -1 \\ -2 & 3 & 2 \\ -1 & 2 & 1 \end{bmatrix} \quad \text{and} \quad \mathbf{B} = \begin{bmatrix} 1 & 2 & 3 \\ 2 & 3 & 4 \\ 3 & 4 & 6 \end{bmatrix}.$$

(*AEM Ref.* p. 355)

Pr.6.12 **(Linear transformations)** With respect to Cartesian coordinates in space, let $\mathbf{y} = \mathbf{Ax}$ and $\mathbf{x} = \mathbf{Bw}$ with $\mathbf{A}$ and $\mathbf{B}$ as in the previous problem. Find the transformation $\mathbf{y} = \mathbf{Cw}$ which transforms $\mathbf{w}$ directly into $\mathbf{y}$. Find the inverse of this transformation. (*AEM Ref.* pp. 316-318)

Pr.6.13 **(Orthogonal vectors)** Show that the following vectors are orthogonal. (*AEM Ref.* p. 365)

$$\mathbf{c} = [3 \quad 2 \quad -2 \quad 1 \quad 0], \qquad \mathbf{d} = [2 \quad 0 \quad 3 \quad 0 \quad 4], \qquad \mathbf{e} = [1 \quad -3 \quad -2 \quad -1 \quad 1]$$

Pr.6.14 **(Extension of an orthogonal system)** Find a vector $\mathbf{x} \neq \mathbf{0}$ orthogonal to the three vectors in Pr.6.13. (*AEM Ref.* p. 365)

Pr.6.15 **(Norm)** Find the Euclidean norm $(c_1^2 + c_2^2 + \ldots + c_n^2)^{1/2}$, etc., of the vectors in Pr.6.13. (*AEM Ref.* p. 362)

Pr.6.16 **(Linear independence)** Check the set of vectors $[-1 \quad 5 \quad 0]$, $[16 \quad 8 \quad -3]$, $[-64 \quad 56 \quad 9]$ for linear independence. (*AEM Ref.* p. 336 (#3))

Pr.6.17 **(Hilbert matrices)** Find the determinant and the inverse of the Hilbert matrix $\mathbf{H} = [h_{jk}]$ with $n = 3$ rows and columns, where $h_{jk} = 1/(j + k - 1)$. Comment on the size of $\det \mathbf{H}$ and of the entries of the inverse. Do the same tasks when $n = 4$ and 5. (*AEM Ref.* p. 358. See also Example 6.2 in this Guide.)

Pr.6.18 **(Experiment on rank)** Find the rank of the $n \times n$ matrix $\mathbf{A} = [a_{jk}]$ with entries $a_{jk} = j + k - 1$ and any n. To understand the reason for the perhaps somewhat surprising result, use the command RowReduce and see what happens for different n. Explain. (*AEM Ref.* pp. 331-336)

Pr.6.19 **(Linear independence)** Are the vectors $[0 \quad 16 \quad 0 \quad -24 \quad 0]$, $[1 \quad 0 \quad -1 \quad 0 \quad 2]$, $[0 \quad -14 \quad 0 \quad 21 \quad 0]$ linearly independent or dependent? (*AEM Ref.* p. 336)

Pr.6.20 **(Row space and column space)** Find a basis of the row space and of the column space of the matrix with the rows $[3 \quad 1 \quad 4]$, $[0 \quad 5 \quad 8]$, $[-3 \quad 4 \quad 4]$, $[1 \quad 2 \quad 4]$. *AEM Ref.* p. 337 (#29))

Chapter 7

Matrix Eigenvalue Problems

Content. Basic commands (Ex. 7.1)
Complex eigenvalues, orthogonal matrices (Exs. 7.2, 7.3, Pr. 7.3)
Complex matrices (Ex. 7.4, Prs. 7.6-7.9)
Similar matrices, diagonalization (Ex. 7.5, Prs. 7.10-7.12)

All matrices in this chapter are square. The eigenvalues of a matrix **A** are obtained by the command `Eigenvalues[A]` and the eigenvectors by the command `Eigenvectors[A]`. Eigenvalues and eigenvectors are simultanously obtained by the command `Eigensystem[A]`.

Accessing parts of the spectrum is explained in Example 7.1.

Examples for Chapter 7

EXAMPLE 7.1 | **EIGENVALUES, EIGENVECTORS,**
ACCESSING SPECTRUM

Find the characteristic determinant, the characteristic polynomial, the eigenvalues, and eigenvectors of the matrix

$$\mathbf{A} = \begin{bmatrix} -2 & 2 & -3 \\ 2 & 1 & -6 \\ -1 & -2 & 0 \end{bmatrix}.$$

Solution. You can obtain all the information about eigenvalues and eigenvectors from the single command `Eigensystem[A]` (see below). or you can obtain eigenvalues and eigenvectors separately. Type

In[1]:= `A = {{-2, 2, -3}, {2, 1, -6}, {-1, -2, 0}}`
Out[1]= {{−2, 2, −3}, {2, 1, −6}, {−1, −2, 0}}

In[2]:= `%//MatrixForm` $\qquad$ (* Out: $\begin{pmatrix} -2 & 2 & -3 \\ 2 & 1 & -6 \\ -1 & -2 & 0 \end{pmatrix}$ *)

In[3]:= `Eigval = Eigenvalues[A]`
Out[3]= {−3, −3, 5} $\qquad$ (* −3 has algebraic multiplicity 2 *)

In[4]:= `Eigval[[1]]` $\qquad$ (* Out: −3 *)

In[5]:= `Eigvec = Eigenvectors[A]` $\quad$ (* Out: ({{3, 0, 1}, {−2, 1, 0}, {−1, −2, 1}} *)

In[6]:= `Eigvec[[2]]` $\qquad$ (* Access the second eigenvector. *)
Out[6]= {−2, 1, 0}

In[7]:= `Eig = Eigensystem[A]`

Out[7]= {{−3, −3, 5}, {{3, 0, 1}, {−2, 1, 0}, {−1, −2, 1}}}

You see that **A** has two eigenvalues, $\lambda_2 = -3$, of multiplicity 2, and $\lambda_1 = 5$, of multiplicity 1. An eigenvector corresponding to 5 is $[-1 \quad -2 \quad 1]^T$. Two linearly independent eigenvectors corresponding to -3 are $[3 \quad 0 \quad 1]^T$ and $[-2 \quad 1 \quad 0]^T$. Keep in mind that Mathematica may sometimes give the eigenvalues and eigenvectors in a different order. You could terminate your work here. But we shall continue with a few further useful commands.

Accessing parts of the spectrum. This can be useful, for instance, in connection with systems of ODE's (Chap. 3). The commands are self-explanatory.

In[8]:= `Eig[[1]]`   (* Out: {−3, −3, 5} *)

In[9]:= `Eig[[1]][[3]]`   (* Out: 5 *)

In[10]:= `Eig[[2]]`   (* Out: {{3, 0, 1}, {−2, 1, 0}, {−1, −2, 1}} *)

In[11]:= `Eig[[2]][[3]]`   (* Out: {−1, −2, 1} *)

In[12]:= `Eig[[2]][[3]][[1]]`   (* Out: −1 *)

Characteristic matrix, determinant, and polynomial; eigenvalues. The characteristic matrix corresponding to **A** is

In[13]:= `charA = A − λ IdentityMatrix[3]; MatrixForm[charA]`

$$\text{Out[13]}= \begin{pmatrix} -2-\lambda & 2 & -3 \\ 2 & 1-\lambda & -6 \\ -1 & -2 & -\lambda \end{pmatrix}$$

The **Greek letter** λ can be entered by typing `\[Lambda]`. Another way of typing λ is the follwing. Click on File. Go down the lines to **Palettes**. Go to the right, to the line of the small list that says **Basic Typesetting**. Your worksheet will again become visible. To the right you will see a list of symbols, including Greek letters. Click on λ, and it will appear wherever you had put your cursor. So you need not type, just click.

The characteristic determinant is the determinant of the matrix $\mathbf{A} - \lambda\mathbf{I}$. By expanding this determinant you obtain the characteristic polynomial

In[14]:= `d = Det[charA]`   (* Out: $45 + 21\,\lambda - \lambda^2 - \lambda^3$ *)

You can also obtain `d` by the command

In[15]:= `cPoly = CharacteristicPolynomial[A, λ]`

Out[15]= $45 + 21\,\lambda - \lambda^2 - \lambda^3$

The characteristic equation `d == 0` gives the eigenvalues

In[16]:= `NSolve[d == 0, λ]`   (* Out: $\lambda \to -3.,\ \lambda \to 5.$ *)

You may confirm this by factorizing the characteristic polynomial and recognizing the algebraic multiplicities of the eigenvalues, in agreement with the results above,

In[17]:= `Factor[cPoly]`   (* Out: $-(-5+\lambda)\,(3+\lambda)^2$ *)

Similar Material in AEM: p. 374

EXAMPLE 7.2   REAL MATRICES WITH COMPLEX EIGENVALUES

A *real* matrix may have *complex* eigenvalues. If $a + ib$ (a, b real) is one of them, the conjugate $a - ib$ must also be an eigenvalue because if a polynomial with *real* coefficients has complex roots, the latter must occur in conjugate pairs. A simple case is

In[1]:= `A = {{0, 1}, {-1, 0}}`   (* Out: {{0, 1}, {-1, 0}} *)

In[2]:= `%//MatrixForm` $\left(* \text{ Out: } \begin{pmatrix} 0 & 1 \\ -1 & 0 \end{pmatrix} *\right)$

In[3]:= `Eigensystem[A]`   (* Out: {{-i, i}, {{i, 1}, {-i, 1}}} *)

You see that the eigenvalues are $-i$ and $i = \sqrt{-1}$. Each eigenvalue has multiplicity 1 because for an $n \times n$ matrix the sum of the algebraic multiplicities of all the eigenvalues must equal n. Eigenvectors are $\begin{bmatrix} 1 & i \end{bmatrix}^T$ and $\begin{bmatrix} 1 & -i \end{bmatrix}^T$.

More generally, consider the matrix

In[4]:= `Clear[a, b]`

In[5]:= `A = {{a, b}, {-b, a}}; MatrixForm[A]` $\left(* \text{ Out: } \begin{pmatrix} a & b \\ -b & a \end{pmatrix} *\right)$

You obtain the eigenvalue $a + ib$ and its conjugate

In[6]:= `Eigenvalues[A]`   (* Out: {a - i b, a + i b} *)

For $a = 0$ and $b = 1$ you get the previous matrix and its eigenvalues. If you want to see the eigenvectors, too, type

In[7]:= `Eigenvectors[A]`   (* Out: {{i, 1}, {-i, 1}} *)

Hence you get $\begin{bmatrix} i & 1 \end{bmatrix}^T$ as an eigenvector for the first eigenvalue and $\begin{bmatrix} -i & 1 \end{bmatrix}^T$ for the second. The previous eigenvectors for the special case $a = 0$, $b = 1$ agree with this result.

This example, solvable almost by inspection, serves to explain relevant Mathematica commands and to show how the computer will handle them.

Similar Material in AEM: p. 375

EXAMPLE 7.3   ORTHOGONAL MATRICES AND TRANSFORMATIONS

The main reason for the importance of orthogonal matrices and transformations results from the fact that these matrices preserve the inner product (the dot product), hence also the length of a vector and the angle between two vectors. Show this for vectors in three-space R^3 in the case of the matrix

$$\mathbf{A} = \begin{bmatrix} \dfrac{2}{3} & \dfrac{1}{3} & \dfrac{2}{3} \\ \dfrac{-2}{3} & \dfrac{2}{3} & \dfrac{1}{3} \\ \dfrac{1}{3} & \dfrac{2}{3} & \dfrac{-2}{3} \end{bmatrix}.$$

Show that this matrix is orthogonal.

Solution. Type the matrix

In[1]:= `A = {{2/3,  1/3,  2/3}, {-2/3,  2/3,  1/3}, {1/3,  2/3, -2/3}}`

Out[1]= $\{\{\frac{2}{3}, \frac{1}{3}, \frac{2}{3}\}, \{-\frac{2}{3}, \frac{2}{3}, \frac{1}{3}\}, \{\frac{1}{3}, \frac{2}{3}, -\frac{2}{3}\}\}$

In[2]:= `%//MatrixForm`   (* Out: $\begin{pmatrix} \frac{2}{3} & \frac{1}{3} & \frac{2}{3} \\ \frac{-2}{3} & \frac{2}{3} & \frac{1}{3} \\ \frac{1}{3} & \frac{2}{3} & \frac{-2}{3} \end{pmatrix}$ *)

In[3]:= `B = Inverse[A]; MatrixForm[B]`   (* Out: $\begin{pmatrix} \frac{2}{3} & -\frac{2}{3} & \frac{1}{3} \\ \frac{1}{3} & \frac{2}{3} & \frac{2}{3} \\ \frac{2}{3} & \frac{1}{3} & -\frac{2}{3} \end{pmatrix}$ *)

You see that the inverse of **A** is the transpose of **A**. To confirm this, type

In[4]:= `B - Transpose[A]`   (* The 3×3 zero matrix *)

Hence **A** is orthogonal, by definition.

Now type two vectors and their inner product (dot product)

In[5]:= `v = {v1, v2, v3};`

In[6]:= `w = {w1, w2, w3};`

In[7]:= `Dot[v, w]`   (* Out: v1 w1 + v2 w2 + v3 w3 *)

Multiply each of the vectors by **A** and take the inner product of the resulting vectors **Av** and **Aw**,

In[8]:= `Dot[A.v, A.w]`

Out[8]= $\left(\frac{v1}{3} + \frac{2\,v2}{3} - \frac{2\,v3}{3}\right)\left(\frac{w1}{3} + \frac{2\,w2}{3} - \frac{2\,w3}{3}\right) +$

$\left(-\frac{2\,v1}{3} + \frac{2\,v2}{3} + \frac{v3}{3}\right)\left(-\frac{2\,w1}{3} + \frac{2\,w2}{3} + \frac{w3}{3}\right) +$

$\left(\frac{2\,v1}{3} + \frac{v2}{3} + \frac{2\,v3}{3}\right)\left(\frac{2\,w1}{3} + \frac{w2}{3} + \frac{2\,w3}{3}\right)$

In[9]:= `Simplify[%]`   (* Out: v1 w1 + v2 w2 + v3 w3 *)

This shows the invariance of the inner product under our (special) orthogonal transformation.

As a special case this includes the invariance of the **length** of a vector under orthogonal transformations; just set $\mathbf{w} = \mathbf{v}$, then you have the square of the Euclidean norm (the square of the usual length of **v**) in the previous formula, and of **Av** in the present formula.

Similarily, the **angle** γ between **v** and **w** is the same as between **Av** and **Aw** because $\mathbf{v} \cdot \mathbf{w} = |\mathbf{v}||\mathbf{w}| \cos \gamma$ shows that γ is expressed in terms of quantities that have just been shown to be invariant.

Similar Material in AEM: pp. 382-384

EXAMPLE 7.4 COMPLEX MATRICES

Three classes of complex matrices are of practical interest (in physics, for instance), mainly because of their spectrum. They are complex generalizations of symmetric, skew-symmetric, and orthogonal real matrices. They are defined as follows.

  A. A matrix is **Hermitian** if it is equal to its conjugate transpose. Its eigenvalues are real.

  B. A matrix is **skew-Hermitian** if it is equal to minus its conjugate transpose. Its eigenvalues are pure imaginary or zero.

  C. A matrix is **unitary** if its inverse is equal to its conjugate transpose. Its eigenvalues have absolute value 1.

A. Show that the following matrix **A** is Hermitian and find its spectrum. Here, I is the Mathematica notation for $i = \sqrt{-1}$.

In[1]:= A = {{4, 1 - 3 I}, {1 + 3 I, 7}}

Out[1]= {{4, 1 − 3 i}, {1 + 3 i, 7}}

In[2]:= %//MatrixForm $\qquad$ (* Out: $\begin{pmatrix} 4 & 1-3\,i \\ 1+3\,i & 7 \end{pmatrix}$ *)

In[3]:= AT = Transpose[A] $\qquad$ (* Out: {{4, 1 + 3 i}, {1 − 3 i, 7}} *)

In[4]:= ACT = Conjugate[AT] $\qquad$ (* The conjugate transpose of **A** *)

Out[4]= {{4, 1 − 3 i}, {1 + 3 i, 7}}

In[5]:= A − ACT $\qquad$ (* The 2 × 2 zero matrix shows that **A** is Hermitian. *)

In[6]:= Eigenvalues[A] $\qquad$ (* Out: {2, 9} *)

B. Show that the following matrix **B** is skew-Hermitian and that its eigenvalues are pure imaginary.

In[7]:= B = {{3 I, 2+I}, {-2+I, -I}}; MatrixForm[B]

Out[7]= $\begin{pmatrix} 3\,i & 2+i \\ -2+i & -i \end{pmatrix}$

In[8]:= BCT = Conjugate[Transpose[B]] $\qquad$ (* The conjugate transpose of **B** *)

Out[8]= {{−3 i, −2 − i}, {2 − i, i}}

In[9]:= B + BCT $\qquad$ (* The 2 × 2 zero matrix *)

This shows that the conjugate transpose BCT of **B** equals minus **B**. Hence **B** is skew-Hermitian, by definition.

In[10]:= Eigenvalues[B] $\qquad$ (* Imaginary eigenvalues *)

Out[10]= {−2 i, 4 i}

C. Show that the following matrix **F** (C is protected - try to use it.) is unitary and its eigenvalues have absolute value 1.

In[11]:= F = {{I/2, Sqrt[3]/2}, {Sqrt[3]/2, I/2}}

Out[11]= {{$\frac{i}{2}$, $\frac{\sqrt{3}}{2}$}, {$\frac{\sqrt{3}}{2}$, $\frac{i}{2}$}}

In[12]:= %//MatrixForm   (* Out: $\begin{pmatrix} \frac{i}{2} & \frac{\sqrt{3}}{2} \\ \frac{\sqrt{3}}{2} & \frac{i}{2} \end{pmatrix}$ *)

In[13]:= FCT = Conjugate[Transpose[F]]   (* The conjugate transpose of **F** *)

Out[13]= $\{\{-\frac{i}{2}, \frac{\sqrt{3}}{2}\}, \{\frac{\sqrt{3}}{2}, -\frac{i}{2}\}\}$

In[14]:= FCT.F   (* Out: {{1, 0}, {0, 1}} *)

In[15]:= F.FCT   (* Out: {{1, 0}, {0, 1}} *)

This shows that **F** is unitary. Note that $\mathbf{F}^T = \mathbf{F}$. The eigenvalues are

In[16]:= Eig = Eigenvalues[F]

Out[16]= $\{\frac{1}{2}(i - \sqrt{3}), \frac{1}{2}(i + \sqrt{3})\}$

In[17]:= Abs[Eig]   (* Out: {1, 1} *)

Hence the eigenvalues have absolute value 1, as expected.

Similar Material in AEM: pp. 385-387

EXAMPLE 7.5   SIMILARITY OF MATRICES. DIAGONALIZATION

A real $n \times n$ matrix **B** is called **similar** to a real $n \times n$ matrix **A** if $\mathbf{B} = \mathbf{P}^{-1}\mathbf{A}\mathbf{P}$ for some nonsingular matrix **P**. Then **A** is similar to **B**, and **A** and **B** are called *similar matrices*. Similar matrices have the same eigenvalues. This accounts for the practical importance of similarity, for instance, in the design of numerical methods for eigenvalues. Furthermore, if **x** is an eigenvector of **A**, then $\mathbf{y} = \mathbf{P}^{-1}\mathbf{x}$ is an eigenvector of the similar matrix **B** corresponding to the same eigenvalue. Verify these statements for the matrices **A** and **P** given by

In[1]:= A = {{10, -3, 5}, {0, 1, 0}, {-15, 9, -10}}; MatrixForm[A]

Out[1]//MatrixForm=

$$\begin{pmatrix} 10 & -3 & 5 \\ 0 & 1 & 0 \\ -15 & 9 & -10 \end{pmatrix}$$

In[2]:= P = {{2, 0, 3}, {0, 1, 0}, {3, 0, 5}}; MatrixForm[P]

Out[2]//MatrixForm=

$$\begin{pmatrix} 2 & 0 & 3 \\ 0 & 1 & 0 \\ 3 & 0 & 5 \end{pmatrix}$$

Solution. In addition to **A** and **P** you need the inverse of **P**, call it **Q**. Then type $\mathbf{B} = \mathbf{P}^{-1}\mathbf{A}\mathbf{P}$.

In[3]:= Q = Inverse[P]

Out[3]= {{5, 0, -3}, {0, 1, 0}, {-3, 0, 2}}

In[4]:= B = Q.A.P; MatrixForm[B]

Out[4]//MatrixForm=

$$\begin{pmatrix} 355 & -42 & 560 \\ 0 & 1 & 0 \\ -225 & 27 & -355 \end{pmatrix}$$

Now compare the eigenvalues. Also obtain eigenvectors to be used afterwards for the diagonalization.

In[5]:= EigA = Eigensystem[A]   (* Ignore the message. *)

General::spell1: Possible spelling error: new symbol name "EigA is similar to existing symbol Eig"

Out[5]= {{-5, 1, 5}, {{-1, 0, 3}, {-1, 2, 3}, {-1, 0, 1}}}

In[6]:= EigB = Eigensystem[B]   (* Ignore the message. *)

General::spell1: Possible spelling error: new symbol name "EigB is similar to existing symbol Eig"

Out[6]= {{-5, 1, 5}, {{-14, 0, 9}, {-14, 2, 9}, {-8, 0, 5}}}

You see that the eigenvalues are the same. They may sometimes come out in a different order, so you have to watch a little when you compare eigenvectors. You get the latter for $\mathbf{A}$, call them $\mathbf{x_1}$, $\mathbf{x_2}$, $\mathbf{x_3}$, and those for $\mathbf{B}$, call them $\mathbf{y_1}$, $\mathbf{y_2}$, $\mathbf{y_3}$, by typing

In[7]:= x1 = EigA[[2, 1]]   (* Out: {-1, 0, 3} *)

In[8]:= x2 = EigA[[2, 2]]   (* Out: {-1, 2, 3} *)

In[9]:= x3 = EigA[[2, 3]]   (* Out: {-1, 0, 1} *)

In[10]:= y1 = Q.x1   (* Out: {-14, 0, 9} *)

In[11]:= y2 = Q.x2   (* Out: {-14, 2, 9} *)

In[12]:= y3 = Q.x3   (* Out: {-8, 0, 5} *)

This agrees with the vectors in EigB.

Diagonalization of A. If $\mathbf{X}$ is a matrix with a basis of eigenvectors of $\mathbf{A}$ as column vectors, then $\mathbf{D} = \mathbf{X^{-1}AX}$ is diagonal with the eigenvalues as diagonal entries. Indeed, write $\mathbf{G}$ for $\mathbf{D}$ (which is protected, that is, it must not be used here) and type

In[13]:= X = Transpose[{x1, x2, x3}]; MatrixForm[X]

Out[13]= $\begin{pmatrix} -1 & -1 & -1 \\ 0 & 2 & 0 \\ 3 & 3 & 1 \end{pmatrix}$

In[14]:= G = Inverse[X].A.X; MatrixForm[G]

Out[14]= $\begin{pmatrix} -5 & 0 & 0 \\ 0 & 1 & 0 \\ 0 & 0 & 5 \end{pmatrix}$

This is the desired diagonal form, the eigenvalues -5, 1, 5 corresponding to the order $\mathbf{x_1}$, $\mathbf{x_2}$, $\mathbf{x_3}$ of the eigenvectors.

Similar Material in AEM: pp. 392-394, 397 (#5)

Problem Set for Chapter 7

Pr.7.1 **(Symmetric and skew-symmetric matrices)** Write a program for representing a square matrix as the sum of a symmetric and a skew-symmetric matrix and apply the program to the matrix in Example 7.1 in this Guide. (*AEM Ref.* p. 381)

Pr.7.2 **(Polynomial matrix)** A *polynomial matrix* is a matrix of the form

$$q(\mathbf{A}) = a_0 \mathbf{I} + a_1 \mathbf{A} + a_2 \mathbf{A}^2 + \ldots + a_m \mathbf{A}^m,$$

where $\mathbf{A}$ is any $n \times n$ matrix and $\mathbf{I}$ is the $n \times n$ unit matrix. If λ is an eigenvalue of $\mathbf{A}$, then $q(\lambda)$ is an eigenvalue of $q(\mathbf{A})$. This is called the **spectral mapping theorem** for polynomial matrices. Verify it for the matrix $\mathbf{A}$ in Example 7.5 in this Guide and the polynomial $q(x) = x^3 + 4x^2 - 10x + 8$. (*AEM Ref.* p. 381)

Pr.7.3 **(Orthogonal transformation, rotation)** Show that the matrix $\mathbf{A}$ given below is orthogonal and the orthogonal transformation $\mathbf{y} = \mathbf{A}\mathbf{x}$ leaves the inner product (the dot product) invariant. (*AEM Ref.* pp. 381-383)

$$\mathbf{A} = \begin{bmatrix} \cos\theta & -\sin\theta \\ \sin\theta & \cos\theta \end{bmatrix}$$

Pr.7.4 **(Cayley's theorem)** Cayley's theorem states that every square matrix $\mathbf{A}$ satisfies its characteristic equation, that is, $p(\mathbf{A}) = \mathbf{0}$, where $p(\lambda)$ is the characteristic polynomial of $\mathbf{A}$. Verify this for the matrix $\mathbf{A}$ in Example 7.1 in this Guide. (*AEM Ref.* p. 373)

Pr.7.5 **(Experiment on Perron's theorem)** Perron's famous theorem states that a square matrix $\mathbf{A}$ all of whose entries are positive has a positive real eigenvalue of multiplicity 1 which is greater than the absolute value of any other eigenvalue of $\mathbf{A}$, and a corresponding real eigenvector whose components are all positive. Verify this theorem for matrices of your own choice. (*AEM Ref.* p. 381. A proof of the theorem is not easy.)

Pr.7.6 **(Complex matrix)** Show that the matrix with rows [2 1−i] and [1+i 3] is Hermitian and find its spectrum. (*AEM Ref.* p. 385)

Pr.7.7 **(Hermitian and skew-Hermitian matrices)** Represent the matrix $\mathbf{A}$ as the sum of a Hermitian and a skew-Hermitian matrix and find the eigenvalues and eigenvectors of $\mathbf{A}$, where $\mathbf{A}$ is as follows. (*AEM Ref.* p. 385)

$$\mathbf{A} = \begin{bmatrix} 4+12i & -12-12i & 12-12i \\ -6+6i & 10-6i & 6+6i \\ 6+6i & -6+6i & -2-6i \end{bmatrix}$$

Pr.7.8 **(Complex matrix)** Let $\mathbf{A}$ be the 3×3 matrix with main diagonal entries 0 and all other entries i. Show that $\mathbf{A}$ is skew-Hermitian and find its spectrum. (Note the multiplicities!) Find the inverse and show that $\mathbf{A} = (1/2)\mathbf{I} - \mathbf{A}^{-1}$, where $\mathbf{I}$ is the 3×3 unit matrix. (*AEM Ref.* p. 385)

Pr.7.9 **(Experiment on unitary matrices)** Consider the powers $\mathbf{F}^m$ (m a positive integer) of the matrix $\mathbf{F}$ in Example 7.4 in this Guide, their eigenvalues, and their inverses. Show that the eigenvalues have absolute value 1. Verify that the inverses are unitary. For what values of m will $\mathbf{F}^m$ be real? For what m will the eigenvalues be 1, 1? Verify that $\mathbf{F}$ satisfies Cayley's theorem (see Pr.7.4). Find other unitary matrices and consider them in a similar fashion. (*AEM Ref.* pp. 385-387)

Pr.7.10 (Similar matrices) Find and discuss the relations between the eigenvalues and eigenvectors of $\mathbf{A}$ and $\mathbf{B} = \mathbf{P}^{-1}\mathbf{A}\mathbf{P}$, where $\mathbf{A}$ and $\mathbf{P}$ are as follows. (*AEM Ref.* p. 397 (#3))

$$\mathbf{A} = \begin{bmatrix} 3 & 4 \\ 4 & -3 \end{bmatrix} \quad \text{and} \quad \mathbf{P} = \begin{bmatrix} -4 & 2 \\ 3 & -1 \end{bmatrix}$$

Pr.7.11 (Similarity transformation) Show that $\mathbf{A}$ and $\mathbf{B} = \mathbf{P}^{-1}\mathbf{A}\mathbf{P}$ have the same eigenvalues and establish the relation between their eigenvectors, where $\mathbf{A}$ and $\mathbf{P}$ are as follows. (*AEM Ref.* p. 392)

$$\mathbf{A} = \begin{bmatrix} -1 & -3 & 3 \\ -6 & 2 & 6 \\ -3 & 3 & 5 \end{bmatrix} \quad \text{and} \quad \mathbf{P} = \begin{bmatrix} 3 & -1 & 1 \\ -15 & 6 & -5 \\ 5 & -3 & 2 \end{bmatrix}$$

Pr.7.12 (Transformation to diagonal form) Show that $\mathbf{D} = \mathbf{X}^{-1}\mathbf{A}\mathbf{X}$ is diagonal with the eigenvalues of $\mathbf{A}$ as diagonal entries. Here $\mathbf{X}$ is a matrix with eigenvectors of $\mathbf{A}$ as column vectors, and $\mathbf{A}$ is as follows. (*AEM Ref.* pp. 392-395)

$$\mathbf{A} = \begin{bmatrix} 5 & 4 \\ 1 & 2 \end{bmatrix}$$

Chapter 8

Vectors in R^2 and R^3.
Dot and Cross Products.
Grad, Div, Curl

Content. Addition of vectors, scalar multiplication (Ex. 8.1, Prs. 8.1-8.4)
Inner product, cross product, triple product (Ex. 8.2, Prs. 8.5-8.13)
Vector fields (Pr. 8.14)
Derivatives, curves (Ex. 8.3, Prs. 8.15-8.18)
Gradient, divergence, curl (Exs. 8.4, 8.5, Prs. 8.19-8.25)

For vectors in R^2 and R^3 various commands are those for vectors in R^n in Chap. 6, for instance, `Dot[u, u]`.

New are the commands for the cross product (Ex. 8.2), gradient (Ex. 8.4), divergence and curl (Ex. 8.5), and the Laplacian (also in Ex. 8.5).

All of these new commands require that you first load the **vector analysis package** by typing

In[1]:= `<< Calculus'VectorAnalysis'`

and then select coordinates, for instance, **Cartesian coordinates**, by typing

In[2]:= `SetCoordinates[Cartesian[x, y, z]]`   `(* Out: Cartesian[x, y, z] *)`

In[3]:= `SetCoordinates[Cartesian[x, y]]`

`Coordinates::invalid :`
   `Cartesian[x, y] is not a valid coordinate system specification.`

Out[3]= `SetCoordinates[Cartesian[x, y]]`

This error message shows you a catch: if you work in the xy-plane, carry along a third coordinate and equate it to zero in your work.

In[4]:= `u = {u1, u2};    v = {v1, v2};`

In[5]:= `CrossProduct[u, v]`

Out[5]= `CrossProduct[u1, u2, v1, v2]`

In[6]:= `u = {u1, u2, 0};    v = {v1, v2, 0};`

In[7]:= `CrossProduct[u, v]`

Out[7]= $\{0, 0, -u2\,v1 + u1\,v2\}$

This calculation illustrates distinctly the reason for carrying along a third coordinate.

Examples for Chapter 8

EXAMPLE 8.1   **VECTORS, ADDITION,**
                              SCALAR MULTIPLICATION

In the applications in this chapter, vectors **v** are often given as arrows from an initial point P to a terminal point Q, where, for instance,

In[1]:= `P = {3, 1, 4}`

Out[1]= {3, 1, 4}

In[2]:= `Q = {1, -2, 4}`

Out[2]= {1, −2, 4}

Then the **vector v** from P to Q is obtained by typing

In[3]:= `v = Q - P`   `(* Out: {−2, −3, 0} *)`

and its **length** is the square root of the dot product **v • v**, typed as

In[4]:= `Sqrt[Dot[v, v]]`   `(* Out: `$\sqrt{13}$` *)`

Note that `P` and `Q` were typed with braces {...}, not with parentheses (...) or brackets [...]. This is quite important. Try it otherwise. **Vector addition and scalar multiplication** proceed as in Chap. 6. For instance, let **v** be as before and **w** = [5, −1, 3]. Calculate **v** + **w**, −4**v**, and 0.5**w**. (The last two commands can be typed with or without an asterisk.)

In[5]:= `w = {5, -1, 3}`   `(* Out: {5, −1, 3} *)`

In[6]:= `v + w`   `(* Out: {3, −4, 3} *)`

In[7]:= `-4*v`   `(* Out: {8, 12, 0} *)`

In[8]:= `-4 v`   `(* Out: {8, 12, 0} *)`

In[9]:= `0.5*w`   `(* Out: {2.5, −0.5, 1.5} *)`

*Similar Material in **AEM:*** pp. 401-406

EXAMPLE 8.2   **INNER PRODUCT. CROSS PRODUCT**

Inner products (dot products) have already occurred in Chap. 6 (see Example 6.1 in this Guide). In 2- and 3-space they can be motivated by the **work done by a force** in a displacement. For instance, find the work W done by the force **p** = [2, 1, 4] in the displacement from $A : (1, 1, 0)$ to $B : (2, 1, 10)$. Also find the angle between **p** and a vector in the direction of the displacement.

Solution. W is defined by $W = $ **p • d**, where **d** is the "displacement vector" from A to B; thus,

In[1]:= `A = {1, 1, 0}`   `(* Out: {1, 1, 0} *)`

In[2]:= `B = {2, 1, 10}`   `(* Out: {2, 1, 10} *)`

In[3]:= `p = {2, 1, 4}`   `(* Out: {2, 1, 4} *)`

In[4]:= `d = B - A`   `(* Out: {1, 0, 10} *)`

In[5]:= W = Dot[p, d]   (* Out: 42 *)

For the **angle** you have $\mathbf{p} \bullet \mathbf{d} = \|\mathbf{p}\|\|\mathbf{d}\| \cos\theta$, hence $\theta = \arccos(\mathbf{p} \bullet \mathbf{d}/(\|\mathbf{p}\|\|\mathbf{d}\|))$. Hence type

In[6]:= theta = ArcCos[Dot[p, d]/(Sqrt[Dot[p, p]] Sqrt[Dot[d, d]])]

Out[6]= Arccos $\left[2\sqrt{\dfrac{21}{101}}\right]$

In[7]:= N[%]

Out[7]= 0.422744   (* Angle in radians *)

In[8]:= %*180/π

Out[8]= 24.2214   (* Angle in degrees *)

Cross products $\mathbf{v} = \mathbf{a} \times \mathbf{b}$ are vectors $\mathbf{v}$ perpendicular to $\mathbf{a}$ and $\mathbf{b}$ and of length $\|\mathbf{v}\|$ equal to the area of the parallelogram with $\mathbf{a}$ and $\mathbf{b}$ as adjacent sides. (If $\mathbf{a}$ and $\mathbf{b}$ are parallel or at least one of them is $\mathbf{0}$, then $\mathbf{v} = \mathbf{0}$ by definition.) Hence, among other applications, cross products can be used to calculate the area of a triangle when its vertices are given, say $A : (1, 1, 1)$, $B : (5, 2, 3)$, $C : (-1, 4, 5)$.

Solution. You need two vectors $\mathbf{a}$ and $\mathbf{b}$ representing two sides of the triangle, say, AB and AC. Accordingly, type

In[9]:= A = {1, 1, 1}   (* Out: {1, 1, 1} *)

In[10]:= B = {5, 2, 3}   (* Out: {5, 2, 3} *)

In[11]:= K = {-1, 4, 5}   (* Out: {-1, 4, 5}   C is protected. Try it. *)

In[12]:= a = B - A   (* Out: {4, 1, 2} *)

In[13]:= b = K - A   (* Out: {-2, 3, 4} *)

To obtain the cross product, you must first load the vector analysis package by typing

In[14]:= <<Calculus`VectorAnalysis`

Then type

In[15]:= v = CrossProduct[a, b]   (* Out: {-2, -20, 14} *)

From this you get the answer (noting that the area of the triangle equals half the area of the parallelogram with $\mathbf{a}$ and $\mathbf{b}$ as sides)

In[16]:= answer = N[Sqrt[Dot[v, v]]/2]   (* Area of the triangle *)

Out[16]= 12.2474

Recalling that a cross product can be written as a symbolical determinant whose first row is $\mathbf{i}, \mathbf{j}, \mathbf{k}$ (unit vectors in the positive directions of the coordinate axes) and whose second and third rows are the two vectors, you can check your cross product by typing

In[17]:= r = {i, j, k}

In[18]:= F = Join[{r}, {a}, {b}]   (* Out: {{i, j, k}, {4, 1, 2}, {-2, 3, 4}} *)

In[19]:= Det[F]   (* Out: $-2\,i - 20\,j + 14\,k$ *)

This is the cross product $[-2, -20, 14]$.

Similar Material in AEM: pp. 410, 414

EXAMPLE 8.3   DIFFERENTIATION OF VECTORS. CURVES AND THEIR PROPERTIES

Investigate the main geometrical properties of the **circular helix**

$$\mathbf{r}(t) = [a\cos t, \; a\sin t, \; ct],$$

which lies on a cylinder of radius a, is right-handed if $c > 0$, and has pitch $2\pi c$.

Solution. You will need the first two derivatives and their inner products (dot products). Type

In[1]:= `Clear[r, a, c]`

In[2]:= `r = {a Cos[t], a Sin[t], c t}`

Out[2]= $\{a\,\mathrm{Cos}\,[t],\, a\,\mathrm{Sin}\,[t],\, c\,t\}$

In[3]:= `r1 = D[r, t]`   `(* Out: {−a Sin [t], a Cos [t], c} *)`

In[4]:= `r2 = D[r1, t]`   `(* Out: {−a Cos[t], −a Sin [t], 0} *)`

In[5]:= `i11 = Dot[r1, r1]`

Out[5]= $c^2 + a^2\,\mathrm{Cos}\,[t]^2 + a^2\,\mathrm{Sin}\,[t]^2$

In[6]:= `i11 = Simplify[%]`   `(* Out: a`^{2}`+ c`^{2}` *)`

In[7]:= `i12 = Dot[r1, r2]`   `(* Out: 0 *)`

In[8]:= `i22 = Simplify[Dot[r2, r2]]`   `(* Out: a`^{2}` *)`

Arc length. The arc length s is the integral of the square root of `i11`,

In[9]:= `s = Integrate[Sqrt[i11], t]`   `(* Out: `$\sqrt{a^2+c^2}\,t$` *)`

Curvature. The curvature κ of a curve represented by $\mathbf{r} = \mathbf{r}(t)$ is given by

In[10]:= `κ = (i11 i22 - i12^2)^(1/2)/i11^(3/2)`

Out[10]= $\dfrac{\sqrt{a^2\,(a^2+c^2)}}{(a^2+c^2)^{3/2}}$

In[11]:= `Simplify[%, a > 0 && c > 0]`   `(* Try without `$a > 0$` && `$c > 0$`. *)`

Out[11]= $\dfrac{a}{a^2+c^2}$

Torsion. You will now also need the third derivative of $\mathbf{r}(t)$. The numerator of the torsion τ (tau) is the determinant of the matrix with $\mathbf{r}', \mathbf{r}'', \mathbf{r}'''$ as rows (or columns), and the denominator is `i11 i22 - i12^2`. Hence type

In[12]:= `r3 = D[r2, t]`   `(* Out: {a Sin [t], −a Cos [t], 0} *)`

In[13]:= `M = Join[{r1}, {r2}, {r3}]; MatrixForm[M]`

Out[13]//MatrixForm=

$$\begin{pmatrix} -a\,\mathrm{Sin}[t] & a\,\mathrm{Cos}[t] & c \\ -a\,\mathrm{Cos}[t] & -a\,\mathrm{Sin}[t] & 0 \\ a\,\mathrm{Sin}[t] & -a\,\mathrm{Cos}[t] & 0 \end{pmatrix}$$

The braces around `r1`, `r2`, `r3` are needed to create a 3×3 matrix. Without them you would obtain a vector with nine components. Try it.

In[14]:= τ = Det[M]/(i11 i22 - i12^2)   (* Torsion τ *)

Out[14]= $\dfrac{a^2\,c\,\text{Cos}\,[t]^2 + a^2\,c\,\text{Sin}\,[t]^2}{a^2\,(a^2 + c^2)}$

In[15]:= Simplify[%]

Out[15]= $\dfrac{c}{a^2 + c^2}$

The command Join has occurred before; here it is again needed for composing a matrix from given vectors to be used as row vectors.

Note that the helix has the remarkable property that both its curvature and its torsion are constant.

If you want to plot the helix, you have to choose specific values for a and c, for instance, $a = 10$ and $c = 1$. Accordingly, type

In[16]:= <<Graphics'Graphics3D'

In[17]:= r0 = {10 Cos[t], 10 Sin[t], t}

In[18]:= P = ParametricPlot3D[Table[{r0[[1]], r0[[2]], r0[[3]]}],
       {t, 0, 8 π}]

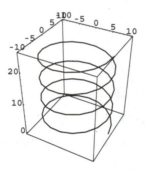

Example 8.3. Helix on a cylinder of radius 10

Similar Material in AEM: pp. 428-435, 440-442

EXAMPLE 8.4   GRADIENT. DIRECTIONAL DERIVATIVE. POTENTIAL

Gradient. To obtain the gradient of a function f, for instance, of the function $f(x, y, z) = 2x^2 + 3y^2 + z^2$, type

In[1]:= <<Calculus'VectorAnalysis'

In[2]:= SetCoordinates[Cartesian[x, y, z]]   (* Out: Cartesian[x, y, z] *)

In[3]:= f = 2 x^2 + 3 y^2 + z^2   (* Out: $2x^2 + 3y^2 + z^2$ *)

In[4]:= v = Grad[f]   (* Out: $\{4x, 6y, 2z\}$ *)

Main applications of the gradient occur in connection with directional derivatives, surface normals, and potentials, as we shall now see.

Directional derivative. The directional derivative of a function f at a point P in the direction of a vector **a** can be expressed in terms of the gradient by

$$\mathrm{D}_a\,f = (1/\|\mathbf{a}\|)\mathbf{a} \bullet \mathrm{grad}\,f.$$

For instance, find the directional derivative of the above f at the point $(2, 1, 3)$ in the direction of the vector $[1, 0, -2]$.

Solution. Type **a** and the corresponding unit vector **b**, then the directional derivative, call it `deriv`. Finally take the value of the derivative at the given point.

In[5]:= `a = {1, 0, -2}`　　　　　　　　　　　　　　　(* Out: $\{1, 0, -2\}$ *)

In[6]:= `b = a/Sqrt[Dot[a, a]]`　　　　　　　　　(* Out: $\left\{\frac{1}{\sqrt{5}}, 0, -\frac{2}{\sqrt{5}}\right\}$ *)

In[7]:= `deriv = Dot[b, v]`　　　　　　　　　　　　(* **v** $= \operatorname{grad} f$; see above. *)

Out[7]= $\dfrac{4\,x}{\sqrt{5}} - \dfrac{4\,z}{\sqrt{5}}$

In[8]:= `% /. {x -> 2, y -> 1, z -> 3}`　　　　　　(* Out: $-\frac{4}{\sqrt{5}}$ *)

In[9]:= `N[%]`　　　　　　　　　　　　　　　　　(* Out: -1.78885 *)

Hence f is decreasing in the direction of **a**.

Surface normal. The gradient $\operatorname{grad} f$ is a normal vector to a level surface $f = const$ of a function $f(x, y, z)$, that is, if $\operatorname{grad} f$ is not the zero vector, it is perpendicular to the tangent plane of the surface $f = const$ passing through the point considered. For instance, find a normal vector to the cone $S: z^2 = 4(x^2 + y^2)$ at the point $(1, 0, 2)$.

Solution. The cone S is the level surface $f = 0$ of the function

In[10]:= `f = 4 (x^2 + y^2) - z^2`　　　　　　(* Out: $4(x^2 + y^2) - z^2$ *)

Thus, a normal vector of S is $\mathbf{N} = \operatorname{grad} f$. (Write `NO` since `N` is protected.)

In[11]:= `NO = Grad[f]`　　　　　　　　　　(* Out: $\{8\,x, 8\,y, -2\,z\}$ *)

In[12]:= `NP = % /. {x -> 1, y -> 0, z -> 2}`　　(* **grad** f at $P: (1, 0, 2)$　*)

Out[12]= $\{8, 0, -4\}$

Hence a unit normal vector of S at the point $(1, 0, 2)$ is (multiply by the reciprocal length of `NP`)

In[13]:= `n = NP/Sqrt[Dot[NP, NP]]`　　　　　(* Out: $\left\{\frac{2}{\sqrt{5}}, 0, -\frac{1}{\sqrt{5}}\right\}$ *)

Potential. A potential is a scalar function f associated with a vector function **v** such that $\mathbf{v} = \operatorname{grad} f$. Not every **v** has a potential. If a **v** does, it is generally more convenient to work with a single scalar function (the potential) than with the triple of component functions of **v**.

Consider the potential $f = c/r$, where $r = \sqrt{x^2 + y^2 + z^2}$ is the distance of a point (x, y, z) from the origin. Type

In[14]:= `r = {x, y, z}`　　　　　　　　　　　　(* Out: $\{x, y, z\}$ *)

In[15]:= `r0 = Sqrt[Dot[r, r]]`　　　　　　　　(* Out: $\sqrt{x^2 + y^2 + z^2}$ *)

In[16]:= `f = c/r0`

Out[16]= $\dfrac{c}{\sqrt{x^2 + y^2 + z^2}}$

This f is the gravitational potential of a mass at the origin, according to Newton's law of gravitation. Here $c > 0$ is the gravitational constant. Indeed, the gradient of f gives the force **v** of attraction,

$\text{In}[17]:= \text{v = Grad[f]}$

$\text{Out}[17]= \left\{ -\dfrac{c\,x}{(x^2+y^2+z^2)^{3/2}},\, -\dfrac{c\,y}{(x^2+y^2+z^2)^{3/2}},\, -\dfrac{c\,z}{(x^2+y^2+z^2)^{3/2}} \right\}$

To see this more distinctly, obtain the length of **v** by typing

$\text{In}[18]:= \text{Sqrt[Dot[v, v]]}$

$\text{Out}[18]= \sqrt{\dfrac{c^2\,x^2}{(x^2+y^2+z^2)^3} + \dfrac{c^2\,y^2}{(x^2+y^2+z^2)^3} + \dfrac{c^2\,z^2}{(x^2+y^2+z^2)^3}}$

$\text{In}[19]:= \text{Simplify[%]}$ $(* \text{ Out: } \sqrt{\dfrac{c^2}{(x^2+y^2+z^2)^2}} \ *)$

Tell the computer that the radicand is positive, so that it can extract the root.

$\text{In}[20]:= \text{Simplify[\%, c > 0 \&\& x\^2 + y\^2 + z\^2 > 0]}$ $(* \text{ Out: } \dfrac{c}{x^2+y^2+z^2} \ *)$

This is $c/r0^2$, well known from **Newtons law of gravitation**. Also, **v** has the direction from (x, y, z) to the origin (note the minus signs!), as it should be for the force with which a mass at (x, y, z) is attracted to a mass at the origin.
       Similar Material in AEM: pp. 446-452

EXAMPLE 8.5   DIVERGENCE, LAPLACIAN, CURL

Divergence. The divergence of a vector function

$$\mathbf{v}(x, y, z) = [v_1(x, y, z),\, v_2(x, y, z),\, v_3(x, y, z)]$$

is obtained by loading the vector analysis package, choosing the kind of coordinates wanted, and typing the command `Div` . That is,

$\text{In}[1]:= \text{<<Calculus`VectorAnalysis`}$

$\text{In}[2]:= \text{SetCoordinates[Cartesian[x, y, z]]}$ $(* \text{ Out: } \text{Cartesian[x, y, z]} \ *)$

$\text{In}[3]:= \text{v = \{v1[x, y, z], v2[x, y, z], v3[x, y, z]\}}$

$\text{Out}[3]= \text{\{v1[x, y, z], v2[x, y, z], v3[x, y, z]\}}$

$\text{In}[4]:= \text{Div[v]}$

$\text{Out}[4]= \text{v3}^{(0,0,1)}\text{[x, y, z]} + \text{v2}^{(0,1,0)}\text{[x, y, z]} + \text{v1}^{(1,0,0)}\text{[x, y, z]}$

$v1^{(1,0,0)}$ means the partial derivative of the first component, v_1, with respect to the first variable, x, and so on. For instance, find the divergence of the following vector function **w**:

$\text{In}[5]:= \text{w = x y z \{x, y, z\}}$ $(* \text{ Out: } \{x^2\,y\,z,\ x\,y^2\,z,\ x\,y\,z^2\} \ *)$

$\text{In}[6]:= \text{Div[w]}$ $(* \text{ Out: } 6\,x\,y\,z \ *)$

Laplacian. The Laplacian of a function f referred to Cartesian coordinates is

$$\nabla^2 f = \operatorname{div}(\operatorname{grad} f) = f_{xx} + f_{yy} + f_{zz}.$$

For instance,

$\text{In}[7]:= \text{f = x y/z}$ $(* \text{ Out: } \dfrac{x\,y}{z} \ *)$

In[8]:= `D[f, x, x] + D[f, y, y] + D[f, z, z]`   (* Out: $\frac{2\,x\,y}{z^3}$ *)

or more quickly,

In[9]:= `Laplacian[f]`   (* Out: $\frac{2\,x\,y}{z^3}$ *)

The Laplacian of a function f equals the divergence of grad f. Indeed,

In[10]:= `Clear[f]`   (* Remove the special assignment from f. *)

In[11]:= `Div[Grad[f[x, y, z]]]`

Out[11]= $f^{(0,0,2)}[x, y, z] + f^{(0,2,0)}[x, y, z] + f^{(2,0,0)}[x, y, z]$

Curl. To obtain the curl of a vector function, for instance, of **w** as given above, type

In[12]:= `Curl[w]`   (* Out: $\{-x\,y^2 + x\,z^2,\, x^2\,y - y\,z^2,\, -x^2\,z + y^2\,z\}$ *)

The curl plays a role in connection with rotations. A **rotation** can be described by a rotation vector **w**, whose direction is that of the axis of rotation and whose length equals the angular velocity. Then the velocity vector **v** at a point P with position vector **r** (whose initial point is any point on the axis of rotation) is obtained as follows.

In[13]:= `w = {w1, w2, w3};   r = {x, y, z};`

In[14]:= `v = Cross[w, r]`

Out[14]= $\{-w3\,y + w2\,z,\, w3\,x - w1\,z,\, -w2\,x + w1\,y\}$

If you now take the curl of **v**, you obtain

In[15]:= `Curl[v]`   (* Out: $\{2\,w1, 2\,w2, 2\,w3\}$ *)

This proves that the curl of the velocity vector of the rotation equals twice the rotation vector. This is a basic relation on rotations which characterizes the nature of the curl in this connection.

If a vector field **v** has a potential f, so that **v** = grad$f$, then **v** is **irrotational**, that is, its curl is the zero vector. To prove this, type

In[16]:= `v = Grad[f]`

Out[16]= $\{0, 0, 0\}$ (* Nonsense *)

In[17]:= `v = Grad[f[x, y, z]]`

Out[17]= $\{f^{(1,0,0)}[x, y, z],\, f^{(0,1,0)}[x, y, z],\, f^{(0,0,1)}[x, y, z]\}$

In[18]:= `Curl[v]`   (* Out: $\{0, 0, 0\}$ *)

Similar Material in AEM: pp. 453-458

Problem Set for Chapter 8

Pr.8.1 (**Components, length**) Find the components and length of the vector **v** with initial point $(1, 2, 3)$ and terminal point $(2, 4, 6)$. (*AEM Ref.* p. 407 (#3))

Pr.8.2 (**Addition, scalar multiplication**) Find $4\mathbf{a} + 8\mathbf{c}$ and $4(\mathbf{a} + 2\mathbf{c})$, where $\mathbf{a} = [3, -2, 1]$ and $\mathbf{c} = [4, 1, -1]$. (*AEM Ref.* p. 407 (#19))

Pr.8.3 (**Resultant force**) Find the resultant of the forces $\mathbf{p} = [4, -2, -3]$, $\mathbf{q} = [8, 8, 1]$, $\mathbf{u} = [-12, -6, 2]$. (*AEM Ref.* p. 407 (#25))

Pr.8.4 **(Equilibrium)** Find **p** such that **p, q** = [3, 2, 0], and **u** = [-2, 4, 0] are in equilibrium. Sketch these forces. (*AEM Ref.* p. 407 (#29))

Pr.8.5 **(Inner product)** Find 2**b**•5**c** and 10**b**•**c,** where **b** = [2, 0, -5] and **c** = [4, -2, 1]. (*AEM Ref.* p. 413 (#5))

Pr.8.6 **(Angle)** Find the angle between **b** and **c** in Pr.8.5. (*AEM Ref.* p. 409)

Pr.8.7 **(Work)** Find the work done by the force **p** = [2, 6, 6] in the displacement from the point $A : (3, 4, 0)$ to the point $B : (5, 8, 0)$. Sketch the vectors involved. (*AEM Ref.* p. 413 (#13))

Pr.8.8 **(Vector product)** Find $\mathbf{a} \times \mathbf{c}$, $\mathbf{c} \times \mathbf{a}$, $|\mathbf{a} \times \mathbf{c}|$, $|\mathbf{c} \times \mathbf{a}|$, **a**•**c**, where **a** = [1, 2, 0] and **c** = [2, 3, 4]. (*AEM Ref.* p. 421 (#3))

Pr.8.9 **(Scalar triple product)** Find $(\mathbf{a} \times \mathbf{b})$•**c** and **a**•$(\mathbf{b} \times \mathbf{c})$, where **b** = $[-3, 2, 0]$ and **a** and **c** are as in Pr.8.8. (*AEM Ref.* p. 421 (#11))

Pr.8.10 **(Tetrahedron)** Find the volume of the tetrahedron with vertices $(1,3,6)$, $(3,7,12)$, $(8,8,9)$, $(2,2,8)$. (*AEM Ref.* p. 422 (#34))

Pr.8.11 **(Linear independence)** Are the vectors [3, 5, 9], [73, −56, 76], [−4, 7, −1] linearly independent? (*AEM Ref.* p. 422 (#37))

Pr.8.12 **(Plane)** Find the plane through the points $(1, 2, 1/4)$, $(4, 2, -2)$, $(0, 8, 4)$. (*AEM Ref.* p. 422 (#31))

Pr.8.13 **(Length of cross product)** Show that for any vectors **a** and **b** the length of $\mathbf{a} \times \mathbf{b}$ equals the square root of $(\mathbf{a}•\mathbf{a})(\mathbf{b}•\mathbf{b}) - (\mathbf{a}•\mathbf{b})^2$ (*AEM Ref.* p. 422 (#38))

Pr.8.14 **(Vector field)** Plot the vector field $\mathbf{v} = [y^2, 1]$. (*AEM Ref.* p. 427 (#19))

Pr.8.15 **(Length of a curve)** Find the length of the **catenary** $\mathbf{r} = [t, \cosh t]$ from $t = 0$ to $t = 1$. Plot this portion of the curve. (*AEM Ref.* p. 434 (#27))

Pr.8.16 **(Lissajous curves)** Plot the special Lissajous curve $[\cos t, \sin 3t]$ for $t = 0 \ldots 2\pi$ (named after the French physicist J. A. Lissajous, 1822-1880). First guess what the curve might look like.

Pr.8.17 **(Torsion)** Find the torsion $\tau = (\mathbf{r}' \ \mathbf{r}'' \ \mathbf{r}''')/[(\mathbf{r}'•\mathbf{r}')(\mathbf{r}''•\mathbf{r}'') - (\mathbf{r}'•\mathbf{r}'')^2]$ of the curve $\mathbf{r}(t) = [t, t^2, t^3]$. (*AEM Ref.* p. 443 (#15). See also Example 8.3 in this Guide.)

Pr.8.18 **(Tangential acceleration)** Find the tangential acceleration of the curve $\mathbf{r}(t) = [\cos t, \sin 2t, \cos 2t]$. Plot the curve. (*AEM Ref.* p. 439 (#8c))

Pr.8.19 **(Gradient)** Find the gradient of the function $f(x, y) = \ln(x^2 + y^2)$ and its value at the point $(2, 0)$. Plot the gradient field in the first quadrant near the origin, say, for x and y from 0.1 to 0.5. (*AEM Ref.* p. 452 (#3))

Pr.8.20 **(Directional derivative)** Find the directional derivative of the function in Pr.8.19 at the point $(4, 0)$ in the direction of the vector $\mathbf{a} = [1, -1]$. (*AEM Ref.* p. 453 (#32))

Pr.8.21 **(Surface normal)** Find a unit normal vector of the cone $z^2 = x^2 + y^2$ at the point $(3, 4, 5)$. (*AEM Ref.* pp. 449, 450)

Pr.8.22 **(Potential)** Using suitable integrations, find out whether the vector function $\cos(xyz)[yz, zx, xy]$ has a potential, and if so, determine it. (*AEM Ref.* p. 450)

Pr.8.23 **(Divergence)** Find the divergence of $(x^2 + y^2)^{-1}[-y, x]$. (*AEM Ref.* p. 456 (#5))

Pr.8.24 **(Divergence)** Prove $\operatorname{div}(f\mathbf{v}) = f\operatorname{div}\mathbf{v} + \mathbf{v} \bullet \operatorname{grad} f$. (*AEM Ref.* p. 456 (#13))

Pr.8.25 **(Curl)** Illustrate the important relations $\operatorname{curl}(\operatorname{grad} f) = \mathbf{0}$ and $\operatorname{div}(\operatorname{curl}\mathbf{v}) = 0$ by examples of your own and list some typical physical applications of these relations. (*AEM Ref.* p. 458)

Chapter 9

Vector Integral Calculus. Integral Theorems

Content. Line integrals (Exs. 9.1, 9.2, Prs. 9.1-9.4)
Double integrals, Green's theorem (Exs. 9.3, 9.4, Prs. 9.5-9.9)
Surface integrals, Gauss and Stokes theorems
(Exs. 9.5-9.7, Prs. 9.10-9.13, 9.16-9.20)
Triple integrals (Prs. 9.14, 9.15)

`Integrate[fC, {t, a, b}]` evaluates line integrals converted to integrals over the parameter interval $a \le t \le b$ of the path of integration C (see Ex. 9.1).
`Integrate[Integrate[...]]` evaluates double integrals (Ex. 9.3).
`D[F, x]`, `D[F, y]` give partial derivatives (Ex. 9.4).

Examples for Chapter 9

EXAMPLE 9.1 │ LINE INTEGRALS

A **line integral** of a function f over a curve $C : \mathbf{r}(t)$, $a \le t \le b$, in space or in the plane (called the **path of integration**) can be defined by

$$\int_C f(\mathbf{r})\, d\mathbf{r} = \int_a^b f(\mathbf{r}(t))\, dt.$$

For example, let C be the helix $\mathbf{r}(t) = [\cos t,\ \sin t,\ 3t]$, $a = 0$, $b = 2\pi$, and

$$f(\mathbf{r}) = f(x, y, z) = (x^2 + y^2 + z^2)^2.$$

Then type

In[1]:= `r = {Cos[t], Sin[t], 3 t}`   (* Out: {Cos [t], Sin [t], 3 t} *)

In[2]:= `f = (x^2 + y^2 + z^2)^2`   (* Out: $(x^2 + y^2 + z^2)^2$ *)

Substituting $x = \cos t$, $y = \sin t$, $z = 3t$ gives $f(\mathbf{r}(t))$, that is, f on C. Denote this by `fC`. Accordingly, type (`r[[1]]` being the first component of `r`, etc.)

In[3]:= `fC = f /. {x -> r[[1]], y -> r[[2]], z -> r[[3]]}`

Out[3]= $(9\,t^2 + \text{Cos } [t]^2 + \text{Sin } [t]^2)^2$

In[4]:= `Simplify[%]`   (* Use $\cos^2 t + \sin^2 t = 1$. *)

Out[4]= $(1 + 9\,t^2)^2$

In[5]:= `answer = Integrate[fC, {t, 0, 2 Pi}]`

Out[5]= $2\,\pi + 48\,\pi^3 + \dfrac{2592\,\pi^5}{5}$

98

In[6]:= N[%]   (* Out: 160135. *)

Work integrals. This is a practical (nonstandard) name for another very useful kind of line integral in which a vector function $\mathbf{F}(\mathbf{r})$ is given and one takes and integrates the tangential component of $\mathbf{F}(\mathbf{r})$ in the tangent direction of the path of integration C. Hence this is the work done by a force $\mathbf{F}$ in a displacement of a body along C. In terms of a formula,

$$\int_C \mathbf{F}(\mathbf{r}) \bullet d\mathbf{r} = \int_a^b \mathbf{F}(\mathbf{r}(t)) \bullet \frac{d\mathbf{r}}{dt}\, dt.$$

For instance, let $\mathbf{F}(\mathbf{r}) = [z,\ x,\ y]$, and let C and $\mathbf{r}(t)$ be as before; thus r = {Cos[t], Sin[t], 3t}.

Denote $\mathbf{F}(\mathbf{r}(t))$ by FC and $\mathbf{r}' = d\mathbf{r}/dt$ by r1. Then

In[7]:= F = {z, x, y}   (* Out: {z, x, y} *)

In[8]:= r1 = D[r, t]   (* Out: {−Sin[t], Cos[t], 3} *)

In[9]:= FC = F /. {x -> r[[1]], y -> r[[2]], z -> r[[3]]}

Out[9]= {3 t, Cos[t], Sin[t]}

In this command, x = r[[1]] is the first component cos t of the position vector $\mathbf{r}$, etc. Similarly, F[[1]] is the first component of $\mathbf{F}$, namely, z, etc. To compute the inner product (the dot product), type

In[10]:= integrand = FC.r1

Out[10]= Cos[t]^2 + 3 Sin[t] − 3 t Sin[t]

In[11]:= answer = Integrate[integrand, {t, 0, 2 Pi}]   (* Out: 7π *)

In[12]:= ParametricPlot3D[Table[{r[[1]], r[[2]], r[[3]]}, {t, 0, 2 Pi}]

Similar Material in AEM: pp. 465-469

| EXAMPLE 9.2 | **INDEPENDENCE OF PATH**

A line integral from a given point A to a given point B will generally depend on the path along which you integrate from A to B. A line integral is called **independent of path in a domain** D in space if for every pair of endpoints A, B in D the integral has the same value for all paths in D that begin at A and end at B.

For instance, show that the work integral of $\mathbf{F} = [2x,\ 2y,\ 4z]$ is independent of path in any domain D in space and find its value if you integrate from $(0,\ 0,\ 0)$ to $(2,\ 2,\ 2)$.

Solution. Necessary and sufficient for path independence in D is that $\mathbf{F}$ (with continuous components) have a potential f in D, that is, that $\mathbf{F}$ be the gradient of a function f in D. Thus type

In[1]:= Clear[x, y, z, f, F]

In[2]:= <<Calculus'VectorAnalysis'

In[3]:= SetCoordinates[Cartesian[x, y, z]]   (* Out: Cartesian[x, y, z] *)

In[4]:= F = {2 x, 2 y, 4 z}   (* Out: {2x, 2y, 4z} *)

You see by inspection that a potential of $\mathbf{F}$ is

In[5]:= `f = x^2 + y^2 + 2 z^2`

In[6]:= `Grad[f]`   `(* Out: {2x, 2y, 4z} *)`

In more complicated cases, instead of inspection you may integrate and see whether you can compose a potential from the three results obtained. Note that in the first integration you have an integration "constant" depending on y and z, and so on.

In[7]:= `Integrate[F[[1]], x]`   `(* Out: `x^2` *)`

In[8]:= `Integrate[F[[2]], y]`   `(* Out: `y^2` *)`

In[9]:= `Integrate[F[[3]], z]`   `(* Out: `$2z^2$` *)`

From this you see that a potential is $x^2 + y^2 + 2z^2$ as claimed before.

Furthermore, if the components of **F** are continuous and have continuous first partial derivatives and if **F** is path independent in a domain D, then its curl is the zero vector. In our case,

In[10]:= `Curl[F]`   `(* Out: {0, 0, 0} *)`

This condition is also sufficient for path independence in D, provided D is simply connected.

Path independence being guaranteed, you may now choose the most convenient path, namely, the straight-line segment C from the origin $A : (0, 0, 0)$ to $B : (2, 2, 2)$, and perform the integration as in the previous example. Let $\mathbf{r} = \mathbf{r}(t)$ be the position vector of C and `r1` its derivative with respect to t. Let `FC` denote **F** on C, that is, $\mathbf{F}(\mathbf{r}(t))$. Type

In[11]:= `r = {t, t, t};`   `r1 = D[r, t]`

Out[11]= $\{1, 1, 1\}$

In[12]:= `FC = F /. {x -> r[[1]], y -> r[[2]], z -> r[[3]]}`

Out[12]= $\{2t, 2t, 4t\}$

In[13]:= `answer = Integrate[FC.r1, {t, 0, 2}]`   `(* Out: 16 *)`

This result is much more quickly obtained by using the potential f and noting that the integral equals $f(B) - f(A)$ (the analog of a well-known formula from calculus) valid in the case of path independence.

In[14]:= `(f /. {x -> 2, y -> 2, z -> 2}) - f /. {x -> 0, y -> 0, z -> 0}`

Out[14]= 16

Of course, this simple example merely serves to explain the basic facts and techniques, and a computer or even a calculator would hardly be needed here.

Similar Material in AEM: pp. 471-475

EXAMPLE 9.3   DOUBLE INTEGRALS. MOMENTS OF INERTIA

Find the moments of inertia I_x and I_y of a mass of density $\sigma = 1$ in the triangle with vertices $(0, 0)$, $(b, 0)$, (b, h) about the coordinate axes, as well as the polar moment $I_0 = I_x + I_y$. (Sketch the triangle.)

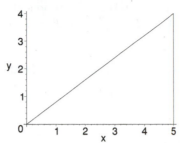

Example 9.3. Region of integration when $b = 5$ and $h = 4$

Solution. Since $\sigma = 1$, the integrand of I_x is y^2. If you integrate stepwise first over y from 0 to xh/b and then over x from 0 to b, you obtain

In[1]:= `Integrate[y^2, {y, 0, h x/b}]` 　　　　　　 (* Out: $\dfrac{h^3 x^3}{3 b^3}$ *)

In[2]:= `Ix = Integrate[%, {x, 0, b}]` 　　　　　　 (* Out: $\dfrac{b h^3}{12}$ *)

If you combine the two steps into one, you obtain the same result,

In[3]:= `Integrate[Integrate[y^2, {y, 0, h x/b}], {x, 0, b}]` 　 (* Out: $\dfrac{b h^3}{12}$ *)

If you integrate first over x, you must integrate y^2 from by/h to b and then thr result over y from 0 to h, obtaining

In[4]:= `Integrate[Integrate[y^2, {x, b y/h, b}], {y, 0, h}]` 　 (* Out: $\dfrac{b h^3}{12}$ *)

Similarly, the integrand of I_y is x^2 and integration gives

In[5]:= `Iy = Integrate[Integrate[x^2, {y, 0, h x/b}], {x, 0, b}]`

Out[5]= $\dfrac{b^3 h}{4}$

The **polar moment of inertia** I_0 is the sum $I_x + I_y$,

In[6]:= `I0 = Ix + Iy` 　　　　　　 (* Out: $\dfrac{b^3 h}{4} + \dfrac{b h^3}{12}$ *)

In[7]:= `Factor[%]` 　　　　　　 (* Out: $\dfrac{1}{12} b h \left(3 b^2 + h^2\right)$ *)

Similar Material in AEM: pp. 481, 484 (#17)

$\boxed{\textbf{EXAMPLE 9.4}}$　　**GREEN'S THEOREM IN THE PLANE**

Green's theorem in the plane transforms a double integral over a region R in the xy-plane into a line integral over the boundary C of R and conversely. The formula is

$$\iint\limits_{R} \left[\left(\frac{\partial}{\partial x} F_2 \right) - \left(\frac{\partial}{\partial y} F_1 \right) \right] dx\, dy = \oint_C (F_1\, dx + F_2\, dy).$$

It is valid under suitable regularity assumptions on R, C, F_1, and F_2, usually satisfied in applications (see AEM, p. 485).

For instance, proceeding first in terms of components (and later repeating the calculation in terms of vectors), let $F_1 = y^2 - 7\,y$, $F_2 = 2\,x\,y + 2\,x$. Type

In[1]:= <<Calculus'VectorAnalysis'

In[2]:= SetCoordinates[Cartesian[x, y, z]]   (* Out: Cartesian[x, y, z] *)

In the following, type space between x and y.

In[3]:= F1 = y^2 - 7 y;   F2 = 2 x y + 2 x;

The integrand of the double integral is

In[4]:= D[F2, x] - D[F1, y]   (* Out: 9 *)

Hence the integrand is constant, so that the integral equals 9 times the area of the region of integration R. For instance, if R is a circular disk of radius 1, the integral equals 9π.

We verify the formula of Green's theorem for this case, assuming that the disk has the center at the origin of the xy-plane. Then its boundary curve C is the circle $x^2 + y^2 = 1$. In polar coordinates, the position vector **r** of C (oriented counterclockwise!) and its derivative **r**$'$ have the components

In[5]:= x = Cos[t];   y = Sin[t];

In[6]:= x1 = D[x, t]   (* Out: −Sin[t] *)

In[7]:= y1 = D[y, t]   (* Out: Cos[t] *)

Hence the line integral on the right-hand side of the formula is

In[8]:= Integrate[F1 x1 + F2 y1, {t, 0, 2 Pi}]   (* Out: 9π *)

This verifies Green's theorem for the present example.

The same example in vectorial notation. Let

In[9]:= Clear[x, y]   (* Unassign x and y. *)

In[10]:= r = {Cos[t], Sin[t], 0}

Out[10]= {Cos [t], Sin [t], 0}

In[11]:= r1 = D[r, t]

Out[11]= {−Sin [t], Cos [t], 0}

In[12]:= F = {y^2 - 7 y, 2 x y + 2 x, 0}

Out[12]= {−7 y + y^2, 2 x + 2 x y, 0}

In[13]:= CU = Curl[F]   (* Out: {0, 0, 9} *)

Hence the integrand (curl **F**) • **k** of the double integral (with **k** the unit vector in the z-direction) equals 9, as before,

In[14]:= CU.{0, 0, 1}   (* Out: 9   Dot product of the curl and **k** *)

For the line integral you get (with r[[1]] and r[[2]] the x and y components of **r**), as before,

In[15]:= FC = F /. {x -> r[[1]], y -> r[[2]]}

Out[15]= {−7 Sin [t] + Sin [t]^2, 2 Cos [t] + 2 Cos [t] Sin [t], 0}

In[16]:= `Integrate[FC.r1, {t, 0, 2 Pi]}`   (* Out: 9π *)

Note that in using the curl, Mathematica requires that you carry along a third component 0 in **r**. (Can you find the reason?)

Similar Material in AEM: pp. 485, 486

$\boxed{\textbf{EXAMPLE 9.5}}$ SURFACE INTEGRALS. FLUX

These are integrals taken over a surface S. We assume S to be given parametrically in the form $\mathbf{r}(u,v)$, where (u,v) varies over a region R in the uv-plane. The integrand is a scalar function given in the form $\mathbf{F}\bullet\mathbf{n}$, where the vector function $\mathbf{F}$ is given and $\mathbf{n}$ is a unit normal vector of S ($-\mathbf{n}$ being the other one). This is very practical in flow problems, where $\mathbf{F} = \rho\mathbf{v}$ (ρ the density, $\mathbf{v}$ the velocity) and $\mathbf{F}\bullet\mathbf{n}$ is the **flux** across S, that is, the mass of fluid crossing S per unit time. The integral is evaluated by reducing it to a double integral over R, namely,

$$\iint_S \mathbf{F}\bullet\mathbf{n}\, dA = \iint_R \mathbf{F}(\mathbf{r}(u,v))\bullet\mathbf{N}(u,v)\, du\, dv,$$

where dA is the element of area, $\mathbf{N} = \mathbf{r}_u\times\mathbf{r}_v$ is a normal vector of S, and $\mathbf{r}_u$ and $\mathbf{r}_v$ are the partial derivatives of $\mathbf{r}(u,v)$.

For instance, let $\mathbf{r} = [u,\ u^2,\ v]$ with u varying from 0 to 2 and v from 0 to 3. You see that $y = x^2$, so that this surface intersects the xy-plane along the parabola $y = x^2$. See the figure on the next page.

Let $\mathbf{F} = [3z^2,\ 6,\ 6xz]$ and $\rho = 1$ gram/cm^3 = 1 ton/meter^3, as for water, speed being measured in meters/sec. Type

In[1]:= `<<Calculus'VectorAnalysis'`

In[2]:= `Clear[u, v]`

In[3]:= `r = {u, u^2, v}`   (* Out: $\{u, u^2, v\}$ *)

In[4]:= `ru = D[r, u]`   (* Out: {1, 2u, 0} *)

In[5]:= `rv = D[r, v]`   (* Out: {0,0,1} *)

In[6]:= `NO = Cross[ru, rv]`   (* Out: {2u, −1, 0}   N is protected *)

In[7]:= `F = {3 z^2, 6, 6 x z}`   (* Out: $\{3z^2, 6, 6xz\}$ *)

On S the vector function $\mathbf{F}$ takes the form $\mathbf{F}(\mathbf{r}(u,v))$. Call it `FS` and type it and the integrand on the right-hand side of the general formula

In[8]:= `FS = F /. {x -> r[[1]], y -> r[[2]], z -> r[[3]]}`

Out[8]= $\{3v^2, 6, 6uv\}$

In[9]:= `integrand = FS.NO`   (* Out: $-6 + 6uv^2$ *)

Now integrate over u from 0 to 2 and over v form 0 to 3,

In[10]:= `flux = Integrate[Integrate[integrand, {u, 0, 2}], {v, 0, 3}]`

Out[10]= 72

Since speed is measured in meters/sec, the flux is measured in tons/sec; this gives the answer 72 tons/sec or 72000 liters/sec.

In the figure the origin is in the back on the right. The x-axis is in the back and points to the left. The y-axis is on the right and points to the front. The z-axis is in the back, pointing upward. The scales should be moved accordingly. The viewpoint was found by experimenting. Try other choices.

In[11]:= `ParametricPlot3D[{r[[1]], r[[2]], r[[3]]}, {u, 0, 2}, {v, 0, 3},`
      `ViewPoint -> {0, 4, 1}]`

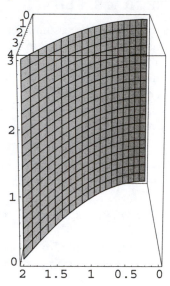

Example 9.5. Surface over which the integral is extended

Similar Material in AEM: pp. 497, 498

EXAMPLE 9.6 **DIVERGENCE THEOREM OF GAUSS**

The **divergence theorem** transforms a triple integral over a region T in space into a surface integral over the boundary surface S of T and conversely. The formula is

$$\iiint_T \operatorname{div} \mathbf{F}\, dV = \iint_S \mathbf{F} \bullet \mathbf{n}\, dA$$

where $\mathbf{F}$ is a given vector function, dV is the volume element, $\mathbf{n}$ is the outer unit normal vector of S, and dA is the element of area of S. The formula is valid under suitable regularity assumptions on T, S, and $\mathbf{F}$ usually satisfied in applications (see AEM, p. 506). Note that if in a flow problem, $\mathbf{F} = \rho\mathbf{v}$ (ρ the density, $\mathbf{v}$ the velocity), then the integral over S gives the flux through S. This is similar to the previous example.

For instance, let T be the solid circular cylinder of radius a with the z-axis as the axis, extending from $z = 0$ to $z = b$ (> 0), and let $\mathbf{F} = [x^3, x^2 y, x^2 z]$. Using the divergence theorem, evaluate the surface integral of $\mathbf{F} \bullet \mathbf{n}$ over the surface S of T (consisting of the upper and lower disks and the lateral cylindrical part).

Solution. Type

In[1]:= `<<Calculus`VectorAnalysis``

In[2]:= `SetCoordinates[Cartesian[x, y, z]]`   (* Out: `Cartesian[x, y, z]` *)

In[3]:= `F = {x^3, x^2 y, x^2 z}`   (* Out: $\{x^3, x^2 y, x^2 z\}$ *)

In[4]:= `divF = Div[F]`   (* Out: $5x^2$ *)

The form of T suggests the use of cylindrical coordinates $x = r\cos\theta$, $y = r\sin\theta$, $z = z$. Accordingly, make the substitution (note that div $\mathbf{F}$ depends only on x)

In[5]:= `Clear[r, `θ`, a, b]`   (* **Unassign** *)

In[6]:= `f = divF /. x -> r Cos[`θ`]`   (* Out: $5r^2\cos[\theta]^2$ *)

This is the integrand of the triple integral in the divergence theorem, which must be multiplied by the volume element $r\,dr\,d\theta\,dz$ and integrated over r from 0 to a, over θ from 0 to 2π, and over z from 0 to b. Type

In[7]:= `answer = Integrate[Integrate[Integrate[f r, {r, 0, a}],`
      `{`θ`, 0, 2 Pi}], {z, 0, b}]`

Out[7]= $\dfrac{5}{4}\mathbf{a}^4\,\mathbf{b}\,\pi$

The direct evaluation of the surface integral should give the same result. It is more elaborate and thus illustrates the usefulness of the divergence theorem. Begin with the disk on the top, whose outer unit normal vector is pointing vertically upward,

In[8]:= `n = {0, 0, 1}`   (* Out: $\{0, 0, 1\}$ *)

You need the dot product (the inner product) $\mathbf{F} \bullet \mathbf{n}$,

In[9]:= `ip = F.n`   (* Out: $x^2 z$ *)

Introduce polar coordinates by setting $x = r\cos\theta$ and then integrate over r from 0 to a and over θ from 0 to 2π. Use $dA = r\,dr\,d\theta$. Call the integral J1. Recall that θ is typed as `\[Theta]` or by the use of **Palettes** (see Ex. 7.1 in this Guide).

In[10]:= `Clear[r]`

In[11]:= `g = ip /. {x -> r Cos[`θ`], z -> b}`

Out[11]= $\mathbf{b}\,\mathbf{r}^2\cos[\theta]^2$

In[12]:= `J1 = Integrate[Integrate[g r, {r, 0, a}], {`θ`, 0, 2 Pi}]`

Out[12]= $\dfrac{1}{4}\mathbf{a}^4\,\mathbf{b}\,\pi$

For the bottom $z = 0$ the inner product is $-x^2 z$, hence it is zero. (The outer normal vector points downward; this gives the minus sign.)

Now turn to the vertical cylinder surface. Represent it by

In[13]:= `r = {a Cos[`θ`], a Sin[`θ`], z}`

Out[13]= $\{\mathbf{a}\cos[\theta],\,\mathbf{a}\sin[\theta],\,\mathbf{z}\}$

Obviously, its normal is horizontal and at each point has the direction of the position vector. Hence the outer unit normal vector n2 of the cylinder is

In[14]:= `n2 = {Cos[`θ`], Sin[`θ`], 0}`   (* Out: $\{\cos[\theta], \sin[\theta], 0\}$ *)

Let `FC` denote **F** on the cylinder. Recall that `r[[1]]`, `r[[2]]`, `r[[3]]` are the components of **r**. Hence type

In[15]:= `FC = F /. {x -> r[[1]], y -> r[[2]], z -> r[[3]]}`

Out[15]= $\{a^3 \, \text{Cos}\,[\theta]^3,\ a^3 \, \text{Cos}\,[\theta]^2 \, \text{Sin}\,[\theta],\ a^2 \, z \, \text{Cos}\,[\theta]^2\}$

In[16]:= `ip2 = FC.n2`

Out[16]= $a^3 \, \text{Cos}\,[\theta]^4 + a^3 \, \text{Cos}\,[\theta]^2 \, \text{Sin}\,[\theta]^2$

In[17]:= `Simplify[%]`   `(* Out: `$a^3 \, \text{Cos}\,[\theta]^2$` *)`

This is the integrand. Multiply it by the element of area $a \, d\theta \, dz$ and integrate over θ from 0 to 2π and over z from 0 to b. Call the integral `J2`. Add it to `J1` to get the answer, which agrees with the previous one.

In[18]:= `J2 = Integrate[Integrate[ip2 a, {`θ`, 0, 2 Pi}], {z, 0, b}]`

Out[18]= $a^4 \, b \, \pi$

In[19]:= `answer = J1 + J2`   `(* Out: `$\frac{5}{4}a^4 \, b \, \pi$` *)`

Similar Material in AEM: pp. 506, 507

EXAMPLE 9.7   STOKES'S THEOREM

Stokes's theorem transforms a surface integral over a surface S into a line integral over the boundary curve C of S and conversely. The formula is

$$\iint\limits_{S} (\text{curl}\,\mathbf{F}) \bullet \mathbf{n}\ dA = \oint_{C} \mathbf{F} \bullet \mathbf{r}'(s)\,ds$$

where **n** is a unit surface normal vector of S and $\mathbf{r}'(s)$ is a unit tangent vector of C and the direction of integration around C appears counterclockwise if one looks from the terminal point of **n** onto the surface. s is the arc length of C. The formula holds under suitable regularity conditions on S, C, and **F** usually satisfied in applications (see AEM, p. 516).

For instance, using Stokes's theorem, evaluate the line integral in the formula when $\mathbf{F} = [2\,y^2,\ x,\ -z^3]$ and C is the circle $x^2 + y^2 = a^2$, $z = b\ (> 0)$.

Solution. Type **F** as given. Then obtain the curl.

In[1]:= `<<Calculus'VectorAnalysis'`

In[2]:= `SetCoordinates[Cartesian[x, y, z]]`   `(* Out: Cartesian[x, y, z] *)`

In[3]:= `F = {2 y^2, x, -z^3}`

Out[3]= $\{2\,y^2,\ x,\ -z^3\}$

In[4]:= `curlF = Curl[F]`   `(* Out: {0, 0, 1 - 4 y} *)`

As surface S choose the disk bounded by C. Obtain the curl on S; denote it by `curlFS`.

In[5]:= `r = {u Cos[v], u Sin[v], b}`

Out[5]= $\{u \, \text{Cos}\,[v],\ u \, \text{Sin}\,[v],\ b\}$

In[6]:= `curlFS = curlF /. y -> r[[2]]`   `(* Out: {0, 0, 1 - 4 u Sin[v]} *)`

A unit normal vector of the disk is

In[7]:= `n = {0, 0, 1}`   (* Out: $\{0,0,1\}$ *)

You can now type the integrand `int` and then integrate over u from 0 to a and over the angle v from 0 to 2π.

In[8]:= `int = curlFS.n`   (* Out: $1 - 4\,u\,\text{Sin}\,[v]$ *)

Recalling that in the polar coordinates u, v the element of area is $u\,du\,dv$, you see that you have to multiply `int` by `u`. This gives

In[9]:= `answer = Integrate[Integrate[int u, {u, 0, a}], {v, 0, 2 Pi}]`

Out[9]= $a^2\,\pi$

Confirmation by direct evaluation of the line integral around the boundary circle $C : x^2 + y^2 = a^2$, $z = b$. A representation of C is the following, in which t goes from 0 to 2π.

In[10]:= `r = {a Cos[t], a Sin[t], b}`   (* Out: $\{a\,\text{Cos}[t], a\,\text{Sin}[t], b\}$ *)

You can use t as a variable of integration because in the integral, by calculus, $\mathbf{r}'(s)\,ds = (d\mathbf{r}/ds)\,ds = (d\mathbf{r}/dt)\,dt$. Denote $\mathbf{F}$ on C by `FC` and obtain it by typing

In[11]:= `FC = F /. {x -> r[[1]], y -> r[[2]], z -> r[[3]]}`

Out[11]= $\{2\,a^2\,\text{Sin}\,[t]^2, a\,\text{Cos}\,[t], -b^3\}$

Denote $d\mathbf{r}/dt$ by `r1` and type

In[12]:= `r1 = D[r, t]`   (* Out: $\{-a\,\text{Sin}[t], a\,\text{Cos}[t], 0\}$ *)

Now obtain the integrand `int2` and finally the integral, which will agree with the previous result.

In[13]:= `int2 = FC.r1`   (* Out: $a^2\,\text{Cos}\,[t]^2 - 2\,a^3\,\text{Sin}\,[t]^3$ *)

In[14]:= `Integrate[int2, {t, 0, 2 Pi}]`   (* Out: $a^2\,\pi$ *)

Similar Material in AEM: pp. 516, 521 (#8)

Problem Set for Chapter 9

Pr.9.1 **(Line integral. Work)** Evaluate the work integral (see Example 9.1 in this Guide) of the force $\mathbf{F} = [2z,\ x,\ -y]$ from $(1, 0, 0)$ to $(1, 0, 4\pi)$ along the helix C : $\mathbf{r} = [\cos t,\ \sin t,\ 2t]$. (*AEM Ref.* p. 470 (#7))

Pr.9.2 **(Path dependence, same endpoints)** Show that the integral in Pr.9.1 changes its value if you integrate from $(1, 0, 0)$ to $(1, 0, 4\pi)$ along the straight-line segment with these endpoints. (*AEM Ref.* p. 470 (#7))

Pr.9.3 **(Independence of path. Potential)** Using a suitable curl, show that the integral $\int_C (3x^2\,dx + 2yz\,dy + y^2\,dz)$ is independent of path in any domain in space. Find a potential and use it to obtain the value of the integral from $A : (0, 1, 2)$ to $B : (1, -1, 7)$. (*AEM Ref.* p. 473)

Pr.9.4 **(Independence of path)** Is the integral of $e^z\,dx + 2y\,dy + xe^z\,dz$ independent of path in space? If this is the case, find a potential by integration. Using the potential, integrate the given form from the origin to the point (a, b, c). (*AEM Ref.* p. 478 (#19))

Pr.9.5 **(Double integral. Center of gravity)** Find the center of gravity of a mass of density $\sigma = 1$ in the portion of the disk $x^2 + y^2 \leq 1$ in the first quadrant of the xy-plane. (*AEM Ref.* pp. 481, 482)

Pr.9.6 **(Double integral. Moment of inertia)** Consider a mass of density $\sigma = 1$ in the portion of the disk $x^2 + y^2 \leq a^2$ in the upper half-plane. Find the polar moment of inertia I_0 of this mass about the origin. (*AEM Ref.* p. 481)

Pr.9.7 **(Green's theorem in the plane)** Using the formula of Green's theorem (see Example 9.4 in this Guide), integrate $\mathbf{F}(\mathbf{r}(t)) \bullet \mathbf{r}'(t)$ counterclockwise around the boundary of the region $R : 1 + x^4 \leq y \leq 2$; here, $\mathbf{F} = [e^y/x, \ e^y \ln x + 2x]$. (*AEM Ref.* p. 490 (#9))

Pr.9.8 **(Green's theorem in the plane)** Using Green's theorem in the plane (see Example 9.4 in this Guide), integrate $\mathbf{F}(\mathbf{r}(t)) \bullet \mathbf{r}'(t)$ with $\mathbf{F} = [x \cosh 2y, \ 2x^2 \sinh 2y]$ counterclockwise around the boundary of the region $R : x^2 \leq y \leq x$. (*AEM Ref.* p. 490 (#10))

Pr.9.9 **(Area)** Choosing $F_1 = 0$, $F_2 = x$ in the formula of Green's theorem in Example 9.4 of this Guide gives $\iint_R dx\,dy = \oint_C x\,dy$. Similarly, $\iint_R dx\,dy = -\oint_C y\,dx$ by choosing $F_1 = -y$, $F_2 = 0$. The double integral is the area A of R. Together,

$$A = \frac{1}{2} \oint_C (x\,dy - y\,dx).$$

Obtain from this the area of an ellipse $x^2/a^2 + y^2/b^2 = 1$. (*AEM Ref.* p. 488)

Pr.9.10 **(Experiment on surface normal)** Find a representation of the **ellipsoid**

$$S : \mathbf{r}(u,v) = [a \cos v \cos u, \ b \cos v \sin u, \ c \sin v]$$

in terms of Cartesian coordinates. Find a normal vector of S. Plot S for some triples a, b, c, for instance, $a = 3$, $b = 2$, $c = 1$. Choose other triples and observe how the surface and its normal change. (*AEM Ref.* p. 495 (#6))

Pr.9.11 **(Surface integral)** Find the flux integral (see Example 9.5 in this Guide for the definition) of $\mathbf{F} = [x^2, \ 0, \ 3y^2]$ over the portion of the plane $x + y + z = 1$ in the first octant in space. (*AEM Ref.* p. 498)

Pr.9.12 **(Surface integral)** Find the flux integral (as defined in Example 9.5 of this Guide) of $\mathbf{F} = [x^2, y^2, z^2]$ over the **helicoid** $\mathbf{r} = [u \cos v, \ u \sin v, \ 3v]$, where $0 \leq u \leq 10$, $0 \leq v \leq 2\pi$. Sketch the surface. Explain its name. (*AEM Ref.* p. 503 (#10))

Pr.9.13 **(Integral over a sphere)** Find the flux integral (as defined in Example 9.5 in this Guide) of $\mathbf{F} = [0, \ x, \ 0]$ over the portion of the sphere $x^2 + y^2 + z^2 = 1$ in the first octant. (*AEM Ref.* pp. 493, 503 (#7))

Pr.9.14 **(Triple integral)** Find the total mass in the box $|x| \leq 1$, $|y| \leq 3$, $|z| \leq 2$ if the mass density is $\sigma = x^2 + y^2 + z^2$. (*AEM Ref.* p. 509 (#1))

Pr.9.15 **(Triple integral. Moment of inertia)** Find the moment of inertia I_x about the x-axis of a mass of density $\sigma = 1$ in the solid circular cylinder of radius a about the x-axis, extending in x-direction from 0 to h. (*AEM Ref.* p. 510 (#11))

Pr.9.16 (**Surface integral. Divergence theorem**) Using the divergence theorem (see Example 9.6 in this Guide), find the integral of the normal component of the vector function $\mathbf{F} = [\,4\,x,\; x^2\,y^2,\; y^2\,z^2\,]$ over the surface of the tetrahedron with vertices $(0,0,0)$, $(1,0,0)$, $(0,1,0)$, $(0,0,1)$. (*AEM Ref.* p. 506)

Pr.9.17 (**Surface integral. Divergence theorem**) Using the divergence theorem (see Example 9.6 in this Guide), integrate the normal component of the vector function $\mathbf{F} = [9x,\; y\cosh^2 x,\; -z\sinh^2 x]$ over the ellipsoid $4\,x^2 + y^2 + 9\,z^2 = 36$. (*AEM Ref.* p. 515 (#5))

Pr.9.18 (**Laplacian, normal derivative**) If $\mathbf{F} = \operatorname{grad} f$ in the formula of the divergence theorem in Example 9.6 in this Guide, show that $\operatorname{div} \mathbf{F} = \nabla^2 f$ and $\mathbf{F} \bullet \mathbf{n} = \partial f / \partial n$ (the normal derivative). Verify the resulting formula

$$\iiint\limits_{T} \nabla^2 f \, dV = \iint\limits_{S} \frac{\partial f}{\partial n} \, dA$$

for $f = (x^2 + y^2 + z^2)^2$ and S a sphere of radius a and center at the origin. *Hint:* Represent the interior of S by $\mathbf{R} = [r\cos u \sin v,\; r\sin u \sin v,\; r\cos v]$ (where r is variable!) and use the corresponding volume element $r^2 \sin v \, dr \, dv \, du$. (*AEM Ref.* pp. 493, 513)

Pr.9.19 (**Surface integral. Divergence theorem**) Using the divergence theorem, find the integral of the normal derivative of $\mathbf{F} = [x^3,\; y^3,\; z^3]$ over the sphere $S : x^2 + y^2 + z^2 = 9$. Represent S as in the previous problem. (*AEM Ref.* p. 510 (#19))

Pr.9.20 (**Stokes's theorem**) Using Stokes's theorem (see Example 9.7 in this Guide), integrate the tangential component of $\mathbf{F} = [e^z,\; e^z \sin y,\; e^z \cos y]$ around the boundary curve of the surface $S : z = y^2$, where $0 \le x \le 4$, $0 \le y \le 2$. (*AEM Ref.* p. 520 (#3))

PART C. FOURIER ANALYSIS AND PARTIAL DIFFERENTIAL EQUATIONS

Content. Fourier series, integrals, and transforms (Chap. 10)
Partial differential equations (PDE's) (Chap. 11)

This part concerns initial and boundary value problems for the "big" PDE's of Applied Mathematics, namely, the wave, heat, and Laplace equations, their solution by separating variables, and the use of Fourier series and integrals for obtaining solutions sufficiently general to satisfy all the given physical conditions.

Chapter 10

Fourier Series, Integrals, and Transforms

Content. Fourier series (period 2π) (Exs. 10.1, 10.4, Prs. 10.1-10.5)
Fourier series (any period) (Exs. 10.2, 10.3, Prs. 10.6-10.10, 10.15)
Half-range expansions (Ex. 10.3, Pr. 10.9)
Trigonometric approximation (Ex. 10.5, Prs. 10.11-10.13)
Fourier integral (Ex. 10.6, Pr. 10.14)

The **Fourier series** of a function $f(x)$ of period $p = 2L$ is obtained by typing (ignore the error message)

In[1]:= `f = a0 + Sum[an Cos[n Pi x/L] + bn Sin[n Pi x/L], {n, 1, Infinity}]`

Out[1]= $a0 + \sum_{n=1}^{\infty} \left(an \cos\left[\frac{n\pi x}{L}\right] + bn \sin\left[\frac{n\pi x}{L}\right] \right)$

with the **Fourier coefficients** a_n and b_n given by the **Euler formulas**

In[2]:= `Clear[f]`

In[3]:= `a[0] = 1/(2 L) Integrate[f[x], {x, -L, L}]`

Out[3]= $\dfrac{\int_{-L}^{L} f[x]\, dx}{2L}$

In[4]:= `a[n] = 1/L Integrate[f[x] Cos[n Pi x/L], {x, -L, L}]`

Out[4]= $\dfrac{\int_{-L}^{L} \cos\left[\frac{n\pi x}{L}\right] f[x]\, dx}{L}$

In[5]:= `b[n] = 1/L Integrate[f[x] Sin[n Pi x/L], {x, -L, L}]`

Out[5]= $\dfrac{\int_{-L}^{L} f[x] \sin\left[\frac{n\pi x}{L}\right] dx}{L}$

110

Functions of period $p = 2\pi$. Replace L by π, obtaining $\cos nx$, $\sin nx$ in all formulas, and $1/(2\pi)$ and $1/\pi$ as factors before the integrals, and integrate from $-\pi$ to π. Write down these (simpler) formulas for yourself.

Examples for Chapter 10

| EXAMPLE 10.1 | FUNCTIONS OF PERIOD 2π. EVEN FUNCTIONS. GIBBS PHENOMENON |

Consider the function $f(x)$ of period 2π which is 0 for $-\pi < x < -\pi/2$, then equals k for $-\pi/2 < x < \pi/2$ and is again 0 for $\pi/2 < x < \pi$ (see the figure, where $k = 1$). You may call this function a **periodic rectangular wave** of period 2π. You obtain the Fourier coefficients from the Euler formulas by integrating from $-\pi/2$ to $\pi/2$ only because $f(x)$ is 0 for the two other parts of the interval of periodicity. Hence type

In[1]:= `a0 = 1/(2 Pi) Integrate[k, {x, -Pi/2, Pi/2}]`   (* Out: $\frac{k}{2}$ *)

Note that this is the mean value of $f(x)$ over the interval from $-\pi$ to π (as it is always the case).

In[2]:= `an = 1/Pi Integrate[k Cos[n x], {x, -Pi/2, Pi/2}]`

Out[2]= $\dfrac{2 k \operatorname{Sin}\left[\frac{n\pi}{2}\right]}{n\pi}$

In[3]:= `bn = 1/Pi Integrate[k Sin[m x], {x, -Pi/2, Pi/2}]`   (* Out: 0 *)

Hence this Fourier series is a **Fourier cosine series**, that is, it has only cosine terms, no sine terms. The reason is that $f(x)$ is an **even function**, that is, $f(-x) = f(x)$. The command `Table` gives you the first few coefficients

In[4]:= `Table[an, {n, 1, 8}]`   (* Out: $\left\{\frac{2k}{\pi}, 0, -\frac{2k}{3\pi}, 0, \frac{2k}{5\pi}, 0, -\frac{2k}{7\pi}, 0\right\}$ *)

For plotting you have to choose a definite value of k, say, $k = 1$,

In[5]:= `An = an /. k -> 1`   (* Out: $\dfrac{2 \operatorname{Sin}\left[\frac{n\pi}{2}\right]}{n\pi}$ *)

Find and plot (on common axes) the first few partial sums

In[6]:= `S1 = 1/2 + Sum[An Cos[n x], {n, 1, 1}]`   (* Out: $\frac{1}{2} + \frac{2\operatorname{Cos}[x]}{\pi}$ *)

In[7]:= `S3 = 1/2 + Sum[An Cos[n x], {n, 1, 3}]`

Out[7]= $\dfrac{1}{2} + \dfrac{2\operatorname{Cos}[x]}{\pi} - \dfrac{2\operatorname{Cos}[3x]}{3\pi}$

In[8]:= `S5 = 1/2 + Sum[An Cos[n x], {n, 1, 5}]`

Out[8]= $\dfrac{1}{2} + \dfrac{2\operatorname{Cos}[x]}{\pi} - \dfrac{2\operatorname{Cos}[3x]}{3\pi} + \dfrac{2\operatorname{Cos}[5x]}{5\pi}$

In[9]:= `P1 = Plot[{S1, S3, S5}, {x, -Pi, Pi}];`

In[10]:= `P2 = Plot[1, {x, -Pi/2, Pi/2}];`

In[11]:= `Show[P1, P2]`

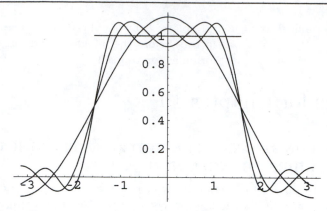

Example 10.1. Given rectangular wave and first three partial sums S1, S3, S5

You see that there are waves near the discontinuities at $-\pi/2$ and $\pi/2$, and you would perhaps expect that these waves would become smaller and smaller if you took partial sums with more and more terms – which could easily be done on the computer. However, these waves do not disappear, but they are shifted closer and closer to those points of discontinuity, as is shown in the figure for the partial sum from $n = 1$ to $n = 50$. This is called the **Gibbs phenomenon**. Experiment with $n = 100$ or even larger n.

In[12]:= Plot[1/2 + Sum[An Cos[n x], {n, 1, 50}], {x, -Pi, Pi}]

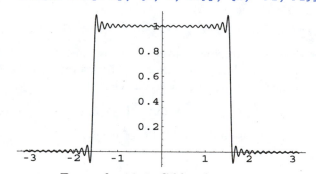

Example 10.1. Gibbs phenomenon

Similar Material in AEM: pp. 529-533

EXAMPLE 10.2   FUNCTIONS OF ARBITRARY PERIOD. ODD FUNCTIONS

For functions $f(x)$ of arbitrary period $p = 2L$ the determination of the Fourier coefficients is the same in principle as for functions of period 2π; just the formulas look a little more complicated.

For instance, consider the function $f(x)$ of period $p = 2L = 2$ in the figure.

In[1]:= P = Line[{{-3,-1},{-1,1}, {-1, -1}, {1,1}, {1,-1}, {3,1}}]

Out[1]= Line[{{-3, -1}, {-1, 1}, {-1, -1}, {1, 1}, {1, -1}, {3, 1}}]

In[2]:= Show[Graphics[P], Axes -> True, AxesLabel -> {x, y},
        AspectRatio -> Automatic]

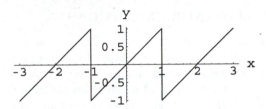

Example 10.2. Odd periodic function of period $p = 2$

Clearly, $f(x) = x$ if $-1 < x < 1$. For this function you obtain from the Euler formulas with $L = 1$ in the chapter opening

In[3]:= a0 = 1/2 Integrate[x, {x, -1, 1}]   (* Out: 0 *)

In[4]:= an = 1/2 Integrate[x Cos[n Pi x], {x, -1, 1}]   (* Out: 0 *)

Hence the series has no cosine terms, only sine terms; it is a **Fourier sine series**, because $f(x)$ is an **odd function,** that is, $f(-x) = -f(x)$. The sine terms have the Fourier coefficients

In[5]:= bn = Integrate[x Sin[n Pi x], {x, -1, 1}]

$$\text{Out[5]} = -\frac{2\cos[n\pi]}{n\pi} + \frac{2\sin n\pi]}{n^2\pi^2}$$

In[6]:= Table[bn, {n, 1, 8}]   (* Out: $\left\{ \frac{2}{\pi}, -\frac{1}{\pi}, \frac{2}{3\pi}, -\frac{1}{2\pi}, \frac{2}{5\pi}, -\frac{1}{3\pi}, \frac{2}{7\pi}, -\frac{1}{4\pi} \right\}$ *)

Now type and then plot some partial sums, including a large one (S50), which will again illustrate the **Gibbs phenomenon** at the points of discontinuity -1 and 1, just as in the preceding example.

In[7]:= S1 = Sum[bn Sin[n Pi x], {n, 1, 1}];

In[8]:= S2 = Sum[bn Sin[n Pi x], {n, 1, 2}];

In[9]:= S3 = Sum[bn Sin[n Pi x], {n, 1, 3}];

In[10]:= S50 = Sum[bn Sin[n Pi x], {n, 1, 50}];

In[11]:= Plot[{S1, S2, S3, S50}, {x, -1, 1}, AspectRatio -> Automatic]

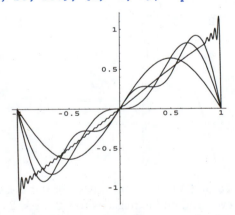

Example 10.2. Approximation of $f(x) = x$ by partial sums of its Fourier series

Similar Material in AEM: pp. 543, 544

EXAMPLE 10.3 | **HALF-RANGE EXPANSIONS**

If you develop a function $f(x)$, given for $0 < x < L$, in a Fourier series, then this series represents a periodic function of period L which for x from 0 to L agrees with $f(x)$. In general, this series contains both cosine and sine terms. In most applications it will be much better to first extend $f(x)$ as an **even** function $f_1(x)$ for $-L < x < L$; then $f_1(x)$ has the period $2L$, and its Fourier series contains only cosine terms because $f_1(x)$ is even. This series is called the **cosine half-range expansion** of $f(x)$ because $f(x)$ is given only over one half of the range (that is, over one half of the interval of periodicity).

Similarly, you may extend the given function $f(x)$ to an **odd** function $f_2(x)$ for $-L < x < L$; then $f_2(x)$ has the period $2L$, and its Fourier series contains only sine terms because $f_2(x)$ is odd. Call this series the **sine half-range expansion** of $f(x)$.

For instance, let $f(x) = x^2$ be given for $0 < x < 2$. Figure (A) shows its even and odd extensions to the interval (the "full range") $-2 < x < 2$. Figure (B) shows its even periodic extension $f_1(x)$ of period $p = 2L = 4$. Figure (C) shows its odd periodic extension $f_2(x)$ of period 4. `AspectRatio -> Automatic` gives equal scales on the axes.

In[1]:= `P1 = Plot[x^2, {x, -2, 2}];`

In[2]:= `P2 = Plot[-x^2, {x, -2, 0}];`

These are combined in the next command to give Fig. (A).

In[3]:= `Show[P1, P2, AspectRatio -> Automatic, PlotRange -> {-4, 4}]`

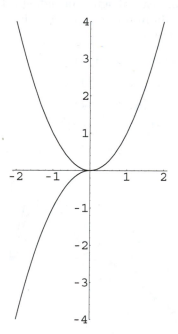

Example 10.3. **(A)** $f(x) = x^2$ (given for $0 < x < 2$)
and its even and odd extensions to $-2 < x < 2$

In[4]:= `P3 = Plot[(x + 4)^2, {x, -6, -2}];`

In[5]:= `P4 = Plot[(x - 4)^2, {x, 2, 6}];`

In[6]:= `P5 = Plot[(x - 8)^2, {x, 6, 10}];`

In[7]:= `Show[P1, P3, P4, P5, AspectRatio -> Automatic]`   (* Fig. (B). *)

In[8]:= `P6 = Plot[-(x + 4)^2, {x, -6, -4}];`

In[9]:= `P7 = Plot[(x + 4)^2, {x, -4, -2}];`

In[10]:= `P8 = Plot[x^2, {x, 0, 2}];`

In[11]:= `P9 = Plot[-(x - 4)^2, {x, 2, 4}];`

In[12]:= `P10 = Plot[(x - 4)^2, {x, 4, 6}];`

In[13]:= `Show[P6, P7, P2, P8, P9, P10]`   (* Fig. (C). *)

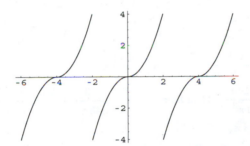

Example 10.3. (**B**) Even periodic extension $f_1(x)$ of $f(x)$

Example 10.3. (**C**) Odd periodic extension $f_2(x)$ of $f(x)$

Now determine the Fourier coefficients a_n of the Fourier cosine series of $f_1(x)$ from the Euler formula. Since $f_1(x)$ and the cosine are both even, so is the product (the integrand). Hence the integral from -2 to 2 equals twice the integral from 0 to 2 (the interval on which $f(x)$ is given). Remembering that $L = 2$, type the commands for a_0, a_n, and for a partial sum S. Figure (D) shows that a good approximation can already be obtained with a small number of terms.

In[14]:= `a0 = 2/(2 2) Integrate[x^2, {x, 0, 2}]`   (* Out: $\frac{4}{3}$ *)

In[15]:= `an = 2/2 Integrate[x^2 Cos[n Pi x/2], {x, 0, 2}]`

$$\text{Out[15]} = \frac{16\,\mathrm{Cos}\,[n\,\pi]}{n^2\,\pi^2} + \frac{2\,(-8 + 4\,n^2\,\pi^2)\,\mathrm{Sin}\,[n\,\pi]}{n^3\,\pi^3}$$

In[16]:= `S = a0 + Sum[an Cos[n Pi x/2], {n, 1, 2}]`

$$\text{Out[16]} = \frac{4}{3} - \frac{16\,\mathrm{Cos}\,\left[\frac{\pi\,x}{2}\right]}{\pi^2} + \frac{4\,\mathrm{Cos}\,[\pi\,x]}{\pi^2}$$

In[17]:= `P11 = Plot[S, {x, -6, 6}];`

In[18]:= `Show[P11, P1, P3, P4, AspectRatio -> Automatic]`   (* Fig. (D). *)

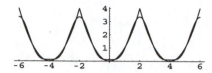

Example 10.3. (D) Approximation of $f_1(x)$ by a partial sum of its Fourier series

For $f_2(x)$ you will need a partial sum consisting of more terms, as Fig. (E) will show, because $f_2(x)$ is discontinuous – and it will be quite interesting to see how a sum of ***continuous*** terms approximates a ***discontinuous*** function. Both $f_2(x)$ and the sine in the Euler formula for b_n are odd, hence the integrand is even, so that the integral from -2 to 2 equals twice the integral from 0 to 2. Hence type

In[19]:= `bn = 2/2 Integrate[x^2 Sin[n Pi x/2], {x, 0, 2}]`

Out[19]= $-\dfrac{16}{n^3\,\pi^3} - \dfrac{2\,(-8+4n^2\,\pi^2)\,\mathrm{Cos}\,[n\,\pi]}{n^3\,\pi^3} + \dfrac{16\,\mathrm{Sin}\,[n\,\pi]}{n^2\,\pi^2}$

In[20]:= `S2 = Sum[bn Sin[n Pi x/2], {n, 1, 7}]`

Out[20]= $\left(-\dfrac{16}{\pi^3} + \dfrac{2\,(-8+4\,\pi^2)}{\pi^3}\right) \mathrm{Sin}\left[\dfrac{\pi\,x}{2}\right] + \left(-\dfrac{2}{\pi^3} - \dfrac{-8+16\,\pi^2}{4\,\pi^3}\right) \mathrm{Sin}\,[\pi\,x] +$

$\left(-\dfrac{16}{27\,\pi^3} + \dfrac{2\,(-8+36\,\pi^2)}{27\,\pi^3}\right) \mathrm{Sin}\left[\dfrac{3\,\pi\,x}{2}\right] + \left(-\dfrac{1}{4\,\pi^3} - \dfrac{-8+64\,\pi^2}{32\,\pi^3}\right) \mathrm{Sin}\,[2\,\pi\,x] +$

$\left(-\dfrac{16}{125\,\pi^3} + \dfrac{2\,(-8+100\,\pi^2)}{125\,\pi^3}\right) \mathrm{Sin}\left[\dfrac{5\,\pi\,x}{2}\right] +$

$\left(-\dfrac{2}{27\,\pi^3} - \dfrac{-8+144\,\pi^2}{108\,\pi^3}\right) \mathrm{Sin}\,[3\,\pi\,x] + \left(-\dfrac{16}{343\,\pi^3} + \dfrac{2\,(-8+196\,\pi^2)}{343\,\pi^3}\right) \mathrm{Sin}\left[\dfrac{7\,\pi\,x}{2}\right]$

In[21]:= `P12 = Plot[S2, {x, -6, 6}];`

In[22]:= `Show[P12, P2, P6, P7, P8, P9, P10, AspectRatio -> Automatic]`

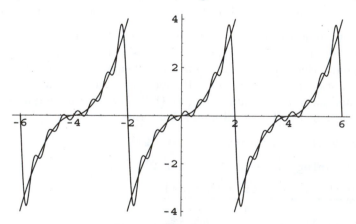

Example 10.3. (E) Approximation of $f_2(x)$ by a partial sum of its Fourier series

Similar Material in AEM: pp. 544-546

EXAMPLE 10.4 · RECTIFIER

A sinusoidal current, for simplicity, $f(t) = \sin t$, where t is time, is passed through a full-wave rectifier which converts the negative half-waves to positive half-waves (and leaves the positive half-waves as they are), resulting in the function $g(t) = |\sin t|$. Find the Fourier series of $g(t)$.

Solution. $g(t)$ is even. Hence you obtain a Fourier cosine series, and the integral for a_n from $t = -\pi$ to π equals twice the integral from 0 to π, an interval in which $g(t) = \sin t$. Using the Euler formulas for a function of period 2π, thus type

In[1]:= `a0 = 1/Pi Integrate[Sin[t], {t, 0, Pi}]`   (* Out: $\frac{2}{\pi}$ *)

In[2]:= `an = Simplify[2/Pi Integrate[Sin[t] Cos[n t], {t, 0, Pi}]]`

Out[2]= $\dfrac{2\,(1 + \text{Cos}\,[\text{n}\,\pi])}{\pi - \text{n}^2\,\pi}$

For $n = 1$ the denominator is zero. Hence type a_1 separately and then partial sums.

In[3]:= `a1 = 2/Pi Integrate[Sin[t] Cos[t], {t, 0, Pi}]`   (* Out: 0 *)

In[4]:= `S3 = a0 + a1 Cos[t] + Sum[an Cos[n t], {n, 2, 3}]`

In[5]:= `S7 = a0 + a1 Cos[t] + Sum[an Cos[n t], {n, 2, 7}]`

Out[5]= $\dfrac{2}{\pi} - \dfrac{4\,\text{Cos}\,[2\,t]}{3\,\pi} - \dfrac{4\,\text{Cos}\,[4\,t]}{15\,\pi} - \dfrac{4\,\text{Cos}\,[6\,t]}{35\,\pi}$

$S7$ is already fairly accurate and can hardly be distinguished from $g(t)$ in the figure.

In[6]:= `P = Plot[{S3, S7}, {t, -Pi, 2 Pi}]`

In[7]:= `Pg = Plot[Abs[Sin[t]], {t, -Pi, 2 Pi}]`

In[8]:= `Show[P, Pg, AspectRatio -> Automatic]`

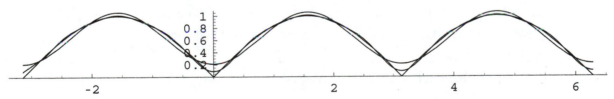

Example 10.4. Full-wave rectification $g(t)$ of $f(t) = \sin t$
and partial sums S_3 and S_7 of its Fourier series

Similar Material in AEM: p. 539

EXAMPLE 10.5 · TRIGONOMETRIC APPROXIMATION. MINIMUM SQUARE ERROR

In approximating a given function f by another function F the quality of the approximation can be measured in various ways, depending on the purpose. A quantity that characterizes the *overall goodness of fit* on the interval of interest, say, $-\pi < x < \pi$, is the so-called **total square error** of F (relative to f) defined by

$$E = \int_{-\pi}^{\pi} (f - F)^2\, dx.$$

If

$$F(x) = A_0 + \sum_{n=1}^{N} (A_n \cos nx + B_n \sin nx),$$

then $F(x)$ is called a **trigonometric polynomial** and the approximation is called a **trigonometric approximation**. In this case the total square error of F relative to f (with fixed N) is minimum if and only if you choose the Fourier coefficients of f as the coefficients of F. It can be shown (see AEM, p. 555) that this **minimum square error** is

$$E^* = \int_{-\pi}^{\pi} f^2 \, dx - \pi \left[2a_0^2 + \sum_{n=1}^{N} (a_n^2 + b_n^2) \right].$$

For instance, find the minimum square errors for the approximation of the given function in Example 10.1 in this Guide with $k = 1$ by the partial sums of its Fourier series (illustrated by the figures in that example).

Solution. The b_n are zero since f is even. Type the formula for the a_n, then the formula for the minimum square error, call it `Emin` , and finally find numerical values of `Emin` for $N = 10, 20, ..., 250$. (Since f is discontinuous, the convergence of the sequence of the minimum square errors to 0 will be slow.)

In[1]:= `an = 1/Pi Integrate[1 Cos[n x], {x, -Pi/2, Pi/2}]`

Out[1]= $\dfrac{2 \sin\left[\frac{n\pi}{2}\right]}{n\pi}$

In[2]:= `Emin = Table[Integrate[1, {x, -Pi/2, Pi/2}] -`
`Pi (2 (1/2)^2 + Sum[N[an^2], {n, 1, 10 M}]), {M, 1, 25}]`

Out[2]= {0.0634527, 0.0318046, 0.0212128, 0.0159122, 0.0127307, 0.0106093,

0.00909395, 0.00795733, 0.00707326, 0.00636599, 0.00578729, 0.00530504,

0.00489698, 0.00454721, 0.00424407, 0.00397882, 0.00374478, 0.00353674,

0.0033506, 0.00318307, 0.0030315, 0.00289371, 0.00276789, 0.00265257,

0.00254647}

Similar Material in AEM: pp. 554-556

EXAMPLE 10.6   **FOURIER INTEGRAL, FOURIER TRANSFORM**

The **Fourier integral representation** of a given function $f(x)$ defined on the x-axis (and satisfying certain regularity conditions; see AEM, p. 559) is

$$f(x) = \int_0^{\infty} [A(w) \cos wx + B(w) \sin wx] \, dw$$

where

$$A(w) = \frac{1}{\pi} \int_{-\infty}^{\infty} f(v) \cos wv \, dv,$$

$$B(w) = \frac{1}{\pi} \int_{-\infty}^{\infty} f(v) \sin wv \, dv.$$

For instance, find the Fourier integral representation of the function $f(x)$ which equals 1 for $-1 < x < 1$ and zero everywhere else on the x-axis.

Solution. The integration in $A(w)$ and $B(w)$ extends from -1 to 1 only because $f(x)$ is zero elsewhere. Thus type

In[1]:= `A = 1/Pi Integrate[1 Cos[w v], {v, -1, 1}]`   (* Out: $\frac{2 \mathrm{Sin}\,[\mathrm{w}]}{\pi\,\mathrm{w}}$ *)

In[2]:= `B = 1/Pi Integrate[1 Sin[w v], {v, -1, 1}]`   (* Out: 0 *)

In[3]:= `f = 2/Pi Integrate[Sin[w]/w Cos[w x], {w, 0, Infinity}]`

Out[3]= $\dfrac{1}{\pi}\left(2\,\mathrm{If}\left[\mathrm{Im}[\mathrm{x}] == 0,\ \dfrac{1}{4}\,\pi\,(\mathrm{Sign}[1-\mathrm{x}] + \mathrm{Sign}[1+\mathrm{x}]),\ \displaystyle\int_{0}^{\infty}\dfrac{\mathrm{Cos}\,[\mathrm{w}\,\mathrm{x}]\,\mathrm{Sin}\,[\mathrm{w}]}{\mathrm{w}}\,\mathrm{dw}\right]\right)$

This shows the desired Fourier integral. It also says that for real x ($\mathrm{Im}\,x = 0$), the integral equals the given function $f(x)$ because $\mathrm{sign}(1-x) + \mathrm{sign}(1+x) = 2$ if $-1 < x < 1$ and 0 otherwise.

It can be shown (see AEM, pp. 569, 570) that the Fourier integral may be converted to complex form, and from it one can derive the **Fourier transform** $F(w)$ of $f(x)$ given by

$$F(w) = \frac{1}{\sqrt{2\pi}} \int_{-\infty}^{\infty} f(x)\, e^{-iwx}\, dx$$

and the **inverse Fourier transform** of $F(w)$ given by

$$f(x) = \frac{1}{\sqrt{2\pi}} \int_{-\infty}^{\infty} F(w)\, e^{iwx}\, dw.$$

For instance, find the Fourier transform of $f(x) = e^{-x^2}$.

Solution. Let F denote the Fourier transform of the given $f(x)$. Type

In[4]:= `F = 1/Sqrt[2 Pi] Integrate[Exp[-x^2] Exp[-I w x],`

   `{x, -Infinity, Infinity}]`

Out[4]= $\dfrac{\mathrm{If}\left[\mathrm{Im}[\mathrm{w}] == 0,\ e^{-\frac{\mathrm{w}^2}{4}}\,\sqrt{\pi},\ \displaystyle\int_{-\infty}^{\infty} e^{-i\,\mathrm{w}\,\mathrm{x}-\mathrm{x}^2}\,\mathrm{dx}\right]}{\sqrt{2\pi}}$

In[5]:= `Simplify[%, Im[w] == 0]`   (* Out: $\dfrac{e^{-\frac{\mathrm{w}^2}{4}}}{\sqrt{2}}$ *)

Hence the computer evaluated the integral under the condition that w is real. Then, since x is real, too, $\exp(-iwx)$ is pure imaginary, and thus has absolute value $(\cos^2 wx + \sin^2 wx)^{1/2} = 1$.

Similar Material in AEM: pp. 570, 571

Problem Set for Chapter 10

Pr.10.1 (Rectangular wave. Gibbs phenomenon) Find the Fourier series of the following function of period 2π and make plots that show the Gibbs phenomenon. (*AEM Ref.* p. 532)

$$f(x) = \begin{cases} -3 & \text{if } -\pi < x < 0 \\ 3 & \text{if } 0 < x < \pi \end{cases}$$

Pr.10.2 (Cosine series) Find the Fourier series of the function (sketch it) given by

$$f(x) = \begin{cases} 1 & \text{if } -\pi/2 < x < \pi/2 \\ -1 & \text{if } \pi/2 < x < 3\pi/2 \end{cases}$$

and periodic with period 2π. (*AEM Ref.* p. 537 (#13))

Pr.10.3 **(Cosine and sine terms)** Find the Fourier series of $f(x)$ given by $f(x) = (x/(2\pi))^4$ if $0 < x < 2\pi$ and of period 2π. (*AEM Ref.* p. 531)

Pr.10.4 **(Half-wave rectifier)** Pass the current $f(t) = \sin t$ through a half-wave rectifier that clips the negative portion of the wave. Find the Fourier series of the resulting periodic function $g(t)$ and plot some of its partial sums. (*AEM Ref.* p. 539)

Pr.10.5 **(Periodic force, resonance)** If a force $f(t) = t^2$ for $0 < t < 2\pi$ and periodic of period 2π is acting as the driving force on a mechanical system, which term of its Fourier series has the greatest coefficient in absolute value (so that the corresponding frequency should be watched for possible resonance effects)? (*AEM Ref.* p. 536 (#8))

Pr.10.6 **(Behavior near a jump)** Find the Fourier series of the periodic function $f(x) = \pi x^3/2$ $(-1 < x < 1)$ of period $p = 2$. Show by plots that at the jumps the partial sums give the arithmetic mean of the right-hand and left-hand limits of $f(x)$. (*AEM Ref.* p. 540 (#12))

Pr.10.7 **(Continuous function)** Find the Fourier series of the periodic function $f(x) = 3x^2$ $(-1 < x < 1)$ of period $p = 2$. Show by a plot that a partial sum of few terms gives a relatively good approximation (except near the cusps). (*AEM Ref.* p. 540 (#7))

Pr.10.8 **(Triangular wave)** Find the Fourier series of the function (sketch it)

$$f(x) = \begin{cases} 1/2 + x & \text{if} \quad -1/2 < x < 0 \\ 1/2 - x & \text{if} \qquad\;\; 0 < x < 1/2 \end{cases}$$

of period $p = 1$. Plot $f(x)$ and some partial sums. (*AEM Ref.* p. 540 (#8))

Pr.10.9 **(Half-range expansion)** Find the Fourier cosine and sine series of $f(x) = x$, where $0 < x < L$ and L is arbitrary. (*AEM Ref.* p. 547 (#21))

Pr.10.10 **(Herringbone wave)** Find the Fourier series of the function (sketch it)

$$f(x) = \begin{cases} x & \text{if} \quad 0 < x < 1 \\ 1 - x & \text{if} \quad 1 < x < 2 \end{cases}$$

of period $p = 2$. Plot $f(x)$ and some partial sums. Observe the Gibbs phenomenon. (*AEM Ref.* p. 540 (#10))

Pr.10.11 **(Error distribution)** Find and plot $f(x) - S_5$ for x from $-\pi$ to π, where $f(x) = x^3$ $(-\pi < x < \pi)$ and periodic with period 2π and S_5 is the partial sum of the Fourier series containing terms of $\sin x$ to $\sin 5x$. (*AEM Ref.* pp. 553-556)

Pr.10.12 **(Minimum square error)** Let $f(x) = x^2$ $(-\pi < x < \pi)$ and periodic with period 2π. Find the minimum square error for the trigonometric approximation of $f(x)$ by polynomials of degree $N = 1, ..., 10$ and $N = 100$. (See Example 10.5 in this Guide. *AEM Ref.* p. 556 (#4))

Pr.10.13 **(Minimum square error)** Do the same task as in Pr.10.12 for the periodic function $f(x) = x$ $(-\pi < x < \pi)$ of period 2π.

Pr.10.14 **(Fourier integral, Fourier cosine integral)** Using the Fourier integral representation in Example 10.6 in this Guide, represent

$$f(x) = \begin{cases} \pi e^{-x} & \text{if} \quad x > 0 \\ \\ 0 & \text{otherwise} \end{cases}$$

by a Fourier integral. (*AEM Ref.* p. 563 (#1))

Pr.10.15 **(Experiment on Gibbs phenomenon)** Study and plot the Gibbs phenomenon for functions of your choice. On what does the height of the "spikes" depend? Or is it always the same? Is the speed of the shift of the waves toward the discontinuity different for different functions? On what does it depend? What about the number of "spikes"?

Chapter 11

Partial Differential Equations (PDE's)

Content. Wave equation, animation (Ex. 11.1, Prs. 11.1-11.3)
Tricomi, Airy equations, vibrating beam (Prs. 11.6, 11.7)
Heat equation, Laplace equation (Exs. 11.2, 11.3, Prs. 11.8-11.12)
Vibrating membrane (Exs. 11.4, 11.5, Prs. 11.13-11.15)

The PDE's considered are solved by separating variables, which reduces them to ODE's, and by the subsequent use of Fourier series and integrals.

Examples for Chapter 11

EXAMPLE 11.1 **WAVE EQUATION. SEPARATION OF VARIABLES. ANIMATION**

The **one-dimensional wave equation** is $u_{tt} = c^2 u_{xx}$. It governs the vertical vibrations of an **elastic string**, such as a violin string. Then $u(x, t)$ is the displacement from rest along the x-axis at a point x and time t. The string of length L is fixed at its ends $x = 0$ and $x = L$. This gives the boundary conditions $u(0, t) = 0$ and $u(L, t) = 0$ for all t. In separating variables you look for solutions of the form $u(x, t) = F(x)G(t)$ satisfying the boundary conditions. Thus, type this $u(x, t)$ and then the wave equation.

In[1]:= `Clear[F, G]`

In[2]:= `u[x, t] = F[x] G[t]`   `(* Out: F[x] G[t] *)`

In[3]:= `pde = D[u[x,t], t, t] == c^2 D[u[x,t], x, x]`

Out[3]= `F[x] G''[t] == c`^{2}` G[t] F''[x]`

Now separate the variables by dividing by $c^2 u(x, t) = c^2 F(x)G(t)$.

In[4]:= `eq = pde[[1]]/(c^2 u[x,t]) == pde[[2]]/(c^2 u[x,t])`

Out[4]= $\dfrac{G''[t]}{c^2\,G[t]} == \dfrac{F''[x]}{F[x]}$

Set each of the the two sides `eq[[1]]` and `eq[[2]]` equal to a constant, which must be negative, say, $-p^2$, in order to avoid ending up with an identically vanishing solution. Begin with the ODE for `F(x)`, that is, `eq[[2]]` $== -p^2$, and solve this ODE.

In[5]:= `sol1 = DSolve[eq[[2]] == -p^2, F[x], x]`

Out[5]= `{{F[x] → C[2] Cos[p x] + C[1] Sin[p x]}}`

Reduce this general solution by the left boundary condition $u(0, t) = 0$.

In[6]:= `s2 = (sol1[[1, 1, 2]] /. x -> 0) == 0`   `(* Out: C[2] == 0 *)`

Hence `C[2] = 0` since the string is fixed at its left end $x = 0$. Also, you may take `C[1] = 1` since arbitrary constants will be provided by $G(t)$. This gives

In[7]:= `s3 = sol1 /. {C[1] -> 1, C[2] -> 0}`

Out[7]= `{{F[x] → Sin[p x]}}`

Now determine p from the right boundary condition.

In[8]:= `s4 = s3 /. x -> L`

Out[8]= `{{F[L] → Sin[L p]}}`

Hence $F(L) = 0$ implies that $pL = n\pi$ with integer n; thus $p = n\pi/L$, $n = 1, 2, \ldots$. (Since $\sin(-\alpha) = -\sin\alpha$, you need not consider $n = -1, -2, \ldots$.) Type

In[9]:= `p = n Pi/L` (* Out: $\frac{n\pi}{L}$ *)

In[10]:= `solF = s3` (* Out: $\left\{\left\{\texttt{F[x]} \to \texttt{Sin}\left[\frac{n\pi x}{L}\right]\right\}\right\}$ *)

Now obtain the ODE for $G(t)$ by setting the left-hand side `eq[[1]]` of `eq` equal to $-p^2$ and solve the ODE.

In[11]:= `solG = DSolve[eq[[1]] == -p^2, G[t], t]`

Out[11]= $\left\{\left\{\texttt{G[t]} \to \texttt{C[2] Cos}\left[\frac{c\,n\,\pi\,t}{L}\right] + \texttt{C[1] Sin}\left[\frac{c\,n\,\pi\,t}{L}\right]\right\}\right\}$

From the solutions $F(x)$ and $G(t)$ in `solF` and `solG` you obtain $u(x, t) = G(t)F(x)$

In[12]:= `u[x,t] = solG[[1, 1, 2]] solF[[1, 1, 2]]`

Out[12]= $\left(\texttt{C[2] Cos}\left[\frac{c\,n\,\pi\,t}{L}\right] + \texttt{C[1] Sin}\left[\frac{c\,n\,\pi\,t}{L}\right]\right) \texttt{Sin}\left[\frac{n\,\pi\,x}{L}\right]$

Taking series of these solutions with $n = 1$ to ∞ and suitable constants, one can determine solutions that satisfy given initial conditions (displacement and velocity), as explained on pp. 590-593 of AEM.

Animation shows the string in motion. The simplest solution $u(x, t)$ is the product $\sin t \sin x$ (when $n = 1$, $c = 1$, $L = \pi$ in the previous formula). Type

In[13]:= `<< Graphics'Animation'`

You must do this **before** you give the command `Animate` (see below). If you violated this order, go out of Mathematica and restart it, loading the above package first.

When you now type and execute the next command, there will appear two coordinate axes directly below the command, a horizontal one marked from 0 to 3, and a vertical one extending from -1 to 1. This gives the string at rest. Below this figure you will also get figures representing the string displaced; ignore these further figures. Go from the coordinate axes to the right-hand margin, where you see a small vertical bracket aligned with the command and then three parallel vertical brackets. Use the mouse to click on the middle one of the three, so that is becomes black. Then simultaneously press Ctrl and y, and you will see the string vibrate as long as you want it, that is, as long as you leave that bracket black. Those further figures may take a while to appear. There will be twenty-four figures altogether (including the first), showing snapshots of the moving string at subsequent times.

In[14]:= `Animate[Plot[Sin[t] Sin[x], {x, 0, Pi}, PlotRange -> {-1, 1}],`
`{t, 0, 2Pi}]`

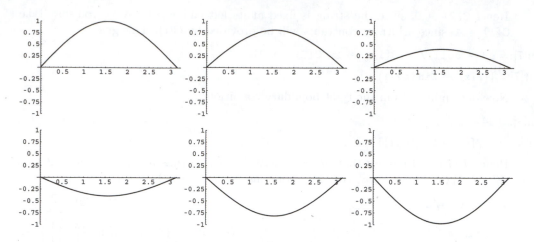

Example 11.1. Fundamental mode of the vibrating string (Animation)

Similar Material in AEM: pp. 587-593

EXAMPLE 11.2　　ONE-DIMENSIONAL HEAT EQUATION

The **one-dimensional heat equation** is $u_t = c^2 u_{xx}$, where $u(x, t)$ is the temperature in a straight bar or wire at a point x and time t. If the ends of a bar are at $x = 0$ and $x = \pi$ (so that the bar has length π and lies along the x-axis) and are kept at temperature 0, the boundary conditions are $u(0, t) = 0$ and $u(\pi, t) = 0$ for all t. Separation of variables leads to solutions $u = u_n$,

$$u_n(x, t) = F_n(x)G_n(t) = B_n \sin nx \exp\left(-(cn)^2 t\right)$$

satisfying the boundary conditions. A series of these will satisfy a given initial condition $f(x) = u(x, 0)$, for instance, a "triangular" initial temperature (see the figure)

$$f(x) = \begin{cases} x & \text{if} & 0 < x < \pi/2 \\ \pi - x & \text{if} & \pi/2 < x < \pi \end{cases}$$

if the coefficients B_n of the series are the Fourier coefficients of the Fourier sine half-range expansion of $f(x)$ (of period 2π). Show this in a plot, which also exhibits the temperatures for various constant values of t. Take $c = 1$ for simplicity.

Solution. Type B_n and then a partial sum of the series of the u_n with these coefficients B_n for various $t = 0, \ 0.1, \ 0.2, \ \dots$. You will see that even for small $t = 0.1, \ 0.2$ you obtain almost a sine curve because the exponential factors of the second and further terms decrease very rapidly to zero. By the Euler formula,

```
In[1]:= Bn = 2/Pi (Integrate[x Sin[n x], {x, 0, Pi/2}] + Integrate[(Pi - x)
          Sin[n x], {x, Pi/2, Pi}])
```

$$\text{Out[1]=} \quad \frac{2\left(\dfrac{2\,\mathrm{Sin}\left[\frac{n\pi}{2}\right]}{n^2} - \dfrac{\mathrm{Sin}\,[n\,\pi]}{n^2}\right)}{\pi}$$

In[2]:= `S = Sum[Bn Sin[n x] Exp[-n^2 t], {n, 1, 13}]`

$$\text{Out[2]=} \quad \frac{4\,e^{-t}\,\mathrm{Sin}\,[x]}{\pi} - \frac{4\,e^{-9\,t}\,\mathrm{Sin}\,[3\,x]}{9\,\pi} + \frac{4\,e^{-25\,t}\,\mathrm{Sin}\,[5\,x]}{25\,\pi} - \frac{4\,e^{-49\,t}\,\mathrm{Sin}\,[7\,x]}{49\,\pi}$$

$$+ \frac{4\,e^{-81\,t}\,\mathrm{Sin}\,[9\,x]}{81\,\pi} - \frac{4\,e^{-121\,t}\,\mathrm{Sin}\,[11\,x]}{121\,\pi} + \frac{4\,e^{-169\,t}\,\mathrm{Sin}\,[13\,x]}{169\,\pi}$$

In[3]:= `S0 = N[S /. t -> 0]`

Out[3]= $1.27324\,\mathrm{Sin}\,[x] - 0.141471\,\mathrm{Sin}\,[3.\,x] + 0.0509296\,\mathrm{Sin}\,[5.\,x] - 0.0259845\,\mathrm{Sin}\,[7.\,x]$

$\quad\quad + 0.015719\,\mathrm{Sin}\,[9.\,x] - 0.0105226\,\mathrm{Sin}\,[11.\,x] + 0.00753396\,\mathrm{Sin}\,[13.\,x]$

In[4]:= `S1 = S /. t -> 0.1;`

In[5]:= `S2 = S /. t -> 0.2;`

In[6]:= `S10 = S /. t -> 1;`

In[7]:= `S20 = S /. t -> 2;`

In[8]:= `S30 = S /. t -> 3;`

In[9]:= `P1 = Plot[{S0, S1, S2, S10, S20, S30}, {x, 0, Pi}]`

In[10]:= `P2 = Plot[t, {t, 0, Pi/2}]`

In[11]:= `P3 = Plot[Pi - t, {t, Pi/2, Pi}]`

In[12]:= `Show[P1, P2, P3]`

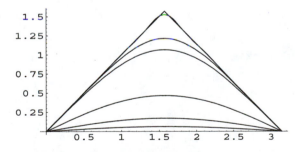

Example 11.2. "Triangular" initial temperature and its decrease with time

Similar Material in AEM: pp. 600-604

EXAMPLE 11.3   **HEAT EQUATION, LAPLACE EQUATION**

Find the steady-state temperature in the rectangular plate $0 \le x \le \pi$, $0 \le y \le \pi/2$ if the upper edge is kept at temperature $u(x, \pi/2) = f(x) = 1$ and the other three edges are kept at temperature 0. Plot the temperature as a surface over the rectangle.

Solution. The two-dimensional heat equation is $u_t = c^2(u_{xx} + u_{yy})$. Since u is assumed to be steady-state (time independent), this equation reduces to **Laplace's equation** $u_{xx} + u_{yy} = 0$. Separation of variables leads to solutions

$$u_n(x, y) = B_n \sin nx \sinh ny$$

that satisfy the three zero boundary conditions. It can be shown that a series of these also satisfies $u(x, \pi/2) = f(x) = 1$ if you choose the $B_n \sinh(n\pi/2)$ to be the coefficients of the Fourier sine half-range expansion of $f(x)$. Thus type the following. By the Euler formula,

In[1]:= `Bn = 2/(Pi Sinh[n Pi/2]) Integrate[1 Sin[n x], {x, 0, Pi}]`

$$\text{Out[1]}= \frac{2\left(\dfrac{1}{n} - \dfrac{\text{Cos}[n\,\pi]}{n}\right)\text{Csch}\left[\dfrac{n\,\pi}{2}\right]}{\pi}$$

In[2]:= `u30 = Sum[Bn Sin[n x] Sinh[n y], {n, 1, 30}]`   `(* 15 terms *)`

In[3]:= `Plot3D[u30, {x, 0, Pi}, {y, 0, Pi/2}, ViewPoint -> {1, 2, 1.5}]`

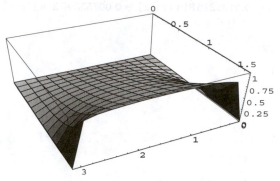

Example 11.3. Temperature given by `u30` shown as a surface over the rectangular plate

To obtain a figure showing the approximation of the temperature along the upper edge, type

In[4]:= `W = u30 /. y -> Pi/2;`

In[5]:= `Plot[W, {x, 0, Pi}, PlotRange -> {0, 1.2}]`

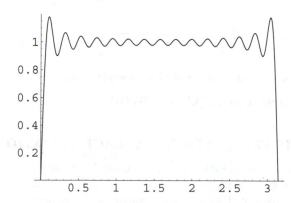

Example 11.3. Temperature given by `u30` on the upper edge
Note the beginning Gibbs phenomenon near 0 and π.

Similar Material in AEM: pp. 605-607

EXAMPLE 11.4   RECTANGULAR MEMBRANE.
DOUBLE FOURIER SERIES

Vibrations of elastic membranes (such as drumheads) are governed by the **two-dimensional wave equation**

$$u_{tt} = c^2(u_{xx} + u_{yy}),$$

where $u(x, y, t)$ is the displacement of the membrane at a point (x, y) and time t from its position at rest in the xy-plane. Let the density of the material and the tension be such that $c^2 = 5$. Let the initial velocity be zero and the initial displacement $f(x, y)$ as typed and as shown in the figure.

In[1]:= `f = 1/10 (4 x - x^2) (2 y - y^2)`

Out[1]= $\dfrac{1}{10}(4\,x - x^2)(2\,y - y^2)$

In[2]:= `Plot3D[f, {x, 0, 4}, {y, 0, 2}]`

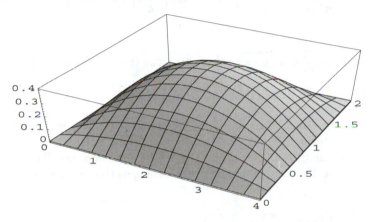

Example 11.4. Initial displacement of the membrane

The solution method is similar to that for the vibrating string. Separation of variables leads to a double sequence of solutions

$$u_{mn} = B_{mn}\cos(\lambda_{mn}t)\sin\left(\frac{m\pi x}{a}\right)\sin\left(\frac{n\pi y}{b}\right) \qquad (m,\ n = 1, 2, \ldots)$$

satisfying the boundary condition $u = 0$ on the edges of the membrane. A double series of these (a "**double Fourier series**") will satisfy the initial conditions $u(x, y, 0) = f(x, y)$ (given initial displacement) and $u_t(x, y, 0) = 0$ (zero initial velocity) if the B_{mn} are the coefficients of the double Fourier series of $f(x, y)$. Accordingly, type the following.

In[3]:= `Clear[m, n]`

In[4]:= `a = 4; b = 2;`   `(* Lengths of the edges of the membrane *)`

The **Euler formulas for a double Fourier series** give

In[5]:= `Bmn = 4/(a b) Integrate[Integrate[f Sin[m Pi x/a] Sin[n Pi y/b],`
     `{x, 0, a}], {y, 0, b}]`

Out[5]= $\frac{1}{20}\left(\frac{128}{m^3\pi^3}-\frac{128\,Cos\,[m\,\pi]}{m^3\,\pi^3}-\frac{64\,Sin\,[m\,\pi]}{m^2\,\pi^2}\right)\left(\frac{16}{n^3\pi^3}-\frac{16\,Cos\,[n\,\pi]}{n^3\,\pi^3}-\frac{8\,Sin\,[n\,\pi]}{n^2\,\pi^2}\right)$

In[6]:= `Simplify[%]`

Out[6]= $\frac{1}{5\,m^3\,n^3\,\pi^6}\,(128\,(-2+2\,Cos\,[m\,\pi]+m\,\pi\,Sin\,[m\,\pi])\,(-2+2\,Cos\,[n\,\pi]+n\,\pi\,Sin\,[n\,\pi]))$

For the time function in u_{mn} you need λ_{mn} as follows.

In[7]:= `Lmn = 5 Pi Sqrt[m^2/a^2 + n^2/b^2]`   (* Out: $5\sqrt{\frac{m^2}{16}+\frac{n^2}{4}}\pi$ *)

In[8]:= `umn = Bmn Cos[Lmn t] Sin[m Pi x/a] Sin[n Pi y/b]`

Out[8]= $\frac{1}{20}\,Cos\left[5\sqrt{\frac{m^2}{16}+\frac{n^2}{4}}\,\pi\,t\right]\left(\frac{128}{m^3\pi^3}-\frac{128\,Cos\,[m\,\pi]}{m^3\,\pi^3}-\frac{64\,Sin\,[m\,\pi]}{m^2\,\pi^2}\right)$
$\left(\frac{16}{n^3\,\pi^3}-\frac{16\,Cos\,[n\pi]}{n^3\,\pi^3}-\frac{8\,Sin\,[n\pi]}{n^2\,\pi^2}\right)Sin\left[\frac{m\,\pi\,x}{4}\right]\,Sin\left[\frac{n\,\pi\,y}{2}\right]$

In[9]:= `Simplify[%]`

Out[9]= $\frac{1}{5\,m^3\,n^3\,\pi^6}\left(128\,Cos\left[\frac{5}{4}\sqrt{m^2+4\,n^2}\,\pi\,t\right]\,(-2+2\,Cos\,[m\,\pi]+m\,\pi\,Sin\,[m\,\pi])\right.$

$\left. (-2+2\,Cos\,[n\,\pi]+n\,\pi\,Sin\,[n\,\pi])\,Sin\left[\frac{m\,\pi\,x}{4}\right]Sin\left[\frac{n\,\pi\,y}{2}\right]\right)$

Now type a partial sum S of these $u_{mn}(x,y,t)$. Very few terms will do. In fact, let us take just a single term. (Try more terms.) Then type S0, which is S for $t = 0$, that is, it is an approximation of the initial shape of the membrane. The corresponding figure looks practically identical to the previous one, so that we need not show it here.

In[10]:= `S = umn /. {m -> 1, n -> 1}`

Out[10]= $\dfrac{2048\,Cos\left[\frac{5}{4}\sqrt{5}\,\pi\,t\right]\,Sin\left[\frac{\pi\,x}{4}\right]\,Sin\left[\frac{\pi\,y}{2}\right]}{5\,\pi^6}$

In[11]:= `S0 = S /. t -> 0`   (* Partial sum for $t = 0$ *)

Out[11]= $\dfrac{2048\,Sin\left[\frac{\pi\,x}{4}\right]\,Sin\left[\frac{\pi\,y}{2}\right]}{5\,\pi^6}$

In[12]:= `Plot3D[S0, {x, 0, 4}, {y, 0, 2}]`

Similar Material in AEM: pp. 619-625

| EXAMPLE 11.5 | **LAPLACIAN. CIRCULAR MEMBRANE. BESSEL EQUATION** |

This concerns the vertical vibrations of a **circular membrane** fixed along its edge $x^2 + y^2 = R^2$ in the xy-plane. Hence you must first obtain the two-dimensional wave equation $u_{tt} = c^2(u_{xx} + u_{yy})$ in polar coordinates.

In[1]:= `<<Calculus'VectorAnalysis'`

In[2]:= `SetCoordinates[Polar[r, theta]]`

```
General::spell1 :  Possible spelling error: new symbol name "theta is similar
to existing symbol Theta"
```

```
Coordinates::invalid :  Polar[r, theta] is not a valid coordinate system
specification.
```

Out[2]= `SetCoordinates[Polar[r, theta]]`

The second error message shows that Mathematica does not want polar coordinates
in the plane, but wants a system of three coodinates in space. Hence add a third
coordinate, z, that is, use cylindrical coordinates (a detour for our purpose).

In[3]:= `Clear[r, z]`

In[4]:= `SetCoordinates[Cylindrical[r, theta, z]]`

Out[4]= `Cylindrical[r, theta, z]`

Now you do not want z, and since we shall consider vibrations that are independent
of the angle (so that at any instant the membrane will look rotationally symmetrical),
you can get rid of the two unnecessary coordinates theta and z simply by typing
u[r] instead of u[r, theta, z]. Actually, you may write u[r, t], anticipating
that time t comes in when you type the wave equation. Also, since you want to solve
the wave equation by separating variables, you may type u[r,t] = W[r]G[t] and
then obtain the Laplacian of this u in polar coordinates.

In[5]:= `Clear[W]`

In[6]:= `u[r,t] = W[r] G[t]`   (* Out: `G[t] W[r]` *)

In[7]:= `lap = Laplacian[u[r, t]]`   (* Out: $\dfrac{\texttt{G[t] W}'\texttt{[r]} + \texttt{r\,G[t] W}''\texttt{[r]}}{\texttt{r}}$ *)

Then type the wave equation

In[8]:= `pde = D[u[r, t], t, t] == c^2 lap`

Out[8]= $\texttt{W[r] G}''\texttt{[t]} == \dfrac{\texttt{c}^2\,\texttt{G[t] W}'\texttt{[r]} + \texttt{r\,G[t] W}''\texttt{[r]}}{\texttt{r}}$

In the next command, pde[[1]] is the left-hand side of pde,

In[9]:= `pde[[1]]`   (* Out: `W[r] G`''`[t]` *)

Similarly pde[[2]]. Type it. Separate the variables in pde by dividing by
$c^2 u = c^2 W G$, obtaining

In[10]:= `eq1 = pde[[1]]/(c^2 u[r,t]) == Simplify[pde[[2]]/(c^2 u[r,t])]`

Out[10]= $\dfrac{\texttt{G}''\texttt{[t]}}{\texttt{c}^2\,\texttt{G[t]}} == \dfrac{\texttt{W}'\texttt{[r]} + \texttt{r\,W}''\texttt{[r]}}{\texttt{r\,W[r]}}$

Equate both sides to a negative constant, say, $-k^2$. For the right-hand side this gives
(take $-k^2$ to the left).

In[11]:= `eq2 = eq1[[2]] + k^2 == 0`

Out[11]= $\texttt{k}^2 + \dfrac{\texttt{W}'\texttt{[r]} + \texttt{r\,W}''\texttt{[r]}}{\texttt{r\,W[r]}} == 0$

This is an ODE for W(r). To obtain it in the usual form, multiply the left-hand side
by $W(r)$ and equate the result to zero,

In[12]:= `eq3 = Together[eq2[[1]] W[r]] == 0`

Out[12]= $\dfrac{k^2\, r\, W[r] + W'[r] + r\, W''[r]}{r} == 0$

In[13]:= `Simplify[%]`

Out[13]= $k^2\, W[r] + \dfrac{W'[r]}{r} + W''[r] == 0$

This is the **Bessel equation** with parameter 0. A general solution is

In[14]:= `DSolve[eq3, W[r], r]`

Out[14]= $\left\{\left\{ W[r] \rightarrow e^{-\sqrt{-k^2}\, r} \right.\right.$

$\left.\left. \left(\dfrac{e^{\sqrt{-k^2}\, r}\, \text{BesselK}\left[0, \sqrt{-k^2}\, r\right] C[1]}{\sqrt{\pi}} + e^{\sqrt{-k^2}\, r}\, \text{BesselI}\left[0, \sqrt{-k^2}\, r\right] C[2] \right)\right\}\right\}$

In[15]:= `sol = FullSimplify[%, k > 0]`

Out[15]= $\left\{\left\{ W[r] \rightarrow \dfrac{\text{BesselK}[0,\, i\, k\, r]\, C[1]}{\sqrt{\pi}} + \text{BesselJ}[0,\, k\, r]\, C[2] \right\}\right\}$

Taking `C[1] = 0` and `C[2] = 1`, you have the Bessel function J_0 of the first kind with parameter 0,

In[16]:= `solJ = sol /. {C[1] -> 0, C[2] -> 1}`

Out[16]= `{{W[r] → BesselJ[0, k r]}}`

In[17]:= `solJ[[1, 1, 2]]`   (* Out: `BesselJ[0, k r]` *)

k must be determined so that $J_0(kr) = 0$ on the edge $r = R$ of the membrane. For plotting, take $R = 1$ and $c = 1$ (the constant in the wave equation). Then k must equal the first positive zero of J_0 (this gives the simplest solution), the second positive zero, etc. (See the graph of J_0 in Example 4.6 in this Guide.) To obtain the first three zeros, type (with $s = kr$)

In[18]:= `zero1 = FindRoot[BesselJ[0, s] == 0, {s, 2}]`

Out[18]= `{s → 2.40483}`

In[19]:= `zero2 = FindRoot[BesselJ[0, s] == 0, {s, 6}]`

Out[19]= `{s → 5.52008}`

In[20]:= `zero3 = FindRoot[BesselJ[0, s] == 0, {s, 10}]`

Out[20]= `{s → 8.65373}`

In[21]:= `zero1[[1, 2]]`   (* Out: 2.40483 *)

You can now plot the three simplest solutions by typing

In[22]:= `ParametricPlot3D[{r Cos[theta], r Sin[theta],`
`        BesselJ[0, zero1[[1, 2]] r]}, {r, 0, 1}, {theta, 0, 2 Pi}]`

In[23]:= `ParametricPlot3D[{r Cos[theta], r Sin[theta],`
`        BesselJ[0, zero2[[1, 2]] r]}, {r, 0, 1}, {theta, 0, 2 Pi}]`

In[24]:= `ParametricPlot3D[{r Cos[theta], r Sin[theta],`
     `BesselJ[0, zero3[[1, 2]] r]}, {r, 0, 1}, {theta, 0, 2 Pi}]`

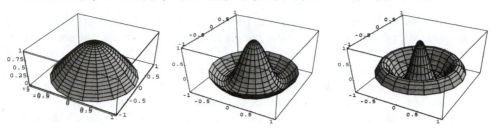

Example 11.5. Circular membrane, simplest forms of vibrations

Similar Material in AEM: pp. 629-634

Problem Set for Chapter 11

Pr.11.1 (Normal modes of the vibrating string) The solutions `F(x)` in `solF` in Example 11.1 in this Guide are called the *normal modes* of the string. Plot them (separately) for $n = 1, 2, 3, 4$ (with $L = \pi$). Multiply them by the corresponding sines of t (with $c = 1$) and apply animation. (See Example 11.1 in this Guide. *AEM Ref.* p. 590)

Pr.11.2 (Animation) Show 5 cycles of the motion

$$u(x,t) = \sin x \cos t - (1/9)\sin 3x \cos 3t + (1/25)\sin 5x \cos 5t$$

(approximating the motion when the initial deflection of the string is "triangular"). (See Example 11.1 in this Guide. *AEM Ref.* p. 593)

Pr.11.3 (Extension of D'Alembert's method) Verify that $u(x,y) = f(x+y) + g(2x-y)$, with arbitrary (twice differentiable) f and g, is a solution of $u_{xx} + u_{xy} = 2u_{yy}$. (*AEM Ref.* p. 598 (#14))

Pr.11.4 (Separation of variables) Solve $u_{xx} - u_{yy} = 0$, choosing the separation constant positive, zero, and negative. (*AEM Ref.* p. 594 (#18))

Pr.11.5 (Checking solutions) Checking is important. The computer may sometimes give you false or insufficient results. Check whether $u = (c_1 e^{kx} + c_2 e^{-kx})(c_3 e^{ky} + c_4 e^{-ky})$ is a solution of $u_{xx} - u_{yy} = 0$. Can you replace k in the functions depending on y by another constant?

Pr.11.6 (Tricomi equation. Airy equation) Find solutions $u(x,y) = F(x)G(y)$ of the Tricomi equation $yu_{xx} + u_{yy} = 0$. Show that for $G(y)$ this gives $G'' + kyG = 0$. (With $k = -1$ or 1 this is called *Airy's equation*.) Find a general solution $G(y)$ involving the Airy functions Ai and Bi (see Ref. [1] in Appendix 1 for information). Obtain Ai$(-y)$ from that general solution and plot it. (*AEM Ref.* p. 598)

Pr.11.7 (Vibrating beam. Vertical vibrations of a horizontal elastic beam of homogeneous material and constant cross section are governed by the equation $u_{tt} = c^2 u_{xxxx}$. Solve this fourth-order PDE by separating variables. (*AEM Ref.* pp. 598, 599)

Pr.11.8 (Heat equation) Solve the heat equation in Example 11.2 in this Guide for a bar of length 10 with $c = 1$ and "parabolic" initial temperature $u(x,0) = x(10 - x)$. (*AEM Ref.* p. 608 (#5), where, however, $c \neq 1$)

Pr.11.9 **(Heat equation)** Derive the solutions $u_n(x,t)$ in Example 11.2 in this Guide by separation of variables. (*AEM Ref.* p. 601)

Pr.11.10 **(Heat flow in a long bar. Fourier integral. Error function)** An infinite bar (the x-axis), practically a very long bar, is heated to 100 degrees at some point, practically, between -1 and 1, the rest being kept at 0. At $t = 0$, heating is terminated and heat begins to flow away to both sides. It can be shown that the temperature is given by $100/\sqrt{\pi}$ times the integral of $\exp(-z^2)$ from $z = -(1+x)/(2c\sqrt{t})$ to $z = (1-x)/(2c\sqrt{t})$. Study the decrease of the temperature by animation, assuming that $c = 1$. The integral of $\exp(-z^2)$ from 0 to v (times $2/\sqrt{\pi}$) is called the **error function** and is denoted by erf v. For animation see the instruction on animation in Example 11.1 in this Guide. (*AEM Ref.* pp. 612, A56)

Pr.11.11 **(Two-dimensional heat equation)** Solve Example 11.3 in this Guide when the upper edge is kept at the temperature $\sin x$, the other data being as before. (*AEM Ref.* pp. 605-607)

Pr.11.12 **(Isotherms)** Find the isotherms (curves of constant temperature) in Pr.11.11 and plot some of them. Do these curves look physically reasonable? (*AEM Ref.* pp. 605-607)

Pr.11.13 **(Animation, rectangular membrane)** Show the motion of the membrane in Example 11.4 in this Guide for u_{22} with $B_{22} = 1$. (*AEM Ref.* p. 625)

Pr.11.14 **(Circular membrane)** Show the motion of the three solutions in the figure of Example 11.5 in this Guide with $c = 1$, $R = 1$, so that $k^2 = z_n{}^2$. Use that $G'' + k^2 G = 0$. (*AEM Ref.* p. 632)

Pr.11.15 **(Circular membrane, vibration depending on angle)** Study the motion of the circular membrane given by $u_{11}(r,\theta,t) = \sin kt\, J_1(kr) \cos \theta$, which has the y-axis ($\theta = \pi/2$ and $3\pi/2$) as a nodal line (line along which the membrane does not move). Here, k is the first positive zero of J_1. (*AEM Ref.* pp. 635, A85)

PART D. COMPLEX ANALYSIS

Content. Complex numbers and functions, conformal mapping (Chap. 12)
Complex integration (Chap. 13)
Power series, Taylor series (Chap. 14)
Laurent series, singularities, residue integration (Chap. 15)
Potential theory (Chap. 16)

The Mathematica notation for $i = \sqrt{-1}$ is I, the response being i. Do not use I otherwise. Symbols $+$, $-$, $*$, $/$, $\hat{}$, are the same in complex as in real. Re, Im, Abs, Argument, and Conjugate give the real part, imaginary part, absolute value, argument, and complex conjugate, respectively. ComplexExpand gives results in the form $a + ib$. This as well as the polar form and the plotting of complex numbers is illustrated in Example 12.1.

Chapter 12

Complex Numbers and Functions.
Conformal Mapping

Content. Complex numbers, polar form, plots (Ex. 12.1, Prs. 12.1-12.5)
Equations, roots, sets (Ex. 12.2, Prs. 12.6-12.11)
Cauchy-Riemann equations, harmonic functions (Ex. 12.3, Pr. 12.12)
Conformal mapping (Ex. 12.4, Prs. 12.13-12.15, 12.17)
Special complex functions (Exs. 12.5, 12.6, Prs. 12.16-12.20)

Examples for Chapter 12

EXAMPLE 12.1   **COMPLEX NUMBERS. POLAR FORM. PLOTTING**

Let us begin with illustrating arithmetic in complex. For $i = \sqrt{-1}$ type I. Let

In[1]:= z1 = 3 + 4 I   (* Out: $3 + 4\,\mathrm{i}$ *)

In[2]:= z2 = 5 - 2 I   (* Out: $5 - 2\,\mathrm{i}$ *)

Then

In[3]:= z1 + z2   (* Out: $8 + 2\,\mathrm{i}$ *)

In[4]:= z1 - z2   (* Out: $-2 + 6\,\mathrm{i}$ *)

In[5]:= z1 z2   (* Out: $23 + 14\,\mathrm{i}$ *)

In[6]:= z1/z2   (* Out: $\frac{7}{29} + \frac{26\,\mathrm{i}}{29}$ *)

In[7]:= N[%]   (* Out: $0.241379 + 0.896552\,\mathrm{i}$ *)

133

In[8]:= (3.0 + 4 I)/(5 - 2 I)   (* Out: $0.241379 + 0.896552\,i$ *)

If a result is not in the form $a+ib$, but you want it in this form, apply ComplexExpand. For instance,

In[9]:= Cos[3 + I]   (* Out: $\mathrm{Cos}[3+i]$ *)

In[10]:= ComplexExpand[%]   (* Out: $\mathrm{Cos}[3]\,\mathrm{Cosh}[1] - i\,\mathrm{Sin}[3]\,\mathrm{Sinh}[1]$ *)

In[11]:= Cos[3.0 + I]   (* Out: $-1.52764 - 0.165844\,i$ *)

Re, Im, Abs, Arg, and Conjugate give the **real part**, **imaginary part**, **absolute value** (**modulus**), **argument**, and **complex conjugate**, respectively. For instance,

In[12]:= z = 16 - 12 I   (* Out: $16 - 12\,i$ *)

In[13]:= Re[z]   (* Out: 16 *)

In[14]:= Im[z]   (* Out: 12 *)

In[15]:= Abs[z]   (* Out: 20 *)

In[16]:= Arg[z]   (* Out: $-\mathrm{Arctan}\left[\frac{3}{4}\right]$ *)

In[17]:= N[%]   (* Out: -0.643501 *)

In[18]:= Conjugate[z]   (* Out: $16 + 12\,i$ *)

Furthermore,

In[19]:= Re[Cos[3 + I]]   (* Out: $\mathrm{Re}[\mathrm{Cos}[3+i]]$ *)

In[20]:= ComplexExpand[%]   (* Out: $\mathrm{Cos}[3]\,\mathrm{Cosh}[1]$ *)

In[21]:= N[%]   (* Out: -1.52764 *)

In[22]:= ComplexExpand[Im[Cos[3 + I]]]   (* Out: $-\mathrm{Sin}[3]\,\mathrm{Sinh}[1]$ *)

In[23]:= Abs[ComplexExpand[Cos[3 + I]]]

Out[23]= $\sqrt{\mathrm{Cos}[3]^2\,\mathrm{Cosh}[1]^2 + \mathrm{Sin}[3]^2\,\mathrm{Sinh}[1]^2}$

In[24]:= Conjugate[-4.8 + 0.3 I]   (* Out: $-4.8 - 0.3\,i$ *)

In[25]:= z = x + I y   (* Out: $x + i\,y$ *)

In[26]:= Conjugate[z]   (* Out: $\mathrm{Conjugate}[x + i\,y]$ *)

In[27]:= ComplexExpand[%]   (* Out: $x - i\,y$ *)

In[28]:= f = z^3   (* Out: $(x + i\,y)^3$ *)

In[29]:= Re[f]   (* Out: $\mathrm{Re}[(x + i\,y)^3]$ *)

In[30]:= ComplexExpand[%]   (* Out: $x^3 - 3\,x\,y^2$ *)

In[31]:= ComplexExpand[Im[f]]   (* Out: $3\,x^2\,y - y^3$ *)

In[32]:= Conjugate[f]   (* Out: $\mathrm{Conjugate}[x + i\,y]^3$ *)

In[33]:= ComplexExpand[Conjugate[f]]   (* Out: $x^3 - 3\,x\,y^2 + i\,(-3\,x^2\,y + y^3)$ *)

The complex conjugate could also be obtained by the following trick.

In[34]:= fbar = f /. Complex[x_, y_] -> Complex[x, -y]   (* Out: $(x - i\,y)^3$ *)

In[35]:= ComplexExpand[%]   (* Out: $x^3 - 3\,x\,y^2 + i\,(-3\,x^2\,y + y^3)$ *)

This can now be used to obtain $\mathrm{Re}\, f$ and $\mathrm{Im}\, f$ in terms of f and $\bar{f}$ as well as the square of the absolute value of f.

In[36]:= `re = Simplify[(f + fbar)/2]`   (* Out: $x^3 - 3\,x\,y^2$ *)

In[37]:= `im = Simplify[(f - fbar)/(2 I)]`   (* Out: $3\,x^2\,y - y^3$ *)

In[38]:= `Simplify[absSquare = f fbar]`   (* Out: $(x^2 + y^2)^3$ *)

Polar form. The polar form, for instance, of $z = 8 + 3i$, is obtained by

In[39]:= `z = 8 + 3 I`   (* Out: $8 + 3i$ *)

In[40]:= `Abs[z] Exp [I Arg[z]]`   (* Out: $\sqrt{73}\,e^{i\,\mathrm{ArcTan}\left[\frac{3}{8}\right]}$ *)

From an expression in polar form, its Cartesian form is obtained by the command `ComplexExpand`. In the present case,

In[41]:= `ComplexExpand[%]`   (* Out: $8 + 3\,I$ *)

Typing may be saved by defining `f[z_]` as shown and then inserting the desired z.

In[42]:= `Clear[z]`

In[43]:= `f[z_] = Abs[z] Exp [I Arg[z]]`   (* Out: $e^{i\,\mathrm{Arg}\,[z]}\,\mathrm{Abs}\,[z]$ *)

In[44]:= `f[1 + I]`   (* Out: $\sqrt{2}\,e^{\frac{i\pi}{4}}$ *)

In[45]:= `f[10 I]`   (* Out: $10\,i$ Should respond $10\,e^{\frac{i\pi}{2}}$ *)

In[46]:= `f[-10]`   (* Out: -10 Should respond $10\,e^{i\pi}$ *)

To obtain the desired results in the last two cases , type

In[47]:= `p[z_] := Abs[z] Exp[HoldForm[Evaluate[I Arg[z]]]]`   (* No response *)

In[48]:= `pol1 = p[1 + I]`   (* Out: $\sqrt{2}\,e^{\frac{i\pi}{4}}$ *)

In[49]:= `pol2 = p[10 I]`   (* Out: $10\,e^{\frac{i\pi}{2}}$ *)

In[50]:= `pol3 = p[-10]`   (* Out: $10\,e^{i\pi}$ *)

The command `HoldForm` keeps `Exp[...]` from being evaluated. In order to convert back to $x + iy$ form, you use

In[51]:= `ComplexExpand[ReleaseHold[pol1]]`   (* Out: $1 + i$ *)

In[52]:= `ComplexExpand[ReleaseHold[pol2]]`   (* Out: $10\,i$ *)

In[53]:= `ComplexExpand[ReleaseHold[pol3]]`   (* Out: -10 *)

The command `ReleaseHold` allows the `Exp[...]` to be evaluated.

Multiplication in polar form is illustrated by the following command.

In[54]:= `Simplify[Abs[r] Exp[I t] Abs[p] Exp[I s]]`

Out[54]= $e^{i\,(s+t)}\,\mathrm{Abs}\,[p]\,\mathrm{Abs}\,[r]$

Plotting complex numbers in the **complex plane.** To obtain $5 + 2i$ and $5 - 5i$ as points, type

In[55]:= `P1 = ListPlot[{{5, 2}, {5, -2}}, PlotRange -> {{0, 6}, {-3, 3}},`
     `Prolog -> AbsolutePointSize[4]]`

To obtain these numbers as vectors (line segments), type (these are line segments with slopes 2/5 and −2/5)

In[56]:= `P2 = Plot[{(2/5) x, -(2/5) x}, {x, 0, 5}]`

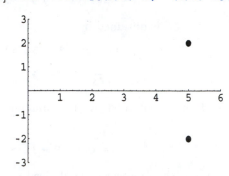
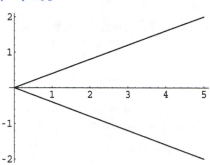

Example 12.1. Complex numbers
plotted as points

Example 12.1. Complex numbers
plotted as vectors (line segments)

To combine both methods of plotting use `Show`, as follows.

In[57]:= `Show[P1, P2]`

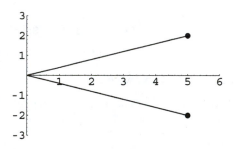

Example 12.1. Combination of the two plots for points

Similar Material in AEM: pp. 652-660

EXAMPLE 12.2 | EQUATIONS. ROOTS. SETS IN THE COMPLEX PLANE

Quadratic and other equations can be solved by `Solve` For instance, solve the equation $z^2 - (5 + i) z + 8 + i = 0$.

Solution.

In[1]:= `Clear[z]`

In[2]:= `eq = z^2 - (5 + I) z + 8 + I == 0`

Out[2]= $(8 + i) - (5 + i) z + z^2 == 0$

In[3]:= `sol = Solve[eq, z]`   `(* Out: {{z → 2 - i}, {z → 3 + 2 i}} *)`

The two solutions can be accessed individually by

In[4]:= `sol[[1, 1, 2]]`   `(* Out: 2 - i *)` and

In[5]:= sol[[2, 1, 2]]   (* Out: 3 + 2i *)

The command FindRoot gives one solution at a time, depending on the choice of an approximation that you have to select. For instance,

In[6]:= FindRoot[eq, {z, 2.5}]   (* Out: {z → 2. − 1.i} *)

In[7]:= FindRoot[eq, {z, 4 + I}]   (* Out: {z → 3. + 2.i} *)

You see that the approximation that you choose can be rather poor. Try $2.5 + 0.5i$ (the point in the middle of the two roots) and see what happens.

Roots of unity. By definition, these are the roots of an equation $z^n = 1$, where n is a given integer. For instance, for $n = 4$ you get 1, i, -1, $-i$. Find and plot the roots of unity for $n = 9$.

Solution. Type

In[8]:= Clear[z, sol, n]

In[9]:= sol = Solve[z^9 == 1.0, z]

Out[9]= {{z → −0.939693 − 0.34202 i}, {z → −0.939693 + 0.34202 i},

{z → −0.5 − 0.866025 i}, {z → −0.5 + 0.866025 i},

{z → 0.173648 − 0.984808 i}, {z → 0.173648 + 0.984808 i},

{z → 0.766044 − 0.642788 i}, {z → 0.766044 + 0.642788 i}, {z → 1.}}

Mathematica cannot plot this *complex* sequence. Convert it to a real sequence of pairs $(\operatorname{Re} z_k, \operatorname{Im} z_k)$, where z_k, $k = 1, \ldots, 9$, are the roots.

In[10]:= S = Table[{Re[sol[[n, 1, 2]]], Im[sol[[n, 1, 2]]]}, {n, 1, 9}]

Out[10]= {{−0.939693, −0.34202}, {−0.939693, 0.34202}, {−0.5, −0.866025},

{−0.5, 0.866025}, {0.173648, −0.984808}, {0.173648, 0.984808},

{0.766044, −0.642788}, {0.766044, 0.642788}, {1., 0}}

You can now plot. AspectRatio -> Automatic in the last command gives equal scalaes on both axes. Without it you would get an ellipse. Try it.

In[11]:= P1 = ListPlot[S, Prolog -> AbsolutePointSize[4]]

In[12]:= P2 = ParametricPlot[{Cos[t], Sin[t]}, {t, 0, 2 Pi}]

In[13]:= Show[P1, P2, AspectRatio -> Automatic]

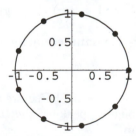

Example 12.2. Roots of unity for $n = 9$

on the unit circle $|z| = 1$

Circles. The circle of radius 1 with center 0 in the figure is called the **unit circle**. Its interior is called the **open unit disk**. From the plotting command you can infer that a circle of radius r and center z_0, for instance, $r = 3$, $z_0 = 2 + i$, can be plotted by the following command.

In[14]:= `P3 = ParametricPlot[{2 + 3 Cos[t], 1 + 3 Sin[t]}, {t, 0, 2 Pi},`
          `AspectRatio -> Automatic]`

Besides disks, other important regions are those bounded by two concentric circles. Such a region is called an **annulus**. For instance, if the center is $2 + i$ and the radii are 3 (this gives the previous circle) and 1.4, you can type (with `P5` for plotting the center)

In[15]:= `P4 = ParametricPlot[{2 + 1.4 Cos[t], 1 + 1.4 Sin[t]}, {t, 0, 2 Pi}]`

In[16]:= `P5 = ListPlot[{{2, 1}}]`

In[17]:= `Show[P3, P4, P5, Prolog -> AbsolutePointSize[4]]`

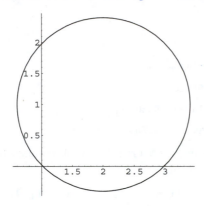

Example 12.2. Circle of radius 3   *Example 12.2.* Annulus with center
and center $2 + i$ $z = 2 + i$ and radii 3 and 1.4

Similar Material in AEM: pp. 660-664

EXAMPLE 12.3   **CAUCHY-RIEMANN EQUATIONS.**
                       HARMONIC FUNCTIONS

A standard notation for complex variables and functions is

$$z = x + iy \quad \text{and} \quad f(z) = u(x, y) + iv(x, y)$$

with real u and v. The **Cauchy-Riemann equations** (involving partial derivatives) are

$$u_x = v_y, \qquad u_y = -v_x$$

or, equivalently, $u_x - v_y = 0$, $u_y + v_x = 0$. These are the most important equations in the whole chapter because, roughly speaking, they are necessary and sufficient for $f(z)$ to be an **analytic function**.

     For instance, find out whether $e^x(\cos y + i \sin y)$ is analytic.

Solution. Type the function and then the partial derivatives needed.

In[1]:= `Clear[f, x, y, u, v]`

In[2]:= `f = Exp[x] (Cos[y] + I Sin[y])`

Out[2]= $e^x (\text{Cos}[y] + i\,\text{Sin}[y])$

In[3]:= `Re[f]`   (* Out: Re[e^x (Cos[y] + i Sin[y])] *)

In[4]:= `u = ComplexExpand[Re[f]]`   (* Out: e^x Cos[y] *)

In[5]:= `v = ComplexExpand[Im[f]]`   (* Out: e^x Sin[y] *)

In[6]:= `ux = D[u, x]`   (* Out: e^x Cos[y] *)

In[7]:= `vy = D[v, y]`   (* Out: e^x Cos[y] *)

Hence the first Cauchy-Riemann equation is satisfied. And so is the second one,

In[8]:= `CR2 = D[u, y] + D[v, x]`   (* Out: 0 *)

Is $|z|^2 = x^2 + y^2$ analytic?

Solution. $f_2 = u + iv = x^2 + y^2$, hence $u = x^2 + y^2 = f_2$ and $v = 0$. Type

In[9]:= `f2 = x^2 + y^2`   (* Out: $x^2 + y^2$ *)

In[10]:= `D[ComplexExpand[Re[f2]], x] - D[ComplexExpand[Im[f2]], y]`

The response is $2x$ (instead of 0). The answer is no. You can stop here. You can also see this result immediately by inspection because f_2 is real, so that the partial derivative of its real part with respect to x is $2x$, whereas its imaginary part is identically zero, and so are its partial derivatives.

Harmonic functions. Solutions of Laplace's equation (with continuous second partial derivatives) are called *harmonic functions*. If u and v are harmonic and such that $f = u + iv$ is analytic, then v is called a **conjugate harmonic function** of u.

Show that $u = x^2 - y^2 - y$ is harmonic and find a conjugate harmonic v.

Solution.

In[11]:= `Clear[x, y, z, h]`

In[12]:= `u = x^2 - y^2 - y`   (* Out: $x^2 - y - y^2$ *)

In[13]:= `<<Calculus'VectorAnalysis'`

In[14]:= `SetCoordinates[Cartesian[x, y, zeta]]`   (* z is used otherwise *)

Out[14]= `Cartesian[x, y, zeta]`

In[15]:= `Laplacian[u]`   (* Out: 0 *)

Hence u is harmonic. Now use the first Cauchy-Riemann equation. Calculate u_x and then integrate $u_x = v_y$ with respect to y, adding an arbitrary "constant" of integration $h(x)$ (not given by Mathematica).

In[16]:= `ux = D[u, x]`   (* Out: 2x *)

In[17]:= `v = Integrate[ux, y] + h[x]`   (* Out: 2xy + h[x] *)

Differentiate this with respect to x, obtaining v_x, and equate it to $-u_y$, thus satisfying the second Cauchy-Riemann equation. Then integrate the result with respect to x.

In[18]:= `vx = D[v, x] == -D[u, y]`

Out[18]= $2\,y + h'[x] == 1 + 2\,y$

In[19]:= v = Integrate[vx[[2]], x] + c   (* Out: $c + x(1 + 2y)$ *)

You can now combine this with the given u, obtaining $f(z) = u(x,y) + iv(x,y)$. Using $x = (z + \bar{z})/2$, $y = (z - \bar{z})/2i$, you can express the result in terms of $z = x + iy$ as $f(z) = z^2 + iz + ic$ (c real) because

In[20]:= f = u + I v   (* Out: $x^2 - y - y^2 + i(c + x(1 + 2y))$ *)

In[21]:= f /. {x -> (z + Conjugate[z])/2, y -> {z - Conjugate[z]}/(2 I)}

Out[21]= $\dfrac{1}{2}$i(z $-$ Conjugate[z]) $+ \dfrac{1}{4}$(z $-$ Conjugate[z])$^2 + \dfrac{1}{4}$(z $+$ Conjugate[z])$^2 +$

         i$\Big($c $+ \dfrac{1}{2}$(1 $-$ i(z $-$ Conjugate[z])) (z $+$ Conjugate[z])$\Big)$

In[22]:= Simplify[%]   (* Out: $ic + z(i + z)$ *)

Similar Material in AEM: pp. 669-673

EXAMPLE 12.4   CONFORMAL MAPPING

The mapping of a region in the z-plane onto a region in the w-plane given by an analytic function $w = f(z) = u(x,y) + iv(x,y)$, $z = x + iy$, is **conformal** (angle-preserving in size and sense), except at points where the derivative $f'(z)$ is zero. To obtain conformal mappings, load the package

In[1]:= <<Graphics`ComplexMap`

A 45-degree rotation of the rectangle $0 \le x \le 2$, $0 \le y \le 1$, combined with a dilatation by a factor $\sqrt{2}$, is obtained by typing

In[2]:= w[z_] = (1 + I) z

In[3]:= P1 = CartesianMap[w, {0, 2}, {0, 1}]

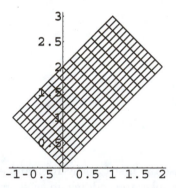

Example 12.4. Conformal mapping by $w = (1 + i)z$

The mapping $w = z^2$ doubles angels at the origin, where $w' = 2z = 0$. For example, it maps the square $0 \le x \le 1$, $0 \le y \le 1$ onto a "gothic window".

In[4]:= w[z_] = z^2

In[5]:= P2 = CartesianMap[w, {0, 1}, {0, 1}]

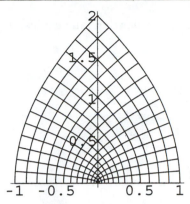

Example 12.4. Conformal mapping of the square $0 \le x \le 1$, $0 \le y \le 1$ by $w = z^2$

Polar coordinates can be employed by using the command `PolarMap`. For example, the semiring $1/2 \le r \le 1$, $0 \le \theta \le \pi$ can be plotted by using the identity map $w = z$; see the figure.

In[6]:= `P3 = PolarMap[Identity, {1/2, 1}, {0, Pi}]`

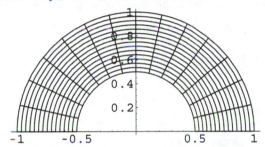

Example 12.4. Semiring $1/2 \le r \le 1$, $0 \le \theta \le \pi$

The function $w = 1/z$ ("**reflection in the unit circle**") maps domains inside the unit circle onto domains outside the circle. For instance, you obtain a domain, say, $1/2 \le r \le 1$, $0 \le \theta \le \pi/4$, and its image under $w = 1/z$, together with the unit circle, on the same figure by typing

In[7]:= `w[z_] = 1/z`   (* Out: $\frac{1}{z}$ *)

In[8]:= `P4 = PolarMap[Identity, {1/2, 1}, {0, Pi/4}]`   (* Given domain *)

In[9]:= `P5 = PolarMap[w, {1/2, 1}, {0, Pi/4}]`   (* Unit circle *)

In[10]:= `P6 = ParametricPlot[{Cos[t], Sin[t]}, {t, 0, 2 Pi}]`

In[11]:= `Show[P4, P5, P6]`

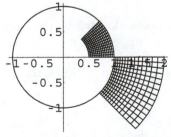

Example 12.4. Mapping by $w = 1/z$ ("Reflection in the unit circle")

The function $w = z + 1/z$ maps a circular ring (an annulus) $0 \leq r \leq 3$ onto an elliptical ring if you type

In[12]:= `w[z_] = z + 1/z`   (* Out: $\frac{1}{z} + z$ *)

In[13]:= `P7 = PolarMap[w, {2, 3}, {0, 2 Pi}]`

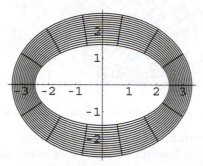

Example 12.4. Mapping of a circular ring onto an elliptical ring by $w = z + 1/z$

$w = z + 1/z$ also gives an airfoil as the image of a special circle. Type

In[14]:= `z = 1/10 (1 + I + Sqrt[122] Exp[I t])`   (* Circle *)

Out[14]= $\dfrac{1}{10}\left((1+i) + \sqrt{122}\, e^{i\,t}\right)$

In[15]:= `w = z + 1/z`   (* Airfoil *)

Out[15]= $\dfrac{10}{(1+i) + \sqrt{122}\, e^{i\,t}} + \dfrac{1}{10}\left((1+i) + \sqrt{122}\, e^{i\,t}\right)$

In[16]:= `Simplify[ComplexExpand[Re[w]]]`

Out[16]= $\dfrac{\left(100 + \mathrm{Abs}\left[(1+i) + \sqrt{122}\, e^{i\,t}\right]^2\right)\left(1 + \sqrt{122}\,\mathrm{Cos}\,[t]\right)}{10\,\mathrm{Abs}\left[(1+i) + \sqrt{122}\, e^{i\,t}\right]^2}$

In[17]:= `ParametricPlot[{Re[w], Im[w]}, {t, 0, 2 Pi}, AspectRatio -> Automatic,`
       `PlotRange -> {-0.5, 0.5}, AxesLabel ->  {"u", "v"},`
       `Ticks -> {{-2, 2}, None}]`

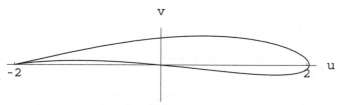

Example 12.4. Joukowski airfoil (the image of a special circle
under $w = z + 1/z$)

Similar Material in AEM: pp. 675-678, 695

EXAMPLE 12.5 | EXPONENTIAL, TRIGONOMETRIC, AND HYPERBOLIC FUNCTIONS

Evaluating functions proceeds as in real, except for the possibility of switching back and forth between z and $x + iy$. To return from $x + iy$ to z, type the command `Clear[z]`. The essential commands will be explained in the following, where they produce the given function $f(z)$, its value at $2 + 3i$, its real part $u(x, y)$, a plot of $u(x, y)$ as a surface, the absolute value $|f(2 + 3i)|$, the unassignment `Clear[z]`, and the derivative $f'(z)$.

In[1]:= `Clear[z, x, y]`

In[2]:= `f = -5 z^3 + (12 - 2 I) z^2 - I z - 200`

Out[2]= $-200 - iz + (12 - 2i)z^2 - 5z^3$

In[3]:= `f /.z -> 2 + 3 I`   (* Out: $-3 + 107i$ *)

In[4]:= `z = x + I y`   (* Out: $x + iy$ *)

In[5]:= `u = ComplexExpand[Re[f]]`

Out[5]= $-200 + 12x^2 - 5x^3 + y + 4xy - 12y^2 + 15xy^2$

In[6]:= `Plot3D[u, {x, 0, 5}, {y, 0, 5}, ViewPoint -> {1.5, 1, 1}]`

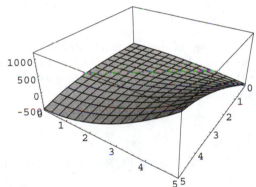

Example 12.5. Real part `u` of `f` plotted as surface over the xy-plane

In[7]:= `Abs[f] /. z -> 2 + 3 I`   (* Out: $\sqrt{11458}$ *)

In[8]:= `N[%]`   (* Out: 107.042 *)

In[9]:= `Clear[z]`

In[10]:= `D[f, z]`   (* Out: $-i + (24 - 4i)z - 15z^2$ *)

Complex exponential function e^z, also written exp z. The command for the exponential function with basis $e = 2.71828...$ is `Exp[z]`. Accordingly,

In[11]:= `Exp[1.4 - 0.6 I]`   (* Out: $3.3469 - 2.28974i$ *)

In[12]:= `Exp[2 + I]`   (* Out: e^{2+i} *)

In[13]:= `ComplexExpand[%]`   (* Out: $e^2 \text{Cos}[1] + i\, e^2 \text{Sin}[1]$ *)

In[14]:= `Re[Exp[2 + I]]`   (* Out: $e^2 \text{Cos}[1]$ *)

In[15]:= `N[Exp[2 + I]]`   (* Out: $3.99232 + 6.21768i$ *)

In[16]:= `N[Exp[1], 50]`   (* Base e *)

Out[16]= 2.7182818284590452353602874713526624977572470937000

In[17]:= `ComplexExpand[Exp[I z]]`   (* Out: `Cos[z] + i Sin[z]` *)

This is **Euler's formula**. Recall that the command `ComplexExpand` regards all variables not included in braces) as real. Hence it regards z as real, but you know that the formula also holds true for complex $z = x + iy$.

Complex trigonometric functions cos z, sin z, tan z, cot z, sec z, cosec z. To obtain values of these functions or their real or imaginary parts, etc., use the commands just illustrated. For instance,

In[18]:= `z = x + I y;`   (* Out: `x + iy` *)

In[19]:= `ComplexExpand[Cos[z]]`   (* Out: `Cos[x] Cosh[y] - i Sin[x] Sinh[y]` *)

In[20]:= `ComplexExpand[Sin[z]]`   (* Out: `Cosh[y] Sin[x] + i Cos[x] Sinh[y]` *)

In[21]:= `N[Cos[2 + 3 I]]`   (* Out: $-4.18963 - 9.10923\,i$ *)

In[22]:= `N[Tan[1 + 3 I]]`   (* Out: $0.00451714 + 1.00205\,i$ *)

To plot the **modular surface** (surface of the absolute value) of sin z, type

In[23]:= `Plot3D[Abs[Sin[x + I y]], {x, 0, 2 Pi}, {y, 0, 2},`
     `ViewPoint -> {-2, -3, 1}]`

The next plot shows the image of the rectangle $-\pi/2 + 0.1 < x < \pi/2 - 0.1$, $-1 < y < 1$ under the mapping by sin z. You see that the two vertical edges of the rectangle are almost folded under the mapping (and would be completely folded if you dropped 0.1 (twice) from the mapping function w).

In[24]:= `Clear[z, w]`

In[25]:= `w[z_] = Sin[z]`

In[26]:= `CartesianMap[w, {-Pi/2 + 0.1, Pi/2 - 0.1}, {-1, 1}]`

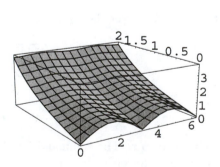 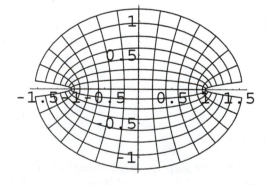

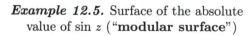

Example 12.5. Surface of the absolute value of sin z ("**modular surface**")   ***Example 12.5.*** Conformal mapping of a rectangle by sin z

cosine and **sine** are defined in terms of exponential functions. To obtain the defining formulas on the computer, type

In[27]:= `Clear[z, w]`

In[28]:= `(Exp[I z] + Exp[-I z])/2`

Out[28]= $\frac{1}{2}\left(e^{-iz} + e^{iz}\right)$

In[29]:= `ComplexExpand[%]`   `(* Out: Cos [z] *)`

Note that the command `ComplexExpand` regards z as real, but the formula also holds true in complex, according to the definition of the complex cosine function.

Hyperbolic functions cosh z, sinh z, tanh z, coth z. In complex, these functions are closely related to the trigonometric functions. To see this for cosh, type the following. For sinh, tanh, and coth the situation is similar. Try it. Numerical values can be obtained as for the other functions just discussed.

In[30]:= `Cosh[I z]`   `(* Out: Cos [z] *)`

In[31]:= `Cos[I z]`   `(* Out: Cosh [z] *)`

In[32]:= `Simplify[Cosh[z]^2 - Sinh[z]^2]`   `(* Out: 1 *)`

In[33]:= `Tanh[1 - 2 I]`   `(* Out: -i Tan[2 + i] *)`

In[34]:= `N[%]`   `(* Out: 1.16674 + 0.243458 i *)`

In[35]:= `h = ComplexExpand[Re[Tanh[x + I y]]]`   `(* Out: `$\frac{Sinh[2x]}{Cos[2y] + Cosh[2x]}$` *)`

This can also be written as $\sinh x \cosh x/(\sinh^2 x + \cos^2 y)$. To verify this claim, type

In[36]:= `h - Sinh[x] Cosh[x]/(Sinh[x]^2 + Cos[y]^2)`

Out[36]= $-\dfrac{Cosh[x]\,Sinh[x]}{Cos[y]^2 + Sinh[x]^2} + \dfrac{Sinh[2x]}{Cos[2y] + Cosh[2x]}$

In[37]:= `Simplify[%]`   `(* Out: 0 *)`

Similar Material in AEM: pp. 679-686

EXAMPLE 12.6   COMPLEX LOGARITHM

In calculus the natural logarithm $\ln x$ is defined for positive real x only and is single-valued (that is, for each such x it has only one value). In complex, $\ln z$ ($z \neq 0$) has infinitely many values $\ln z = \ln |z| + i\,\text{Arg}\,z \pm 2n\pi i$ ($n = 0, 1, 2, ...$). These values all have the same real part $\ln |z|$. Their imaginary parts differ by integer multiples of 2π. Mathematica gives the **principal value** of $\ln z$, by definition corresponding to $n = 0$ and $-\pi < \text{Arg}\,z \leq \pi$ (called the **principal value** of arg z and denoted by Arg z, as shown). For example,

In[1]:= `ComplexExpand[Log[x + I y]]`

Out[1]= $i\,\text{Arg}[x + iy] + \text{Log}[\text{Abs}[x + iy]]$

In[2]:= `Arg[x + I y]`   `(* Out: Arg[x + i y] *)`

In[3]:= `ComplexExpand[Log[4 - 3 I]]`   `(* Out: -i ArcTan`$\left[\frac{3}{4}\right]$` + Log [5] *)`

In[4]:= `Re[Log [3 - 4 I]]`   `(* Out: Re[Log[3 - 4 i]] *)`

In[5]:= `ComplexExpand[%]`   `(* Out: Log[5] *)`

In[6]:= `ComplexExpand[Im[Log[3 - 4 I]]]`   `(* Out: -ArcTan`$\left[\frac{4}{3}\right]$` *)`

In[7]:= N[Log[3 - 4 I]]   (* Out: $1.60944 - 0.927295\,i$ *)

In[8]:= Log[-1]   (* Out: $i\pi$ *)

In[9]:= N[%, 50]   (* This approximates πi. *)

Out[9]= 3.1415926535897932384626433832795028841971693993751 i

Derivative. The derivative of the complex logarithm is

In[10]:= D[Log[z], z]   (* Out: $\frac{1}{z}$ *)

Plotting. Type $\ln(3 - 4i)$ in the form $a + bi$. Type all the values you want to plot; this is the complex sequence S .

In[11]:= L = Log[3 - 4 I] + 2 n Pi I

Out[11]= $2\,i\,n\,\pi + \text{Log}\,[3 - 4\,i]$

In[12]:= S = Table[N[L], {n, -2, 3}]

Out[12]= {$1.60944 - 13.4937\,i$, $1.60944 - 7.21048\,i$, $1.60944 - 0.927295\,i$,

        $1.60944 + 5.35589\,i$, $1.60944 + 11.6391\,i$, $1.60944 + 17.9223\,i$}

Recall from Example 12.1 that for plotting you need a real sequence (the sequence of the real and imaginary parts). Note that in the new sequence you renumber the six terms from $n = 1$ to $n = 6$. (Try the numbering in S and see what you get.)

In[13]:= S2 = Table[{Re[S[[n]]], Im[S[[n]]]}, {n, 1, 6}]

Out[13]= {{1.60944, −13.4937}, {1.60944, −7.21048}, {1.60944, −0.927295},

        {1.60944, 5.35589}, {1.60944, 11.6391}, {1.60944, 17.9223}}

In[14]:= ListPlot[S2, Prolog -> AbsolutePointSize[4]]

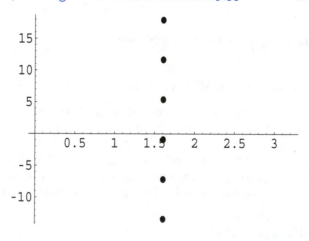

Example 12.6. Some values of $\ln(3 - 4i)$

Similar Material in AEM: pp. 687-690

Problem Set for Chapter 12

Pr.12.1 **(Complex arithmetic)** Let $z_1 = 8 + 3I$, $z_2 = 9 - 2I$. Find $z_1 + z_2$, $z_1 - z_2$, $z_1 z_2$, z_1/z_2, $|z_1/z_2|$, $|z_1|/|z_2|$, $\operatorname{Re} z_1$, $\operatorname{Im}(z_1{}^2)$, $\operatorname{Arg} z_1$. (*AEM Ref.* pp. 653, 654)

Pr.12.2 **(Complex arithmetic)** Let $z_1 = 4 + 3i$, $z_2 = 2 - 5i$. Find $z_1 \bar{z}_2$, $\bar{z}_1 z_2$ (why must these two products be conjugate?), $1/|z_1|$, $|z_1| + |z_2| - |z_1 + z_2|$ (why must this be nonnegative?), $\operatorname{Re}(z_1{}^3)$, $(\operatorname{Re} z_1)^3$, $\operatorname{Im}((z_1 - z_2)/(z_1 + z_2))$. (*AEM Ref.* pp. 655-657)

Pr.12.3 **(Real and imaginary parts)** Obtain the important formulas $x = (z + \bar{z})/2$ and $y = (z - \bar{z})/2i$ on the computer. (*AEM Ref.* p. 656)

Pr.12.4 **(Polar form)** Using the computer, convert the following complex numbers to polar form and back. $2 + 2i$, $-2 - 2i$, $(1 + i)(-2 - 2i)$, -20, i, $((6 + 8i)/(4 - 3i))^2$. (*AEM Ref.* pp. 657-660)

Pr.12.5 **(Plotting complex numbers)** Plot $(0.9 + 0.4i)^n$ for integer $n = -20, ..., 20$. Also plot the unit circle on the same axes. Can you find n for each of the 41 points that you see in the plot? (For plotting points see Example 12.6 in this Guide.)

Pr.12.6 **(Quadratic equation in z^2)** Solve $z^4 - (3 + 6i)z^2 - 8 + 6i = 0$. (*AEM Ref.* p. 662 (#30))

Pr.12.7 **(Roots)** Find and plot all cube roots of $1 + i$, as well as the circle on which these roots lie. (*AEM Ref.* p. 662 (#21))

Pr.12.8 **(Roots of unity)** Find and plot the 16th roots of unity. Can you visualize the values before you plot them? (*AEM Ref.* pp. 660-662)

Pr.12.9 **(Annulus)** Plot the annulus with center $2 + 2i$ whose outer circle passes through the origin and whose inner circle touches the coordinate axes. (*AEM Ref.* p. 664)

Pr.12.10 **(Complex plane)** Find the curve satisfying $|z + i|/|z - i| = 1$ (a) by a geometric argument without calculation, (b) on the computer. (*AEM Ref.* p. 655)

Pr.12.11 **(Cauchy-Riemann equations)** Is $z|z|^2$ analytic? (*AEM Ref.* pp. 669)

Pr.12.12 **(Cauchy-Riemann equations)** Is $f(z) = \operatorname{Re}(z^2) - i \operatorname{Im}(z^2)$ analytic? (*AEM Ref.* p. 673 (#12))

Pr.12.13 **(Experiment on conformal mapping)** Make plots of the images of rectangles under $w = z^n$, $n = 2, 3, 4$, etc. How do these images change as functions of the exponent n? What happens to the image if you change the position of a rectangle in the plane? (*AEM Ref.* pp. 674-677)

Pr.12.14 **(Experiment on conformal mapping)** Experiment with the image of a square $a \le x \le a + 1$, $b \le y \le b + 1$ under the mapping $w = z^3$ and characterize the position and form of the image for various $a \ge 0$ and $b \ge 0$. (*AEM Ref.* pp. 674-678)

Pr.12.15 **(Inversion in the unit circle)** Plot the square $1/2 \le x \le 3/2$, $1/2 \le y \le 3/2$ and its image under $w = 1/z$ as well as the unit circle on common axes. Discuss how the image curves correspond to the straight-line segments in the given square. (*AEM Ref.* p. 692)

Pr.12.16 **(Exponential function)** Find e^z (in the form $u + iv$) and $|e^z|$ if z equals $2 + 3\pi i$, $1 + i$, $2\pi(1 + i)$, $0.95 - 1.6i$, and $-\pi i/2$. (*AEM Ref.* p. 682 (##1-5))

Pr.12.17 (Conformal mapping by cos z) Find the image of the rectangle $0 \leq x \leq 2\pi$, $1/2 \leq y \leq 1$ under the mapping $w = \cos z$. What are the images of the vertical sides $x = 0$ and $x = 2\pi$? (*Hint* : What will happen if you shorten the x-interval slightly? (*AEM Ref.* p. 687 (#18))

Pr.12.18 (Hyperbolic functions) Obtain the basic formulas $\cosh^2 z - \sinh^2 z = 1$ and $\cosh^2 z + \sinh^2 z = \cosh 2z$ on the computer. (*AEM Ref.* p. 686 (#3))

Pr.12.19 (Natural logarithm ln z) Find the principal value $\mathrm{Ln}\, z$ for $z = -5, -12 - 16i, 1 + i$, $1 - i, -10 + 0.1i, -10 - 0.1i$. What does a comparison of the last two values illustrate? (*AEM Ref.* p. 691 (##5-8))

Pr.12.20 (General powers) Find (in the form $u + iv$) $(2i)^{2i}$, 3^{4-i}, $(1 + 3i)^i$, $i^{1/2}$. (*AEM Ref.* p. 691)

Chapter 13

Complex Integration

Content. Indefinite integration (Ex. 13.1)
Use of path (Ex. 13.2, Prs. 13.1-13.3, 13.5, 13.7-13.9)
Cauchy theorem and formula (Ex. 13.3, Pr. 13.6, 13.10)
Use of partial fractions (Prs. 13.4, 13.5)

Further integration methods follow in Chap. 15.

Examples for Chapter 13

EXAMPLE 13.1   INDEFINITE INTEGRATION OF ANALYTIC FUNCTIONS

Let $f(z)$ be analytic in a domain D that is **simply connected** (that is, every closed curve in D encloses only points of D). Then there exists an analytic function $F(z)$ such that $F'(z) = f(z)$ everywhere in D, and for all paths in D from any point z_0 to any point z_1 in D,

$$\int_{z_0}^{z_1} f(z)\, dz = F(z_1) - F(z_0).$$

$F(z)$ is called an **antiderivative** or **indefinite integral** of $f(z)$. This is the analog of a known formula from calculus, but note well that it applies to *analytic* functions only. For instance,

In[1]:= Integrate[3 z^2, {z, 0, 1 + I}]   (* Out: $-2 + 2\,i$ *)

In[2]:= Integrate[Cos[z], {z, -Pi I, Pi I}]   (* Out: $2\,i\,\text{Sinh}[\pi]$ *)

In[3]:= N[%]   (* Out: $0. + 23.0975\,i$ *)

In[4]:= Integrate[1/z, {z, -I, I}]   (* Out: $i\,\pi$ *)

In the last integral an antiderivative is the principal value $\text{Ln}\, z$ of $\ln z$, which is not analytic on the negative real axis $x \leq 0$, and the paths from $-i$ to i must be restricted accordingly.

    Similar Material in AEM: pp. 707, 708

EXAMPLE 13.2   INTEGRATION: USE OF PATH. PATH DEPENDENCE

If the function $f(z)$ you want to integrate is not analytic, the integral will generally depend on path, and you have to use a representation of the path C that is given. Let C be represented by $z = z(t)$, $a \leq t \leq b$, and piecewise smooth. Let $f(z)$ be continuous on C. Then

$$\int_C f(z)\, dz = \int_a^b f(z(t))\, \dot{z}(t)\, dt \qquad \dot{z} = dz/dt.$$

For instance, if you want to integrate $f(z) = 1/z^n$ with any given $n = 1, 2, \ldots$ counterclockwise around the unit circle C, there is no simply connected domain containing C in which $1/z^n$ is analytic. ($1/z^n$ is analytic, for instance, in the annulus $D : 1/2 < |z| < 3/2$, but D is not simply connected.) Represent C, obtain $\dot{z}$, and then integrate from $a = 0$ to $b = 2\pi$.

In[1]:= `z = Exp[I t]`   (* Out: e^{it} Type a space between `I` and `t` ! *)

In[2]:= `zdot = D[z, t]`   (* Out: ie^{it} *)

In[3]:= `Clear[n]`

In[4]:= `J = Integrate[1/z^n zdot, {t, 0, 2 Pi}]`

Out[4]= $\mathtt{i}\left(-\dfrac{\mathtt{i}\,(-1)^{-\mathtt{n}}}{-1+\mathtt{n}} - \dfrac{\mathtt{i}\,\mathtt{Conjugate}[(-1)^{1-\mathtt{n}}]}{-1+\mathtt{n}}\right)$

You see that $n - 1 = 0$ when $n = 1$; hence this formula holds true for any integer $n \neq 1$.

In[5]:= `Simplify[ComplexExpand[%]]`   (* Out: $\dfrac{2\,\mathtt{i}\,\mathtt{Sin}[\mathtt{n}\,\pi]}{-1+\mathtt{n}}$ *)

Hence $J = 0$ for $n = 2, 3, \ldots$ (and also for $n = 0, -1, -2, \ldots$, that is, for $1, z, z^2, \ldots$). Indeed,

In[6]:= `Table[J, {n,-10, 10}]`

`Power::infy: Infinite expression` $\frac{1}{0}$ `encountered`

Out[6]= $\{0, 0, 0, 0, 0, 0, 0, 0, 0, 0;\, 0,$ `Indeterminate`$, 0, 0, 0, 0, 0, 0, 0, 0, 0\}$

However, for $n = 1$ you have $f(z) = 1/z$ and will obtain the **very important result**

$$\oint_C \frac{1}{z}\, dz = 2\pi i \qquad \text{(counterclockwise around the unit circle)}$$

that is frequently needed throughout complex analysis. In fact,

In[7]:= `Integrate[1/z zdot, {t, 0, 2 Pi}]`   (* Out: $2\,\mathtt{i}\,\pi$ *)

Path dependence of the integral can be illustrated by many examples. For instance, integrate $f(z) = |z|^2$ from 0 to $1 + i$ over two different paths, (A) over the straight segment joining these points, (B) over the parabolic arc $y = x^2$ (see the figure).

Solution. (A) Over the segment you obtain the value $(2 + 2i)/3$. Indeed,

In[8]:= `z1 = t + I t`   (* Straight segment - Space between `I` and `t` ! *)
Out[8]= $(1 + \mathtt{i})\,\mathtt{t}$

In[9]:= `z1dot = D[z1, t]`   (* Out: $1 + \mathtt{i}$ *)

In[10]:= `f1 = Abs[z1]^2`   (* Function f on the segment *)
Out[10]= $2\,\mathtt{Abs}[\mathtt{t}]^2$

In[11]:= `ComplexExpand[%]`   (* Out: $2\,\mathtt{t}^2$ *)

In[12]:= `Integrate[f1 z1dot, {t, 0, 1}]`

Out[12]= $\dfrac{2}{3} + \dfrac{2\,\mathtt{i}}{3}$

(B) The value of the integral taken over the parabolic arc will differ from the value just obtained.

In[13]:= `z2 = t + I t^2`   (* Out: $t + i t^2$ *)

In[14]:= `z2dot = D[z2, t]`   (* Out: $1 + 2 i t$ *)

In[15]:= `z2bar = ComplexExpand[Conjugate[z2]]`   (* Out: $t - i t^2$ *)

In[16]:= `z2 z2bar`   (* Out: $(t - i t^2)(t + i t^2)$ *)

In[17]:= `f2 = ComplexExpand[%]`   (* Out: $t^2 + t^4$ *)

In[18]:= `Integrate[f2 z2dot, {t, 0, 1}]`

Out[18]= $\dfrac{8}{15} + \dfrac{5\,i}{6}$

In[19]:= `P1 = Plot[t, {t, 0, 1}]`

In[20]:= `P2 = Plot[t^2, {t, 0, 1}]`

In[21]:= `Show[P1, P2, AspectRatio -> Automatic, Ticks -> {{1}, {1}},`
        `AxesLabel -> {x, y}]`

Example 13.2. Paths of the integrals in (A) and (B)

Similar Material in AEM: pp. 708-710

| **EXAMPLE 13.3** | **CONTOUR INTEGRATION BY CAUCHY'S INTEGRAL THEOREM AND FORMULA** |

Cauchy's integral theorem. Let $f(z)$ be analytic in a simply connected domain D. Let C be any closed path in D. Then the integral of $f(z)$ around C is zero.

Cauchy's integral formula and derivative formulas. Let $f(z)$ and D be as before. Let C be as before and *simple* (that is, without self-intersections). Let z_0 be any point inside C. Then the integral

$$(1) \qquad \oint_C \frac{f(z)}{(z - z_0)^{n+1}}\, dz$$

(taken counterclockwise) equals $2\pi i f(z_0)$ for $n = 0$ (**"Cauchy's integral formula"**) and ($f^{(n)}$ the nth derivative of f with respect to z)

$$(2) \qquad \frac{2\pi i}{n!} f^{(n)}(z_0) \qquad \text{for } n = 1, 2, \dots .$$

For instance, taking $f(z) = 1$ and $z_0 = 0$, you see without any calculation that for $n = 0$ (that is, for $1/z$) you get $2\pi i$, and for $n = 1, 2, \dots$ the integrals (of $1/z^2$, $1/z^3, \dots$) are all zero.

As a second application, find the integral of $\dfrac{e^z}{z\,e^z - 2\,i\,z}$ counterclockwise around the circle $|z| = 0.5$.

Solution. From Cauchy's integral formula with $f(z) = \dfrac{e^z}{e^z - 2i}$ and $z_0 = 0$ you obtain

In[1]:= `Clear[z]`

In[2]:= `f = Exp[z]/(Exp[z] - 2 I)`   (* Out: $\dfrac{e^z}{-2\,i + e^z}$ *)

In[3]:= `2 Pi I f /. z -> 0`   (* Out: $\left(-\dfrac{4}{5} + \dfrac{2\,i}{5}\right)\pi$ Space between `Pi I f`! *)

You still have to make sure that $f(z)$ is analytic everywhere inside and on C. Now the only points at which something could happen are those at which the denominator of $f(z)$ is zero; that is, $e^z = 2i$, $z = \ln 2i = \ln 2 + \pi i/2 \pm 2\,n\,\pi\,i$, but these points lie outside C, as can be seen from

In[4]:= `Solve[Exp[z] - 2 I == 0, z]`

> `Solve::ifun : Inverse functions are being used by Solve, so some solutions`
>   `may not be found`

Out[4]= $\left\{\left\{z \to \dfrac{1}{2}\,i\,(\pi - 2\,i\,\mathrm{Log}\,[2])\right\}\right\}$

In[5]:= `Simplify[%]`

Out[5]= $\left\{\left\{z \to \dfrac{i\,\pi}{2} + \mathrm{Log}\,[2]\right\}\right\}$

In[6]:= `Abs[%[[1, 1, 2]]]`   (* Out: $\sqrt{\dfrac{\pi^2}{4} + \mathrm{Log}[2]^2}$ *)

In[7]:= `N[%]`   (* Out: 1.71693 *)

and the fact that the other solutions are even greater in absolute value. Your result is now established.

As a third application, find the integral of $(\tan \pi z)/z^6$ counterclockwise around the circle $|z| = 1/4$.

Solution. From (1) and (2) with $f(z) = \tan \pi z$, $z_0 = 0$, and $n + 1 = 6$, hence $n = 5$, you obtain

In[8]:= `Clear[z]`

In[9]:= `2 Pi I/5! D[Tan [Pi z], z, z, z, z, z]`

Out[9]= $\dfrac{1}{60}\,i\,\pi\,(16\,\pi^5\,\mathrm{Sec}\,[\pi\,z]^6 + 88\,\pi^5\,\mathrm{Sec}\,[\pi\,z]^4\,\mathrm{Tan}\,[\pi\,z]^2 + 16\,\pi^5\,\mathrm{Sec}\,[\pi\,z]^2\,\mathrm{Tan}\,[\pi\,z]^4)$

In[10]:= `% /. z -> 0`   (* Out: $\dfrac{4\,i\,\pi^6}{15}$ *)

A factor π results from $2\pi i$, and π^5 from the chain rule in the five differentiations.

Note further that the circle of integration is so small that the points $\pm 1/2$, $\pm 3/2 \ldots$ where $\tan \pi z$ is not analytic lie outside the circle.

Similar Material in AEM: pp. 714, 722, 725 (#14), 726

Problem Set for Chapter 13

Pr.13.1 **(Use of path)** Integrate $\operatorname{Re} z$ over the shortest path from $1 + i$ to $3 + 2i$. (*AEM Ref.* p. 712 (#15))

Pr.13.2 **(Contour integral)** Using a representation of the path, integrate $3/(z-i) - 6/(z-i)^2$ clockwise around the circle $|z-i| = 5$. Confirm the answer by the method in Example 13.3 in this Guide. (*AEM Ref.* p. 712 (#26))

Pr.13.3 **(Use of path)** Integrate $\operatorname{Re} z$ from $1 + i$ vertically upward to $1 + 2i$ and then horizontally to $3 + 2i$. (*AEM Ref.* p. 712 (#16))

Pr.13.4 **(Partial fractions)** Integrate $(2z^3 + z^2 + 4)/(z^4 - 4z^2)$ clockwise around the circle of radius 5 and center 2 first as given and then by using partial fractions. (*AEM Ref.* pp. 722, 726)

Pr.13.5 **(Partial fractions)** Integrate $(4z^2 + 17z - 68)/(z^3 - 12z + 16)$ counterclockwise around the circle $|z| = 3$ by using partial fractions. *AEM Ref.* pp. 722, 726)

Pr.13.6 **(Derivative formulas)** Integrate $(\tanh z)/z^4$ counterclockwise around the unit circle. (See Example 13.3 in this Guide.)

Pr.13.7 **(Path dependence)** Integrate $\bar{z}$ from 0 to $1 + i$ (A) along the shortest path, (B) along the parabola $y = x^2$. (*AEM Ref.* p. 712 (#18))

Pr.13.8 **(Experiment on integrand)** Integrate $(e^z \sin z)/z^{n+1}$ counterclockwise around the circle $|z - 1| = 2$ for $n = 1, \ldots, 10$. (*AEM Ref.* p. 726)

Pr.13.9 **(Use of contour)** Integrate $\operatorname{Re} z^2$ clockwise around the boundary of the square with vertices $0, i, 1 + i, 1$. (*AEM Ref.* p. 712 (#20))

Pr.13.10 **(Derivative formulas)** Integrate $(z^3 + \sin z)/(z - i)^3$ counterclockwise around the boundary of the square with vertices ± 2 and $\pm 2i$. (*AEM Ref.* p. 729 (#12))

Power Series, Taylor Series

Content. Sequences (Ex. 14.1, Prs. 14.1, 14.2)
Convergence tests (Ex. 14.2, Prs. 14.3-14.5)
Power series (Ex. 14.3, Prs. 14.6, 14.7, 14.11)
Taylor series (Ex 14.4, Prs. 14.8-14.14)
Uniform convergence (Ex. 14.5, Pr. 14.15)

Complex sequences and series are obtained by the same commands as in real, namely, `Table`, `Series` and `Sum`.

Coefficients of z^n in a power series p are accessed as before, by typing the command `Coefficient[p, z^n]`.

Plotting complex sequences (z_n). Convert (z_n) to a sequence of $[\text{Re}\,z_n, \text{Im}\,z_n]$. Then plot.

Examples for Chapter 14

EXAMPLE 14.1   SEQUENCES AND THEIR PLOTS

To obtain a sequence (z_n), you may first type z_n and then apply the command `Table`

In[1]:= `zn = ((9 + I)/10)^n`

Out[1]= $\left(\dfrac{9}{10} + \dfrac{i}{10}\right)^n$

In[2]:= `S1 = Table[zn, {n, 1, 5}]`

Out[2]= $\left\{\dfrac{9}{10} + \dfrac{i}{10}, \dfrac{4}{5} + \dfrac{9i}{50}, \dfrac{351}{500} + \dfrac{121i}{500}, \dfrac{1519}{2500} + \dfrac{36i}{125}, \dfrac{12951}{25000} + \dfrac{7999i}{25000}\right\}$

or you may do this with one command, inserting the given expression for z_n directly into `Table`. Try it. If you want the terms as decimal fractions, type

In[3]:= `N[%]`

Out[3]= $\{0.9 + 0.1\,i,\ 0.8 + 0.18\,i,\ 0.702 + 0.242\,i,\ 0.6076 + 0.288\,i,\ 0.51804 + 0.31996\,i\}$

To plot the sequence, you must convert each term $a + ib$ to the form $\{a, b\}$. (See also Example 12.6 in this Guide, where the same was done.) Then give the plotting command.

In[4]:= `Table[{Re[zn], Im[zn]}, {n, 1, 5}]`

Out[4]= $\left\{\left\{\dfrac{9}{10}, \dfrac{1}{10}\right\}, \left\{\dfrac{4}{5}, \dfrac{9}{50}\right\}, \left\{\dfrac{351}{500}, \dfrac{121}{500}\right\}, \left\{\dfrac{1519}{2500}, \dfrac{36}{125}\right\}, \left\{\dfrac{12951}{25000}, \dfrac{7999}{25000}\right\}\right\}$

In[5]:= `S2 = Table[{Re[zn], Im[zn]}, {n, 1, 100}];`

In[6]:= `ListPlot[S2, Prolog -> AbsolutePointSize[2],`
    `AspectRatio -> Automatic, PlotRange -> {{-0.1, 1}, {-0.1, 0.4}}]`

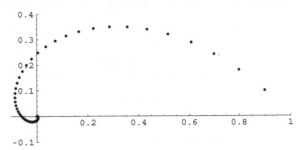

Example 14.1. Plot of the sequence (z_n), where $z_n = (0.9 + 0.1\,i)^n$, 100 terms

Function. You can obtain the sequence S2 of real pairs (here denoted by S3) from the complex expression zn by a function of z that gives the real and imaginary parts, namely,

$\text{In}[7]:=$ `S[z_, M_] := Table[{Re[z], Im[z]}, {n, 1, M}]`

$\text{In}[8]:=$ `S3 = S[zn, 5]`

$\text{Out}[8]= \left\{ \left\{ \dfrac{9}{10}, \dfrac{1}{10} \right\}, \left\{ \dfrac{4}{5}, \dfrac{9}{50} \right\}, \left\{ \dfrac{351}{500}, \dfrac{121}{500} \right\}, \left\{ \dfrac{1519}{2500}, \dfrac{36}{125} \right\}, \left\{ \dfrac{12951}{25000}, \dfrac{7999}{25000} \right\} \right\}$

This function computes the real and imaginary parts of M complex numbers zn .
    Similar Material in AEM: pp. 732-735

| EXAMPLE 14.2 | **RATIO CONVERGENCE TEST FOR COMPLEX SERIES** |

Ratio test. This is the practically most important test for convergence of a series $z_1 + z_2 + \dots$. It uses the ratios $|z_{n+1}/z_n|$, $n = 1, 2, \dots$. If for all n greater than some N these ratios do not exceed a fixed $q < 1$, the series converges absolutely. If for every $n > N$ the ratios are 1 or greater, the series diverges.

    Very often the sequence of the test ratios will have a limit L. Then if $L < 1$, the series converges absolutely. If $L > 1$, it diverges. If $L = 1$, the test fails, that is, no conclusion is possible.

    Guess whether the following series converges or diverges (consider how fast the terms increase). Then apply the test.

$$\sum_{n=0}^{\infty} \frac{(100 + 75\,i)^n}{n!} = 1 + (100 + 75i) + \frac{1}{2!}(100 + 75\,i)^2 + \dots$$

Solution. The test ratio is

$\text{In}[1]:=$ `ratio = ((100 + 75 I)^(n + 1)/(n + 1)!)/((100 + 75 I)^n/n!)`

$\text{Out}[1]= \dfrac{(100 + 75i)\, n!}{(1 + n)!}$

$\text{In}[2]:=$ `Abs[FullSimplify[ratio]]`   (* Out: $\dfrac{125}{\text{Abs}[1 + n]}$ *)

$\text{In}[3]:=$ `Limit[%, n -> Infinity]`   (* Out: 0 *)

    The limit is 0. Hence the series converges.
    Similar Material in AEM: pp. 737-739

EXAMPLE 14.3   POWER SERIES. RADIUS OF CONVERGENCE

A **power series** is a series of the form

$$(1) \qquad \sum_{n=0}^{\infty} a_n (z - z_0)^n.$$

It may converge for all z (the nicest case) or no $z \neq z_0$ (the useless case) or in a circle

$$(2) \qquad |z - z_0| = R.$$

If this is the *smallest* circle with center z_0 that includes all the points at which the series (1) converges, then its radius is called the **radius of convergence** of (1), and is given by the **Cauchy-Hadamard formula**

$$(3) \qquad R = \lim_{n \to \infty} \left| \frac{a_n}{a_{n+1}} \right|.$$

For instance, find the radius of convergence of the series $\displaystyle\sum_{n=0}^{\infty} \frac{(2n)!}{(n!)^2}(z - 3i)^n$.

Solution. Use (3). You can save work by typing a_n and then getting the quotient by using the substitution `/.  n -> n + 1` as shown. The radius of convergence will be 1/4.

In[1]:= `an = (2 n)!/(n!)^2`   (* Out: $\frac{(2n)!}{n!^2}$ *)

In[2]:= `an1 = an /. n -> n + 1`   (* Out: $\frac{(2(1+n))!}{(1+n)!^2}$ *)

In[3]:= `FullSimplify[an/an1]`   (* Out: $\frac{1+n}{2+4n}$ *)

In[4]:= `Limit[%, n -> Infinity]`   (* Out: $\frac{1}{4}$ *)

Similar Material in AEM: pp. 741-745

EXAMPLE 14.4   TAYLOR SERIES

A **Taylor series** of a given function $f(z)$ is a power series (1), Example 14.3, with coefficients $a_n = f^{(n)}(z_0)/n!$, where $f^{(n)}$ is the nth derivative. Mathematica knows practically all elementary and higher real and complex functions of general interest. If such an $f(z)$ has a Taylor series, you can obtain it by the command `Series`. For instance, if $f(z) = \cos \pi z$ and $z_0 = 1/2$, type

In[1]:= `ser = Series[Cos[Pi z], {z, 1/2, 7}]`

Out[1]= $-\pi \left(z - \frac{1}{2}\right) + \frac{1}{6}\pi^3 \left(z - \frac{1}{2}\right)^3 - \frac{1}{120}\pi^5 \left(z - \frac{1}{2}\right)^5 + \frac{\pi^7 \left(z - \frac{1}{2}\right)^7}{5040} + 0\left[z - \frac{1}{2}\right]^8$

7 is the highest power of $z - 1/2$ which you wish to have in the series. Try other values. For computing values or for plotting, drop first the error term by the command `Normal`. Decimal values for the coefficients and the center are obtained by the command `N[%]` and a value `f0` of f, for instance, at $0.4 + 0.1i$, by the command for substitution as shown.

In[2]:= `poly = Normal[%]`

Out[2]= $-\pi\left(-\frac{1}{2}+z\right)+\frac{1}{6}\pi^3\left(-\frac{1}{2}+z\right)^3-\frac{1}{120}\pi^5\left(-\frac{1}{2}+z\right)^5+\frac{\pi^7\left(-\frac{1}{2}+z\right)^7}{5040}$

In[3]:= `N[%]`

Out[3]= $-3.14159\,(-0.5+z)+5.16771\,(-0.5+z)^3-2.55016\,(-0.5+z)^5+$

$\qquad 0.599265\,(-0.5+z)^7$

In[4]:= `f0 = poly /. z -> 0.4 + 0.1 I`

Out[4]= $0.324392-0.303722\,i$

Confirm your result by the coefficient formula for the Taylor series.

In[5]:= `f[z_] = Cos[Pi z]`   `(* Out: Cos[π z] *)`

In[6]:= `S = Table[D[f[z], {z, n}]/n! /. z -> 1/2, {n, 0, 11}]`

Out[6]= $\left\{0,\,-\pi,\,0,\,\dfrac{\pi^3}{6},\,0,\,-\dfrac{\pi^5}{120},\,0,\,\dfrac{\pi^7}{5040},\,0,\,-\dfrac{\pi^9}{362880},\,0,\,\dfrac{\pi^{11}}{39916800}\right\}$

In this command, {z, n} means z, z, ..., z (n times), so that you obtain the nth derivative, as needed. Individual coefficients are obtained from S by typing `S[[1]]`, `S[[2]]`, ... Try it. Accordingly, you get a Taylor polynomial, in agreement with the previous result, by typing

In[7]:= `s7 = Sum[S[[n]] (z - 1/2)^(n - 1), {n, 1, 9}]`

Out[7]= $-\pi\left(-\frac{1}{2}+z\right)+\frac{1}{6}\pi^3\left(-\frac{1}{2}+z\right)^3-\frac{1}{120}\pi^5\left(-\frac{1}{2}+z\right)^5+\frac{\pi^7\left(-\frac{1}{2}+z\right)^7}{5040}$

You can now plot this, together with other Taylor polynomials, for real z, to create a figure as you may have seen it in your calculus book, illustrating how Taylor polynomials approximate given functions.

In[8]:= `s5 = Sum[S[[n]] (z - 1/2)^(n - 1), {n, 1, 7}]`

In[9]:= `s9 = Sum[S[[n]] (z - 1/2)^(n - 1), {n, 1, 11}]`

In[10]:= `Plot[{s5, s7, s9}, {z, 0, 2.5}]`

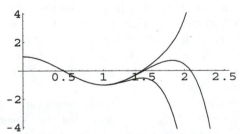

Example 14.4. Taylor polynomials approximating $\cos \pi z$ for real $z = x$
(s_7 goes upward.)

Maclaurin series. You can now easily obtain familiar Maclaurin series, which have the same form as those for real functions in calculus. In particular, this holds for the geometric series and the series for e^z, $\cos z$, and $\sin z$.

In[11]:= `S1 = Series[1/(1 - z), {z, 0, 10}]`   `(* Geometric series *)`

Out[11]= $1 + z + z^2 + z^3 + z^4 + z^5 + z^6 + z^7 + z^8 + z^9 + z^{10} + O[z]^{11}$

In[12]:= S2 = Series[Exp[z], {z, 0, 10}]

Out[12]= $1 + z + \dfrac{z^2}{2} + \dfrac{z^3}{6} + \dfrac{z^4}{24} + \dfrac{z^5}{120} + \dfrac{z^6}{720} + \dfrac{z^7}{5040} + \dfrac{z^8}{40320} + \dfrac{z^9}{362880} + \dfrac{z^{10}}{3628800} + O[z]^{11}$

In[13]:= S3 = Series[Cos[z], {z, 0, 10}]

Out[13]= $1 - \dfrac{z^2}{2} + \dfrac{z^4}{24} - \dfrac{z^6}{720} + \dfrac{z^8}{40320} - \dfrac{z^{10}}{3628800} + O[z]^{11})$

In[14]:= S4 = Series[Sin[z], {z, 0, 10}]

Out[14]= $z - \dfrac{z^3}{6} + \dfrac{z^5}{120} - \dfrac{z^7}{5040} + \dfrac{z^9}{362880} + O[z]^{11}$

Differentiation and integration of Taylor series is permissible. You can easily do it on the computer. For example, obtain series agreeing with $(\sin z)' = \cos z$. Using series, illustrate other relations of your choice.

In[15]:= D[S4, z]   (* The derivative of $\sin z$ is $\cos z$. *)

Out[15]= $1 - \dfrac{z^2}{2} + \dfrac{z^4}{24} - \dfrac{z^6}{720} + \dfrac{z^8}{40320} + O[z]^{10}$

In[16]:= Integrate[S3, z]   (* Agrees with $\int \cos z \, dz = \sin z + C$, where $C = 0$. *)

Out[16]= $z - \dfrac{z^3}{6} + \dfrac{z^5}{120} - \dfrac{z^7}{5040} + \dfrac{z^9}{362880} - \dfrac{z^{11}}{39916800} + O[z]^{12}$

Similar Material in AEM: pp. 751-757

EXAMPLE 14.5   UNIFORM CONVERGENCE

Let the series $f_0(z) + f_1(z) + f_2(z) + \ldots$ be convergent for all z in some region G in the z-plane. Let $s(z)$ be its sum and $s_n(z)$ its nth partial sum. Then because of convergence, for a given $\varepsilon > 0$ you can find an N such that for all $n > N$ you have $|s(z) - s_n(z)| < \varepsilon$, where z is in G. Here N will generally depend on both ε and z. If for every given $\varepsilon > 0$ you can find an N, **depending only on epsilon** such that **for all z in G** and all $n > N$ that inequality holds, then the convergence of that series is called **uniform convergence** in G. This concept is important because in the case of uniform convergence in G, if the terms $f_n(z)$ are continuous in G, so is the sum $s(z)$, and termwise integration of the series over any path C in G is permissible, that is, it gives a series whose sum is the integral of $s(z)$ over C. This is not generally true in the case of ordinary convergence.

Show that the following series of continuous terms has a discontinuous sum, the discontinuity being at $x = 0$. (It can be shown that the convergence of the series in any interval on the real x-axis containing $x = 0$ is not uniform.) Plot the partial sums $s_1, s_4, s_{16}, s_{64}, s_{256}, s_{1024}$, illustrating the approach to the sum $s(x)$.

$$x^2 + \frac{x^2}{1 + x^2} + \frac{x^2}{(1 + x^2)^2} + \ldots$$

Solution. Apart from the factor x^2 this is a geometric series with the quotient $q = 1/(1 + x^2)$ and nth partial sum $(1 - q^{n+1})/(1 - q)$. Accordingly, type

In[1]:= Clear[q, n, x]

In[2]:= q = 1/(1 + x^2)   (* Out: $\dfrac{1}{1+x^2}$ *)

In[3]:= sn = x^2 (1 - q^(n+1))/(1 - q)   (* Out: $\dfrac{x^2\left(1-\left(\dfrac{1}{1+x^2}\right)^{1+n}\right)}{1-\dfrac{1}{1+x^2}}$ *)

In[4]:= sn = Simplify[%]   (* Out: $1+x^2-\left(\dfrac{1}{1+x^2}\right)^n$ *)

For $x = 0$ this is $1 - 1 = 0$ for all n. Hence $s(0) = 0$. For $x \neq 0$ the limit as $n \to \infty$ is $s(x) = 1 + x^2$. Hence the sum of the series is discontinuous at $x = 0$. To plot those partial sums, replace n by 4^n in **sn** and then plot.

In[5]:= s4n = sn /. n -> 4^n   (* Out: $1+x^2-\left(\dfrac{1}{1+x^2}\right)^{4^n}$ *)

In[6]:= S = Table[s4n, {n, 0, 5}]

Out[6]= $\left\{1+x^2-\dfrac{1}{1+x^2},\ 1+x^2-\dfrac{1}{(1+x^2)^4},\ 1+x^2-\dfrac{1}{(1+x^2)^{16}},\right.$
$\left.1+x^2-\dfrac{1}{(1+x^2)^{64}},\ 1+x^2-\dfrac{1}{(1+x^2)^{256}},\ 1+x^2-\dfrac{1}{(1+x^2)^{1024}}\right\}$

In[7]:= Plot[{S[[1]], S[[2]], S[[3]], S[[4]], S[[5]], S[[6]]}, {x, -1, 1}]

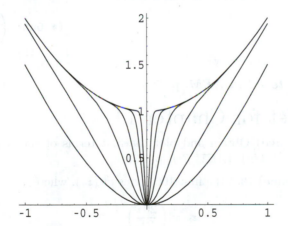

Example 14.5. Partial sums of a convergent series
not converging uniformly in any interval containing 0

Termwise integration giving false results. This is not the fault of the software, but a consequence of mathematics. It can happen if convergence is not uniform. Show that this is the case for the series with the general term $f_m(x) = u_m(x) - u_{m-1}(x)$, where $u_m = mx \exp\left(-mx^2\right)$, and integration over x from 0 to 1. (It can be shown that the convergence of this series on the interval of integration is not uniform.)

Solution. Type u_m, u_{m-1}, then f_m, then the partial sum s_n.

In[8]:= Clear[m, n, x]

In[9]:= um = m x Exp[-m x^2]   (* Out: $e^{-m x^2} m x$ Space after **m** and **x**! *)

In[10]:= um1 = um /. m -> m - 1   (* Out: $e^{-(-1+m) x^2} (-1 + m) x$ *)

In[11]:= fm = um - um1   (* Out: $-e^{-(-1+m) x^2} (-1 + m) x + e^{-m x^2} m x$ *)

In[12]:= `Table[Sum[fm, {m, 1, n}], {n, 1, 10}]`

Out[12]= $\left\{ e^{-x^2}\,x,\, 2\,e^{-2\,x^2}\,x,\, 3\,e^{-3\,x^2}\,x,\, 4\,e^{-4\,x^2}\,x,\, 5\,e^{-5\,x^2}\,x,\, 6\,e^{-6\,x^2}\,x,\, 7\,e^{-7\,x^2}\,x,\, 8\,e^{-8\,x^2}\,x, \right.$

$\left. 9\,e^{-9\,x^2}\,x,\, 10\,e^{-10\,x^2}\,x \right\}$

In[13]:= `sn = n x Exp[-n x^2]`   (* Out: $e^{-n\,x^2}\,n\,x$ *)

(Do you see that this is $u_n(x)$? Can you explain why?) The limit of `sn` as $n \to \infty$ is 0 because the exponential function decreases much faster than the factor n increases. L'Hôpital's rule applied to $n\,x\,e^{-n\,x^2}$ shows this even more distinctly.

In[14]:= `D[n x, x]/D[Exp[n x^2], x]`   (* Out: $\dfrac{e^{-n\,x^2}}{2\,x}$ *)

(Direct application of the command `Limit`, even along with conditions, such as `Im[x] == 0` or $0 \le x \le 1$, does not seem to work.) Hence the sequence of the partial sums has the limit 0 for any fixed x. This is the sum of the series, by definition. Integration of 0 from 0 to 1 gives 0. Now show that by first integrating the partial sums and then taking the limit of the sequence of the integrals you get a different result, namely, 1/2 instead of 0. This shows that termwise integration of the series is not permissible, that is, it leads to a false result.

In[15]:= `Integrate[sn, {x, 0, 1}]`   (* Out: $\left(\dfrac{1}{2\,n} - \dfrac{e^{-n}}{2\,n}\right)\,n$ *)

In[16]:= `Limit[%, n -> Infinity]`   (* Out: $\frac{1}{2}$ *)

Similar Material in AEM: p. 763

Problem Set for Chapter 14

Pr.14.1 **(Complex sequence)** Obtain and plot the first terms of the sequence $\{z_n\}$ with $z_n = 1 - 1/n^2 + i(2 + 4/n)$. (*AEM Ref.* p. 732)

Pr.14.2 **(Complex sequence)** Plot the first 100 terms of (z_n), where

$$z_n = \left(\frac{20\,i}{21}\right)^{0.1\,n}.$$

(*AEM Ref.* p. 732)

Pr.14.3 **(Ratio test)** Is the following series

$$\sum_{n=0}^{\infty} \frac{(20 + 30\,i)^n}{n!}$$

convergent or not? Use the ratio test. (*AEM Ref.* p. 737)

Pr.14.4 **(Ratio test)** Test the series with general term $(n!)^2/(2n)!$ for convergence. (*AEM Ref.* p. 740 (#16))

Pr.14.5 **(Ratio test)** Is the following series

$$\sum_{n=1}^{\infty} \frac{(3\,i)^n\,n!}{n^n}$$

convergent or not? (*AEM Ref.* p. 740 (#17))

Pr.14.6 (Radius of convergence) Find the radius of convergence of the series

$$\sum_{n=0}^{\infty} n\,(n-1)\,2^n\,z^{3n}.$$

(See Example 14.3 in this Guide. *AEM Ref.* p. 745 (#8))

Pr.14.7 (Radius of convergence) Find the radius of convergence of the series

$$\sum_{n=0}^{\infty} \frac{(3n)!}{2^n\,(n!)^3}z^n.$$

(See Example 14.3 in this Guide. *AEM Ref.* p. 745 (#11))

Pr.14.8 (Maclaurin series) The general term $a_n\,z^n$ of the Maclaurin series of a function $f(z)$ has the coefficient $a_n = f^{(n)}(0)/n!$. Using this formula, obtain the Maclaurin series of e^z (9 terms) and confirm the result by the command `Series`.
(*AEM Ref.* p. 755)

Pr.14.9 (Taylor series) Find the Taylor series of Ln z with center $z_0 = 1$ from the coefficient formula $a_n = f^{(n)}(z_0)/n!$ (5 terms) and confirm the result by the command `Series`.
(*AEM Ref.* p. 756)

Pr.14.10 (Taylor series) Find the Taylor series of $f(z) = \cos^2 z$ with center $z_0 = \pi/2$ (4 terms). Confirm the result by expressing $f(z)$ in terms of $\cos 2z$ and developing in a series. (*AEM Ref.* p. 759 (# 26))

Pr.14.11 (Integration of power series) The integral of $1/(1+z^2)$ is arctan z (plus a constant). Using this, obtain the Maclaurin series of arctan z (powers up to z^{19}) by integrating a suitable series. (*AEM Ref.* p. 756)

Pr.14.12 (Experiment on sine integral) The *sine integral* Si (z) is the integral of $(\sin z)/z$ (integration from 0 to z). It is one of the more important integrals occurring in applications that cannot be integrated by the usual methods of calculus. Obtain its Maclaurin series (up to the power z^{15}) and use it to plot Si (z) for real z from 0 to 8. Then plot Si (z) by using the command `SinIntegral[z]`. Plot jointly. Finally, find experimentally a value of z at which the Maclaurin series still gives values whose first two decimals are accurate. (*AEM Ref.* pp. 758 (#10), A57)

Pr.14.13 (Bernoulli numbers) The Bernoulli numbers B_n are defined by the Maclaurin series of $z/(e^z - 1)$, written in the form $1 + B_1 z + B_2 z^2/2! + B_3 z^3/3! + \dots$. Find the first seven nonzero Bernoulli numbers. (*AEM Ref.* p. 758 (#13))

Pr.14.14 (Bessel functions) Find the Maclaurin series of $J_0(z)$. From the first five partial sums of the series find five approximate values of $J_0(0.5 - 0.5i)$. Note the rapid approach to the 10S-value obtained directly by the command `BesselJ[0, 0.5 - 0.5 I]`.
(*AEM Ref.* p. 220)

Pr.14.15 (Lack of uniform convergence) Plot the partial sums S_m, $m = 2^n$, $n = 1, 2, ..., 10$, of the series

$$x^4 \sum_{m=1}^{\infty} (1+x^4)^{-m}$$

on common axes. (*AEM Ref.* pp. 762, 763)

Chapter 15

Laurent Series. Residue Integration

Content. Laurent series (Ex. 15.1, Prs. 15.1-15.3)
Singularities, zeros (Ex. 15.2, Prs. 15.4-15.7)
Residue integration (Ex. 15.3, Prs. 15.8-15.11)
Real integrals (Exs. 15.4, 15.5, Prs. 15.12-15.15)

Laurent series are infinite series of positive and negative integer powers of $z - z_0$ of the form

$$(1) \qquad a_0 + a_1 (z - z_0) + a_2 (z - z_0)^2 + \cdots + b_1 (z - z_0)^{-1} + b_2 (z - z_0)^{-2} + \cdots.$$

They represent functions $f(z)$ analytic in an annulus with center z_0. The series of the negative powers is called the **principal part** of (1). A function $f(z)$ can have several Laurent series with the same center z_0 but valid in different annuli. Particularly important is the series (1) of a given $f(z)$ that converges immediately near z_0 (except at z_0 itself), say, for $0 < |z - z_0| < R$, where $R > 0$. For this series the coefficient b_1 of $1/(z - z_0)$ is called the **residue** of $f(z)$ at z_0. Residues can be used for evaluating contour integrals by the elegant method of **residue integration**. See Example 15.3.

Examples for Chapter 15

> **EXAMPLE 15.1**   **LAURENT SERIES**

Find all Laurent series of $f(z) = 1/(z^3 - z^4)$ with center $z_0 = 0$.

Solution. The first series is S1 (below), which is obtained without difficulty. It converges for $0 < |z| < 1$. (This is a "degenerate" annulus.) Its principal part has three terms. The **residue** of $f(z)$ at 0 is 1 because $1/z$ has the coefficient 1.

In[1]:= f = 1/(z^3 - z^4)　　　　　　　　　　　　　(* Out: $\frac{1}{z^3 - z^4}$ *)

In[2]:= S1 = Series[f, {z, 0, 10}]

Out[2]= $\frac{1}{z^3} + \frac{1}{z^2} + \frac{1}{z} + 1 + z + z^2 + z^3 + z^4 + z^5 + z^6 + z^7 + z^8 + z^9 + z^{10} + O[z]^{11}$

The second Laurent series S2 is obtained by the **standard trick** of setting $z = 1/w$ in the function $f(z)$, developing $f(1/w) = g(w)$ in a series of powers of w, and resubstituting $w = 1/z$ into the series, which in the present case converges for $|z| > 1$.

In[3]:= `Series[f, {1/z, 0, 10}]`

General:: ivar : $\frac{1}{z}$ is not a valid variable

Out[3]= $\text{Series}\left[\frac{1}{z^3 - z^4}, \left\{\frac{1}{z}, 0, 10\right\}\right]$

In[4]:= `g = Simplify[f /. z -> 1/w]`

Out[4]= $\dfrac{w^4}{-1 + w}$

In[5]:= `Series[g, {w, 0, 10}]`

Out[5]= $-w^4 - w^5 - w^6 - w^7 - w^8 - w^9 - w^{10} + \text{O}[w]^{11}$

In[6]:= `S2 = % /. w -> 1/z`

Out[6]= $-\left(\frac{1}{z}\right)^4 - \left(\frac{1}{z}\right)^5 - \left(\frac{1}{z}\right)^6 - \left(\frac{1}{z}\right)^7 - \left(\frac{1}{z}\right)^8 - \left(\frac{1}{z}\right)^9 - \left(\frac{1}{z}\right)^{10} + \text{O}\left[\frac{1}{z}\right]^{11}$

Similar Material in AEM: p. 774

| EXAMPLE 15.2 | **SINGULARITIES AND ZEROS**

A function $f(z)$ has a **singularity** at a point z_0 if $f(z)$ is not analytic (perhaps not even defined) at z_0, but every neighborhood of z_0 contains points at which $f(z)$ is analytic. For instance, to find the singularities of $f(z) = e^z/z^3$, type

In[1]:= `f = Exp[z]/z^3`   (* Out: $\frac{e^z}{z^3}$ *)

In[2]:= `S1 = Series[f, {z, 0, 5}]`

Out[2]= $\dfrac{1}{z^3} + \dfrac{1}{z^2} + \dfrac{1}{2z} + \dfrac{1}{6} + \dfrac{z}{24} + \dfrac{z^2}{120} + \dfrac{z^3}{720} + \dfrac{z^4}{5040} + \dfrac{z^5}{40320} + \text{O}[z]^6$

S1 is the Laurent series of $f(z)$ with center 0. It converges for $|z| > 0$ (thus for $z \neq 0$). It shows that $f(z)$ has a **singularity** (more precisely, a **pole** of third order) at $z = 0$, with residue $1/2$. To see whether $f(z)$ is singular at infinity, develop it in a series in powers of $w = 1/z$.

In[3]:= `S2 = S1 /. z -> 1/w`

Out[3]= $\dfrac{1}{\left(\frac{1}{w}\right)^3} + \dfrac{1}{\left(\frac{1}{w}\right)^2} + \dfrac{1}{\frac{2}{w}} + \dfrac{1}{6} + \dfrac{1}{24\,w} + \dfrac{1}{120}\left(\frac{1}{w}\right)^2 + \dfrac{1}{720}\left(\frac{1}{w}\right)^3 + \dfrac{\left(\frac{1}{w}\right)^4}{5040} + \dfrac{\left(\frac{1}{w}\right)^5}{40320} + \text{O}\left[\frac{1}{w}\right]^6$

This shows that $g(w) = f(1/z)$ has a singularity at $w = 0$ (because of the negative powers of w). Hence $f(z)$ has a **singularity at** ∞, by definition.

Zeros. A *zero* of an analytic function $f(z)$ in a domain D is a z_1 in D such that $f(z_1) = 0$. For instance, to find the zeros of $\cosh z$, type

In[4]:= `Solve[Cosh[z] == 0, z]`

Solve:: ifun : Inverse functions are being used by Solve, so some solutions
   may not be found

Out[4]= $\left\{\left\{z \to -\dfrac{i\,\pi}{2}\right\}, \left\{z \to \dfrac{i\,\pi}{2}\right\}\right\}$

$\cosh z$ has infinitely many zeros. To see this, type

In[5]:= `Cosh[z] /. z -> I Z`   (* Out: `Cos[Z]` *)

You know the familiar real zeros of the cosine at $\pi/2 + n\pi$. Hence they give the pure imaginary zeros of cosh z, as follows, where n is any integer.

In[6]:= `Cosh[Pi I/2 + n Pi I]`   (* Out: $-\text{Sin}[n\pi]$ *)

If a function $f(z)$ is a quotient of two functions, you may find the **singularities** of $f(z)$ from the zeros of the denominator. For instance, type

In[7]:= `f = (-z^2 - 22 z + 8)/(z^3 - 5 z^2 + 4 z)`

Out[7]= $\dfrac{8 - 22z - z^2}{4z - 5z^2 + z^3}$

In[8]:= `den = Denominator[f]`

Out[8]= $4z - 5z^2 + z^3$

In[9]:= `sol = Solve[den == 0, z]`   (* Out: $\{\{z \to 0\}, \{z \to 1\}, \{z \to 4\}\}$ *)

Make sure that the numerator and the denominator have no common factors (in which case a zero might cancel out and give no singularity). Show that the numerator `num` is not zero at 0, 1, or 4. In the last command below, `sol[[n]]` gives you 0, 1, 4 for $n = 1, 2, 3$, respectively. Try it separately.

In[10]:= `num = Numerator[f]`   (* Out: $8 - 22z - z^2$ *)

In[11]:= `Table[num /. z -> sol[[n, 1, 2]], {n, 1, 3}]`   (* Out: $\{8, -15, -96\}$ *)

Similar Material in AEM: pp. 776-780

EXAMPLE 15.3   RESIDUE INTEGRATION

Residue theorem. Let $f(z)$ be analytic inside and on a simple closed path C, except for k points $z_1, z_2, ..., z_k$ inside C. Then the integral of $f(z)$ taken counterclockwise around C equals $2\pi i$ times the sum of the k residues of $f(z)$ at $z_1, ..., z_k$.

Hence this elegant integration method concerns contour integrals. You can obtain residues from Laurent series of $f(z)$ as in Example 15.2 in this Guide, or by formulas without using a series. There are two such formulas for the residue of $f(z)$ at a **pole of first order** (**"simple pole"**), that is, a singularity at a z_j for which the principal part of the Laurent series that converges near z_j has just one term, $b_1/(z - z_j)$. The first formula is

(1) $$\lim_{z \to z_j} (z - z_j) f(z).$$

The second formula holds for functions of the form $f(z) = p(z)/q(z)$. Then the residue of f at a simple pole $z = z_j$ is

(2) $$p(z_j)/q'(z_j) \qquad\qquad q' = dq/dz.$$

For instance, integrate $f(z) = (4 - 3z)/(z^2 - z)$ counterclockwise over a simple closed path C for which 0 and 1 lie inside C.

Solution. $z^2 - z = z(z - 1) = 0$ for $z = 0$ and 1. At these points, $f(z)$ has simple poles with residues obtained from (2) by typing

In[1]:= `res = (4 - 3 z)/D[z^2 - z, z]`   (* Out: $\dfrac{4 - 3z}{-1 + 2z}$ *)

In[2]:= `answer = 2 Pi I ((res /. z -> 0) + res /. z -> 1)`   (* Out: $-6\,\mathrm{i}\,\pi$ *)

The parentheses (...) around `res /. z -> 0` are quite important. Without them you get a false result. Try it. You may confirm this by typing

In[3]:= `f = (4 - 3 z)/(z^2 - z)`   (* Out: $\dfrac{4-3\,z}{-z+z^2}$ *)

In[4]:= `Residue[f, {z, 0}]`   (* Out: -4 *)

In[5]:= `Residue[f, {z, 1}]`   (* Out: 1 *)

Pole of mth order at z_j. This is a singularity for which the principal part of the Laurent series, convergent near z_j, has $1/(z - z_j)^m$ as the highest negative power. At such a pole the residue of $f(z)$ is

$$(3) \qquad \frac{1}{(m-1)!} \lim_{z \to z_j} \left(\frac{d^{m-1}}{dz^{m-1}} \Big[(z - z_j)^m\, f(z_j) \Big] \right).$$

For instance, integrate $f(z) = (\tan 4z)/z^2$ clockwise around the circle $C : |z| = 1/4$.

Solution. $z = 0$ is the only point inside C at which $f(z)$ has a singularity, because $4z = \pi/2$ gives $z = \pi/8 > 1/4$. This is a pole of second order. From (3) with $m = 2$ you obtain the residue 4, so that the answer is $-8\pi i$, with the minus sign resulting from *clockwise* integration.

In[6]:= `res = D[Tan [4 z], z] /. z -> 0`   (* Out: 4 *)

Essential singularity. This is a singularity of $f(z)$ at z_0 such that the Laurent series of $f(z)$, convergent for $0 < |z - z_0| < R$, has infinitely many negative powers. Here we assume that singularity to be **isolated**, that is, it is the only singularity of $f(z)$ in a (possibly very small) neighborhood of z_0. In this case use the Laurent series with center z_0 just mentioned. For instance, integrate $f(z) = \exp(1/(z - 1))$ counterclockwise around the circle $|z| = 2$.

Solution. $f(z)$ has a singularity at $z = 1$, where $z - 1 = 0$. Obtain the Laurent series by setting $1/(z - 1) = w$, thus $z = (w + 1)/w$.

In[7]:= `f = Exp[1/(z - 1)]`   (* Out: $e^{\frac{1}{-1+z}}$ *)

In[8]:= `Series[f, {z, 1, 10}]`

Series:: esss : Essential singularity encountered in $e^{\frac{1}{z-1}} + O[z-1]^{11}$

Out[8]= $\mathtt{Series}\left[e^{\frac{1}{-1+z}}, \{z, 1, 10\} \right]$

In[9]:= `g = Simplify[f /. z -> (w + 1)/w]`   (* Out: e^w *)

In[10]:= `S1 = Series[g, {w, 0, 5}]`

Out[10]= $1 + w + \dfrac{w^2}{2} + \dfrac{w^3}{6} + \dfrac{w^4}{24} + \dfrac{w^5}{120} + O[w]^6$

In[11]:= `S2 = S1 /. w -> 1/(z - 1)`

Out[11]= $1 + \dfrac{1}{-1+z} + \dfrac{1}{2}\left(\dfrac{1}{-1+z}\right)^2 + \dfrac{1}{6}\left(\dfrac{1}{-1+z}\right)^3 + \dfrac{1}{24}\left(\dfrac{1}{-1+z}\right)^4 + \dfrac{1}{120}\left(\dfrac{1}{-1+z}\right)^5 + O\left[\dfrac{1}{-1+z}\right]^6$

You see that the residue of $f(z)$ at $z = 1$ (the coefficient of $1/(z - 1)$) is 1. Hence the answer is $2\pi i$. To confirm this, type

In[12]:= `2 Pi I Residue[f, {z, 1}]`   (* Out: $2\,\mathrm{i}\,\pi$ *)

The singularity of $f(z)$ at $z = 1$ is an essential singularity. Type $\{w, 0, 20\}$ or $\{w, 0, 100\}$ in S1 or whatever you want. Then S2 will have as many negative powers as you want.

Similar Material in AEM: pp. 781-786

EXAMPLE 15.4 | REAL INTEGRALS OF RATIONAL FUNCTIONS OF COS AND SIN

Certain real integrals can be evaluated by complex contour integration by converting a real interval of integration to a complex contour. For instance, evaluate the integral of the function $1/(\sqrt{2} - \cos t)$ from $t = 0$ to 2π (t real) by complex contour integration.

Solution. The interval of integration is transformed to the unit circle by introducing the new variable of integration $z = e^{it}$. Thus $t = -i \operatorname{Ln} z$ and $dt = dz/iz$, so that the integrand int is $(dt/dz)/(\sqrt{2} - \cos t)$. Accordingly, type

In[1]:= t = Solve[z == Exp[I T], T]

Solve:: ifun : Inverse functions are being used by Solve, so some solutions may not be found

Out[1]= $\{\{\texttt{T} \to -i \operatorname{Log}[\texttt{z}]\}\}$

In[2]:= t[[1, 1, 2]]   (* Out: $-i \operatorname{Log}[z]$ *)

In[3]:= c = Cos[t[[1, 1, 2]]]   (* Out: $\frac{1+z^2}{2z}$ *)

In[4]:= Expand[%]   (* Out: $\frac{1}{2z} + \frac{z}{2}$ *)

In[5]:= d = D[t[[1, 1, 2]], z]   (* Out: $-\frac{i}{z}$ *)

In[6]:= int = Simplify[d/(Sqrt[2] - c)]   (* Out: $\frac{2i}{1-2\sqrt{2}z+z^2}$ *)

The integrand int has singularities at the roots of the quadratic polynomial in the denominator, which you obtain by the command

In[7]:= zeros = Solve[Denominator[int] == 0, z]

Out[7]= $\left\{\left\{\texttt{z} \to -1+\sqrt{2}\right\}, \left\{\texttt{z} \to 1+\sqrt{2}\right\}\right\}$

You need the residue of int at the first zero, at which int has a pole of first order. The second zero is greater than 1, so that it lies outside the unit circle and is of no interest here. Formula (2) in the previous example gives this residue if you type

In[8]:= quot = Numerator[int]/D[Denominator[int], z]   (* Out: $\frac{2i}{-2\sqrt{2}+2z}$ *)

In[9]:= zeros[[1, 1, 2]]   (* Out: $-1+\sqrt{2}$ *)

In[10]:= res = Simplify[quot /. z -> zeros[[1, 1, 2]]]   (* Out: $-i$ *)

Confirm this by typing

In[11]:= Residue[int, {z, zeros[[1, 1, 2]]}]   (* Out: $-i$ *)

In[12]:= answer = 2 Pi I res   (* Out: 2π *)

Mathematica can evaluate this kind of integral directly. In the present case, confirm the answer by using the command Integrate.

In[13]:= `Clear[t]`

In[14]:= `Simplify[Integrate[1/(Sqrt[2] - Cos[t]), {t, 0, 2 Pi}]]`

Out[14]= 2π

       Similarly,

In[15]:= `Integrate[(1 + Sin[t])/(3 + Cos[t]), {t, 0, 2 Pi}]`   (* Out: $\frac{\pi}{\sqrt{2}}$ *)

In[16]:= `Integrate[Cos[t]/(13 - 12 Cos[2 t]), {t, 0, 2 Pi}]`   (* Out: 0 *)

Similar Material in AEM: pp. 787, 793

    EXAMPLE 15.5   **IMPROPER REAL INTEGRALS OF RATIONAL FUNCTIONS**

This example concerns the evaluation of **improper integrals** of rational functions $f(x)$ from $-\infty$ to ∞ by complex contour integration under the assumption that the denominator of $f(x)$ is nowhere zero and its degree is at least 2 units higher than that of the numerator. Then the integral of $f(x)$ equals $2\pi i$ times the sum of the residues at the poles of the ***complex*** function $f(z)$ in the upper half-plane. For instance, show that

$$\int_{-\infty}^{\infty} \frac{1}{1+x^4}\,dx = \frac{\pi}{\sqrt{2}}.$$

Solution. $f(z) = 1/(1+z^4)$ has poles where $1 + z^4$ has zeros. You obtain them as follows. (See also the figure on the next page.)

In[1]:= `Z0 = Solve[ z^4 == -1, z]`

Out[1]= $\{\{z \to -(-1)^{1/4}\}, \{z \to (-1)^{1/4}\}, \{z \to -(-1)^{3/4}\}, \{z \to (-1)^{3/4}\}\}$

In[2]:= `z0 = Solve[z^4 == -1.0, z]`   (* -1.0 gives decimal fractions. *)

Out[2]= $\{\{z \to -0.707107 - 0.707107\,i\}, \{z \to -0.707107 + 0.707107\,i\},$

       $\{z \to 0.707107 - 0.707107\,i\}, \{z \to 0.707107 + 0.707107\,i\}\}$

In[3]:= `z0[[3, 1, 2]]`   (* Do the same for the other roots. *)

Out[3]= $0.707107 - 0.707107\,i$

In[4]:= `S = Table[{Re[z0[[n, 1, 2]]], Im[z0[[n, 1, 2]]]}, {n, 1, 4}]`

Out[4]= $\{\{-0.707107, -0.707107\}, \{-0.707107, 0.707107\}, \{0.707107, -0.707107\},$

       $\{0.707107, 0.707107\}\}$

In[5]:= `P1 = ListPlot[S, AspectRatio -> Automatic,`
         `Prolog -> AbsolutePointSize[4]]`

In[6]:= `P2 = ParametricPlot[{Cos[t], Sin[t]}, {t, 0, 2 Pi}]`   (* Circle *)

In[7]:= `Show[P1, P2]`

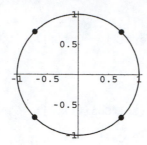

Example 15.5. Poles of $f(z)$ on the unit circle

The four poles are simple. The two poles at `S[[2]]` and `S[[4]]` lie in the upper half-plane because their imaginary parts are positive. From (2) in Example 15.3 in this Guide you obtain the residues and the answer by typing

In[8]:= `res = 1/D[1 + z^4, z]`   (* Out: $\frac{1}{4\,z^3}$ *)

In[9]:= `2 Pi I ((res /. z -> z0[[2, 1, 2]]) + res /. z -> z0[[4, 1, 2]])`

Out[9]= $2.22144 + 5.2318 \times 10^{-16}\,\text{i}$

Integrating a real function over the real axis must give a real value; hence the very small imaginary part of the response is due to round-off. Also, the result is $\pi/\sqrt{2}$ because

In[10]:= `N[Pi/Sqrt[2]]`   (* Out: 2.22144 *)

To confirm this, use `Z0` in the very first command and type

In[11]:= `Simplify[2 Pi I ((res /. z -> Z0[[2, 1, 2]]) +`
        `res /. z -> Z0[[4, 1, 2]])]`

Out[11]= $\dfrac{\pi}{\sqrt{2}}$

This shows that the very small imaginary part in the previous result is indeed due to round-off.

Mathematica can evaluate the integral directly. Thus, you may confirm your result by typing

In[12]:= `Integrate[1/(1 + x^4), {x, -Infinity, Infinity}]`   (* Out: $\frac{\pi}{\sqrt{2}}$ *)

Integrals as before but with singularities on the real axis. Then under the degree condition the **Cauchy principal value** (see AEM, pp. 788, 791) of the integral equals $2\pi i$ times the sum of the residues at the poles of $f(z)$ in the upper half-plane (as before) plus πi times the sum of the residues at the poles of $f(z)$ on the x-axis.

For instance, you will see that

$$f(x) = \frac{1}{(x^2 - 3x + 2)(x^2 + 1)}$$

is of this kind. Type the denominator den with z instead of x and determine its zeros.

In[13]:= `den = (z^2 - 3 z + 2) (z^2 + 1)`   (* Out: $(1 + z^2)\,(2 - 3z + z^2)$ *)

In[14]:= `z0 = Solve[den == 0, z]`   (* Out: $\{\{z \to -i\}, \{z \to i\}, \{z \to 1\}, \{z \to 2\}\}$ *)

These are the locations of four simple poles. You need the residues of those at i, 1, and 2. The pole at $-i$ lies in the lower half-plane and is not needed. Use again formula (2) in Example 15.3. Type

In[15]:= res = Simplify[1/D[den, z]]   (* Out: $\dfrac{1}{-3 + 6z - 9z^2 + 4z^3}$ *)

In[16]:= z0[[2, 1, 2]]   (* Out: i *)

In[17]:= res /. z -> z0[[2, 1, 2]]   (* Out: $\dfrac{3}{20} - \dfrac{i}{20}$ *)

In[18]:= res /. z -> z0[[3, 1, 2]]   (* Out: $-\dfrac{1}{2}$ *)

In[19]:= res /. z -> z0[[4, 1, 2]]   (* Out: $\dfrac{1}{5}$ *)

In[20]:= answer = 2 Pi I (res /. z -> z0[[2, 1, 2]]) +
        Pi I ((res /. z -> z0[[3, 1, 2]]) + res /. z -> z0[[4, 1, 2]])

Out[20]= $\dfrac{\pi}{10}$

You may confirm this result by using the command Residue, typing

In[21]:= 2 Pi I Residue[1/den, {z, z0[[2, 1, 2]]}] +
        Pi I (Residue[1/den, {z, z0[[3, 1, 2]]}] +
               Residue[1/den, {z, z0[[4, 1, 2]]}])

Out[21]= $\dfrac{\pi}{10}$

Similar Material in AEM: pp. 788-793

Problem Set for Chapter 15

Pr.15.1 (**Laurent series**) Find the Laurent series of $(z - \pi i)^{-2} \cosh z$ with center $z_0 = -\pi i$ (4 nonzero terms). (*AEM Ref.* pp. 770-776)

Pr.15.2 (**Laurent series**) Find all Laurent series of $f(z) = 1/(z^2 + 1)$ with center $z_0 = i$ as well as the residue of $f(z)$ at z_0. (*AEM Ref.* p. 776 (#9))

Pr.15.3 (**Laurent series**) Find the Laurent series of $f(z) = \exp(-1/z^2)/z^2$ with center 0 (eight nonzero terms). What kind of singularity does $f(z)$ have at 0 and what is the corresponding residue? (*AEM Ref.* p. 776 (#7))

Pr.15.4 (**Singularities**) Find the location and kind of the singularities of $f(z) = \tan(\pi z/2)$ and the corresponding residues. (*AEM Ref.* p. 780 (#10))

Pr.15.5 (**Singularities**) Find the location and kind of the singularities and the residues of the function
$$f(z) = \frac{3z^4 - 18z^3 + 36z^2 - 24z + 1}{z^6 - 10z^5 + 40z^4 - 80z^3 + 80z^2 - 32z}.$$
(*AEM Ref.* p. 777)

Pr.15.6 (**Zeros**) Find the zeros of $(z^4 - z^2 - 6)^3$. (*AEM Ref.* p. 780 (#7))

Pr.15.7 (**Poles**) Find the residues at the poles of $(-z^2 - 22z + 8)/(z^3 - 5z^2 + 4z)$. (*AEM Ref.* pp. 782, 783)

Pr.15.8 (**Residue**) Find the residues of $f(z) = 50z/[(z + 4)(z - 1)^2]$ at all singularities. (*AEM Ref.* p. 784)

Pr.15.9 **(Residue)** Find the residues of $f(z) = 1/(1 - e^z)$ at all singularities. (*AEM Ref.* p. 786 (#5))

Pr.15.10 **(Residue integration)** Integrate $f(z) = \exp{(-z^2)}/\sin 4z$ counterclockwise around the unit circle. (*AEM Ref.* p. 786 (#20))

Pr.15.11 **(Residue integration)** Integrate $f(z) = (z \cosh \pi z)/(z^4 + 13 z^2 + 36)$ counterclockwise around the circle $|z| = \pi$. (*AEM Ref.* p. 786 (#19))

Pr.15.12 **(Real integral by residue integration)** Integrate $(1 + 4 \cos t)/(17 - 8 \cos t)$ from 0 to 2π by residue integration. Confirm the result by the command `Integrate`. (*AEM Ref.* p. 793 (#8))

Pr.15.13 **(Real integral by residue integration)** Integrate $(\cos t)/(13 - 12 \cos 2t)$ from 0 to 2π by the residue method. Confirm the result by the command `Integrate`. Could you get the result by jointly plotting the numerator and the reciprocal of the denominator and reasoning geometrically? (*AEM Ref.* p. 793 (#7))

Pr.15.14 **(Improper real integral)** Integrate $1/(x^2 - 2x + 5)^2$ from $-\infty$ to ∞ by the residue method. (*AEM Ref.* p. 794 (#14))

Pr.15.15 **(Poles on the real axis)** Integrate $x/(8 - x^3)$ from $-\infty$ to ∞, using the method in Example 15.5 in this Guide. (*AEM Ref.* p. 794 (#23))

Chapter 16

Complex Analysis in Potential Theory

Content. Complex potential (Ex. 16.1, Prs. 16.1-16.3)
Conformal mapping (Ex. 16.2, Prs. 16.4, 16.5)
Fluid flow (Ex. 16.3, Prs. 16.6, 16.7)
Series for potential (Ex. 16.4, Pr. 16.8)
Mean value theorems (Ex. 16.5, Prs. 16.9, 16.10)

Complex potential. $F(z) = \Phi(x, y) + i\Psi(x, y)$ permits the use of complex analysis for two-dimensional potential problems. Its advantage is that it simultaneously gives equipotential lines $\Phi(x, y) = const$ and lines of force (streamlines of flows) $\Psi(x, y) = const$.

Examples for Chapter 16

EXAMPLE 16.1 | **COMPLEX POTENTIAL. RELATED PLOTS**

Find and plot the equipotential lines and the lines of force of the **complex potential** $F(z) = -i z^2$.

Solution. Write $F(z) = \Phi(x, y) + i\Psi(x, y)$. Type F and Φ. Then solve the equation $\Phi(x, y) = k = const$ for y. Use y for plotting a sequence S of **equipotential lines** (hyperbolas).

In[1]:= `Clear[x, y, z]`

In[2]:= `F = -I z^2`   (* Out: $-i z^2$ *)

In[3]:= `z = x + I y`   (* Out: $x + i y$ *)

In[4]:= `Phi = ComplexExpand[Re[F]]`   (* Out: $2xy$ *)

In[5]:= `y0 = Solve[Phi == k, y]`

Out[5]= $\{\{y \to \frac{k}{2x}\}\}$

Warning! Do not type `y` (instead of `y0`) on the left; Mathematica would refuse it.

In[6]:= `S = Table[y0[[1, 1, 2]], {k, 0, 10}]`

Out[6]= $\{0, \frac{1}{2x}, \frac{1}{x}, \frac{3}{2x}, \frac{2}{x}, \frac{5}{2x}, \frac{3}{x}, \frac{7}{2x}, \frac{4}{x}, \frac{9}{2x}, \frac{5}{x}\}$

In[7]:= `N[%]`   (* Out: $\{0., \frac{0.5}{x}, \frac{1}{x}, \frac{1.5}{x}, \frac{2.}{x}, \frac{2.5}{x}, \frac{3.}{x}, \frac{3.5}{x}, \frac{4.}{x}, \frac{4.5}{x}, \frac{5.}{x}\}$ *)

In[8]:= `P1 = Plot[Evaluate[S], {x, 0, 2.5},`
`AspectRatio -> Automatic, PlotRange -> {0, 2.5}]`

Lines of force can be plotted in a similar fashion. Plotting the sequence `S` and a sequence `S2` of lines of force on the same axes, and using `Show`, you obtain a rectangular net of curves.

In[9]:= `Psi = ComplexExpand[Im[F]]`   (* Out: $-x^2 + y^2$ *)

In[10]:= `y2 = Solve[Psi == k, y]`

Out[10]= $\{\{y \to -\sqrt{k + x^2}\}, \{y \to \sqrt{k + x^2}\}\}$

In[11]:= `y2[[2, 1, 2]]`   (* Out: $\sqrt{k + x^2}$ *)

You see that the equation $\Psi = const$ has two solutions. The second, `y2[[2, 1, 2]]`, gives curves in the first quadrant. (`y2[[1, 1, 2]]` gives curves in the fourth; try it.) For positive (negative) `k` you get curves above (below) the line $y = x$. (Try it.)

In[12]:= `S2 = Table[y2[[2, 1, 2]], {k, -10, 10}]`

Out[12]= $\Big\{ \sqrt{-10 + x^2}, \sqrt{-9 + x^2}, \sqrt{-8 + x^2}, \sqrt{-7 + x^2}, \sqrt{-6 + x^2}, \sqrt{-5 + x^2}, \sqrt{-4 + x^2},$

$\sqrt{-3 + x^2}, \sqrt{-2 + x^2}, \sqrt{-1 + x^2}, \sqrt{x^2}, \sqrt{1 + x^2}, \sqrt{2 + x^2}, \sqrt{3 + x^2},$

$\sqrt{4 + x^2}, \sqrt{5 + x^2}, \sqrt{6 + x^2}, \sqrt{7 + x^2}, \sqrt{8 + x^2}, \sqrt{9 + x^2}, \sqrt{10 + x^2} \Big\}$

In[13]:= `P2  = Plot[Evaluate[S2], {x, 0, 2.5}, AspectRatio -> Automatic,`
           `PlotRange ->  {0, 2.5}]`

```
Plot::plnr : √-10+x²  is not a machine-size real number
at x  = 1.0416666666666667`*^-7
```

(And several similar messages will occur due to the square root of negative numbers.)

In[14]:= `Show[P1, P2]`

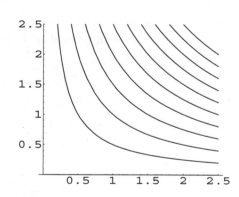 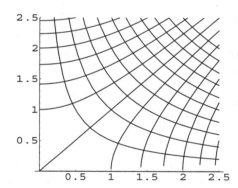

Example 16.1. Equipotential lines   **Example 16.1.** Equipotential lines
$\Phi = 2xy = const$ (hyperbolas)   and lines of force

Similar Material in AEM: pp. 799-802

| **EXAMPLE 16.2** | USE OF CONFORMAL MAPPING |

Conformal mapping provides another approach to the plotting problem. For instance, in the previous example proceed as follows. Solving $w = F(z) = -iz^2$ (see before) you have $z = \sqrt{iF} = \sqrt{iw}$. Interchanging the notations gives the inverse $w = \sqrt{iz}$. Now load the package (***always do this first!***).

In[1]:= `<<Graphics'ComplexMap'`

and type the mapping function in the form

In[2]:= `Clear[z]`

In[3]:= `w[z_] = Sqrt[I z]`   (* Out: $\sqrt{iz}$ *)

Then give the mapping command

In[4]:= `CM = CartesianMap[w, {0, 12.5}, {-12.5, 12.5}]`

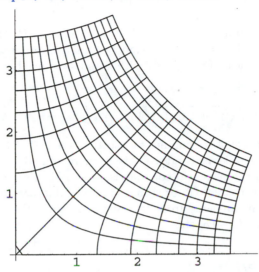

Example 16.2. Plotting equipotential lines and lines of force
by conformal mapping

To explain: Multiplication by i rotates the rectangle $0 \leq x \leq 12.5$, $-12.5 \leq y \leq 12.5$ in
the mapping command into the rectangle $-12.5 \leq x \leq 12.5$, $0 \leq y \leq 12.5$ in the upper
half-plane. The root maps the segment $0 \leq y \leq 12.5$ on the imaginary axis onto a
segment of length $\sqrt{12.5} = 3.55$ on the bisecting line of the first quadrant. Similarly,
the segment $-12.5 \leq x \leq 0$ maps into the imaginary axis. The image in the figure
contains the square of side 2.5 (whose diagonal has length $2.5\sqrt{2} = 3.55$) shown in a
figure of the previous example, and the present curves agree with those in that figure
in their form (but there are now more in number).

Similar Material in AEM: pp. 799-806

EXAMPLE 16.3   FLUID FLOW

Flow around a cylinder. Consider the complex potential $F(z) = z + 1/z$. Find
the **stream function** Ψ and the **streamlines** $\Psi = const.$ In particular, show that
the streamline $\Psi = 0$ consists of the x-axis and the unit circle, so that the flow can
be regarded as a flow around a cylinder (intersecting the z-plane in the unit circle),
which far enough from the cylinder is practically parallel because $1/z$ in $F(z)$ becomes
negligible compared to z when $|z|$ is large. Plot streamlines in the z-plane.

Solution. The physical situation suggests polar coordinates, at least to some extent.
Type z, then the complex potential F, and then its imaginary part, which is the

stream function $\Psi(r, \theta)$. Write t instead of θ, for simplicity. (Time t occurs nowhere in this chapter.)

In[1]:= `z = r Exp[I t]` $\qquad\qquad$ (* Out: $e^{it} r$ *)

In[2]:= `F = z + 1/z` $\qquad\qquad$ (* Out: $\dfrac{e^{-it}}{r} + e^{it} r$ *)

In[3]:= `Psi = ComplexExpand[Im[F]]` $\qquad$ (* Out: $-\dfrac{\mathrm{Sin}[t]}{r} + r\,\mathrm{Sin}[t]$ *)

From this formula you can see that the x-axis ($t = 0$ and π) and the unit circle ($r = 1$) give $\Psi = 0$, as claimed.

Now get the streamlines by solving $\Psi = const$ for $r = r(t)$. Denote the constant by k times a numerical factor (0.25) found by trial and error to have a reasonably dense field of streamlines. (Experiment with k and see what happens.)

In[4]:= `sol = Solve[Psi == k, r]`

Out[4]= $\left\{\left\{r \to \dfrac{1}{2}\,\mathrm{Csc}\,[t]\left(k - \sqrt{k^2 + 4\,\mathrm{Sin}[t]^2}\right)\right\}, \left\{r \to \dfrac{1}{2}\,\mathrm{Csc}\,[t]\left(k + \sqrt{k^2 + 4\,\mathrm{Sin}[t]^2}\right)\right\}\right\}$

In making up a sequence of streamlines for plotting, csc t, that is, sin t in the denominator would cause trouble at 0 and π. This is the reason why you stay away from 0 and π, as you see in the next two commands. Try otherwise, and you will see that the figure is no longer usable. Observe that the streamlines are practically horizontal and parallel in regions far away from the cylinder. 0.5 in the command lets k take the values $-2, -1.5, -1, -0.5$, etc.

In[5]:= `S1 = Table[{sol[[1, 1, 2]] Cos[t], sol[[1, 1, 2]] Sin[t]},`
`        {k, -2, 2, 0.5}];`

In[6]:= `S2 = Table[{sol[[2, 1, 2]] Cos[t], sol[[2, 1, 2]] Sin[t]},`
`        {k, -2, 2, 0.5}];`

In[7]:= `ParametricPlot[Evaluate[Join[S1, S2]],`
`        {t, 0.01, Pi - 0.01}, AspectRatio -> Automatic]`

To obtain a larger number of streamlines, type

In[8]:= `sol2 = Solve[Psi == k/5, r]`

Out[8]= $\left\{\left\{r \to \dfrac{1}{10}\,\mathrm{Csc}\,[t]\left(k - \sqrt{k^2 + 100\,\mathrm{Sin}[t]^2}\right)\right\},\right.$
$\qquad\left.\left\{r \to \dfrac{1}{10}\,\mathrm{Csc}\,[t]\left(k + \sqrt{k^2 + 100\,\mathrm{Sin}[t]^2}\right)\right\}\right\}$

In[9]:= `S3 = Table[{sol2[[1, 1, 2]] Cos[t], sol2[[1, 1, 2]] Sin[t]},`
`        {k, -5, 5, 0.5}];`

In[10]:= `S4 = Table[{sol2[[2, 1, 2]] Cos[t], sol2[[2, 1, 2]] Sin[t]},`
`        {k, -5, 5, 0.5}];`

In[11]:= `P1 = ParametricPlot[Evaluate[Join[S3, S4]],`
`        {t, 0.01, Pi - 0.01}, AspectRatio -> Automatic]`

To cover the (physically meaningless) streamlines inside the circle (the cylinder), type

In[12]:= `P2 = Show[Graphics[Disk[{0, 0}, 1]]]`

In[13]:= `Show[P1, P2]`

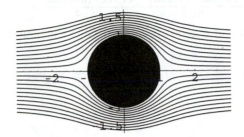

Example 16.3. Flow around a cylinder

Similar Material in AEM: pp. 814, 818

| EXAMPLE 16.4 | **SERIES REPRESENTATION OF POTENTIAL**

This concerns the potential in a disk $|z| \leq R$ for a given boundary potential whose Fourier series is

$$\Phi(R, \theta) = a_0 + \sum_{n=1}^{\infty} (a_n \cos n\theta + b_n \sin n\theta).$$

The series for the potential inside the disk is

$$\Phi(r, \theta) = a_0 + \sum_{n=1}^{\infty} \left(\frac{r}{R}\right)^n (a_n \cos n\theta + b_n \sin n\theta).$$

For instance, find the series corresponding to the boundary potential $-\theta/\pi$ if $-\pi < \theta < 0$ and θ/π if $0 < \theta < \pi$ on the unit circle $|z| = 1$. (Make a sketch of that potential.)

Solution. Since the boundary potential is an even function, its sine terms are zero, so that you get a Fourier cosine series. The constant term is $a_0 = 1/2$ (the mean value of the boundary potential). The other Fourier cosine coefficients are

In[1]:= `an = 2/Pi Integrate[theta/Pi  Cos[n theta], {theta, 0, Pi}]`

$$\text{Out[1]} = \frac{2\left(-\frac{1}{n^2} + \frac{\cos[n\pi]}{n^2} + \frac{\pi \sin[n\pi]}{n}\right)}{\pi^2}$$

In[2]:= `Simplify[%]`   (* Out: $\dfrac{2(-1 + \cos[n\pi] + n\pi \sin[n\pi])}{n^2 \pi^2}$ *)

In[3]:= `Table[an, {n, 1, 10}]`

$$\text{Out[3]} = \left\{ -\frac{4}{\pi^2}, 0, -\frac{4}{9\pi^2}, 0, -\frac{4}{25\pi^2}, 0, -\frac{4}{49\pi^2}, 0, -\frac{4}{81\pi^2}, 0 \right\}$$

Hence the corresponding partial sum of the series representing the potential inside the disk is

In[4]:= `Phi = 1/2 + Sum[an r^n Cos[n theta], {n, 1, 10}]`

$$\text{Out[4]} = \frac{1}{2} - \frac{4\,r\,\text{Cos[theta]}}{\pi^2} - \frac{4\,r^3\,\text{Cos[3 theta]}}{9\,\pi^2} - \frac{4\,r^5\,\text{Cos[5 theta]}}{25\,\pi^2} - \frac{4\,r^7\,\text{Cos[7 theta]}}{49\,\pi^2} - \frac{4\,r^9\,\text{Cos[9 theta]}}{81\,\pi^2}$$

Setting $r = 1$, you can plot a good approximation of the boundary potential by typing

```
In[5]:= Plot[Phi /. r -> 1, {theta, -Pi, Pi},
    Ticks -> {{-Pi, -Pi/2, Pi/2, Pi}, Automatic}]
```

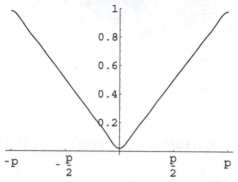

Example 16.4. Boundary potential approximated by the partial sum Phi
(with $r = 1$) (where $p = \pi$)

You may also plot the potential Φ as a surface over the disk $|z| \leq 1$ by the command

```
In[6]:= ParametricPlot3D[{r Cos[theta], r Sin[theta], Phi}, {r, 0, 1},
    {theta, 0, 2 Pi}]
```

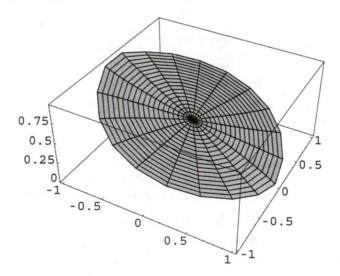

Example 16.4. Potential as a surface over the xy-plane

Similar Material in AEM: p. 821

| EXAMPLE 16.5 | **MEAN VALUE THEOREM FOR ANALYTIC FUNCTIONS** |

Mean Value Theorem. If $F(z)$ is analytic in a simply connected domain D, then
its value at any z_0 in D equals the mean value of $F(z)$ over any circle of radius r in

D with center z_0. As a formula,

$$F(z_0) = \frac{1}{2\pi} \int_0^{2\pi} F(z_0 + r\, e^{i\,t})\, dt.$$

Verify this for $F(z) = (z+2)^2$ and a circle of radius 1 and center 1.

Solution. Type a representation of the circle, then F on the circle, then the integral. Clearly, it should have the value $F(1) = 9$.

In[1]:= z = 1 + Exp[I t]   (* Out: $1 + e^{i\,t}$ *)

In[2]:= F = (z + 2)^2   (* Out: $\left(3 + e^{i\,t}\right)^2$ *)

In[3]:= 1/(2 Pi) Integrate[F, {t, 0, 2 Pi}]

Out[3]= $\dfrac{\frac{13\,i}{2} + \frac{1}{2}\left(-13\,i + 36\,\pi\right)}{2\,\pi}$

In[4]:= Simplify[%]   (* Out: 9 *)

Remark. The present theorem has applications to potential problems (see Pr.16.9).

Similar Material in AEM: p. 822

Problem Set for Chapter 16

Pr.16.1 **(Complex potential)** Find and plot the equipotential lines of $F(z) = z^2$. (*AEM Ref.* p. 803 (#16))

Pr.16.2 **(Equipotential surfaces)** Plot the equipotential surfaces $\ln |z| = k$, thus $|z| = e^k$, of $F(z) = \ln z$ for $k = 1, 2, 3$. Use the package << Graphics'Shapes'. (*AEM Ref.* pp. 800, 803)

Pr.16.3 **(Orthogonal net, maximum principle)** Find and plot equipotential lines and lines of force of $F(z) = \cos z$ jointly in the z-plane. From the plot find the points at which this net is not orthogonal. Plot the real potential Φ as a surface over a square in the z-plane and convince yourself that Φ satisfies the *maximum principle*, that is, the maximum and minimum values of Φ are taken on the boundary of the square. (*AEM Ref.* pp. 799-803, 824)

Pr.16.4 **(Use of conformal mapping)** Obtain a portion of the orthogonal net in Pr.16.3, say, the portion in the rectangle $0 \le x \le \pi/2$, $0 \le y \le 4$, by conformal mapping. See Example 16.2 in this Guide. (*AEM Ref.* pp. 799-807)

Pr.16.5 **(Pair of opposite electrical charges)** The field of two charges, $+1$ at $z = -1$ and -1 at $z = 1$, has the complex potential $F = \text{Ln}\left(-(z+1)/(z-1)\right)$ (the first minus amounting to adding an imaginary constant πi to F). Plot a net of equipotential lines and lines of force by conformal mapping. (*AEM Ref.* p. 802)

Pr.16.6 **(Arbitrary cylinder)** Find the complex potential of a flow around a cylinder of radius r. Plot the streamlines for a cylinder of radius 2. (*AEM Ref.* p. 817 (#9))

Pr.16.7 **(Complex potential)** Find and discuss the streamlines of $F(z) = z^3$. (*AEM Ref.* pp. 799-807)

Pr.16.8 **(Series representation of potential)** Find a series of the potential inside the unit circle if the upper half of the circle is kept at 1kV (1000 volts) and the lower half is grounded (is kept at 0 volt). Plot the potential as a surface. Also plot the boundary potential as a curve, to see how accurately the partial sums used will approximate the actual potential. (*AEM Ref.* p. 822 (#10))

Pr.16.9 **(Mean value of harmonic functions)** By taking the real part on both sides of the formula in Example 16.5 of this Guide you obtain a corresponding mean value theorem for harmonic functions. Verify this theorem for $\Phi = (x-1)(y-1)$, the point $(3,-3)$, and a circle of radius 1. (*AEM Ref.* p. 825 (#7))

Pr.16.10 **(Mean value of harmonic functions)** If Φ is harmonic in a simply connected domain D, its value at a point $P_0 : (x_0, y_0)$ in D equals the mean value of Φ over any disk D_0 in D with center P_0 (defined as the integral over D_0 divided by the area of D_0). Verify this for the function in Pr.16.9 and a disk of radius 1 with center $(3,-3)$. (*AEM Ref.* p. 825 (#7))

PART E.   NUMERICAL METHODS

Content. Basic concepts, solving equations, interpolation, integration (Chap. 17)
  Gauss, Gauss-Seidel methods, LU, matrix inversion, methods for
    eigenvalues (Chap. 18)
  Euler, Runge-Kutta, Adams-Moulton methods for ODE's and systems,
    methods for PDE's (Chap. 19)

Chapter 17

Numerical Methods in General

Content. Significant digits, quadratic equations (Ex. 17.1, Pr. 17.1)
  Solution of equations (Newton's method, etc.; Exs. 17.2-17.5,
      Prs. 17.2-17.6)
  Interpolation (Lagrange, Newton, splines; Exs. 17.6, 17.7,
      Prs. 17.7-17.11)
  Integration (Ex. 17.8, Prs. 17.12-17.15)

`SetPrecision[expression, k]` gives all numbers in expression with kS (k significant digits).

Examples for Chapter 17

EXAMPLE 17.1   LOSS OF SIGNIFICANT DIGITS. QUADRATIC EQUATION

Find the roots of the quadratic equation $x^2 - 40x + 2 = 0$ to 4S (4 significant digits).

Solution. Type a general quadratic equation. Then obtain the usual formula by the command `Solve`.

In[1]:= `Clear[a, b, c, x]`

In[2]:= `eq = a x^2 + b x + c == 0`   (* Out: `c + b x + a x^2 == 0` *)

In[3]:= `sol = Solve[eq, x]`

Out[3]= $\left\{\left\{x \to \dfrac{-b - \sqrt{b^2 - 4ac}}{2a}\right\}, \left\{x \to \dfrac{-b + \sqrt{b^2 - 4ac}}{2a}\right\}\right\}$

Now substitute the given $a = 1$, $b = -40$, $c = 2$ into the solution formula, obtaining

In[4]:= `Sol = {sol[[1, 1, 2]], sol[[2, 1, 2]]} /. {a -> 1, b -> -40, c -> 2}`

Out[4]= $\left\{\dfrac{1}{2}\left(40 - 2\sqrt{398}\right), \dfrac{1}{2}\left(40 + 2\sqrt{398}\right)\right\}$

Since you are supposed to do your calculations with 4S (4 significant digits), use the command `SetPrecision`. For the larger root `Sol[[2]]` you then obtain

179

In[5]:= x2 = SetPrecision[Sol[[2]], 4]   (* Out: 39.95 *)

From this and the fact that the sum of the two roots equals the coefficient 40 of the equation you may conclude that the second root equals 40 - 39.95 = 0.05. This shows the loss of two significant digits. This loss can be avoided by the formula $x_1 = (c/a)/x_2$ resulting from the fact that the product of the roots equals c/a, in our case, $c/a = 2/1 = 2$. For the smaller root you thus obtain

In[6]:= x1 = (2/1)/x2   (* Out: 0.05006 *)

This is the 4S-value, and due to the use of the improved formula the loss of significant digits no longer occurs. Confirm this result by typing

In[7]:= SetPrecision[Sol[[1]], 4]   (* Out: 0.05006 *)

To avoid misunderstandings: 4S was chosen for convenience, but for more digits the situation is the same in principle. For instance, you get 30S and 27S for the roots, respectively, from the the usual formula, but 30S for the smaller root by the improved formula. Indeed,

In[8]:= x2b = SetPrecision[Sol[[2]], 30]   (* Out: 30S−value *)

Out[8]= 39.9499373432600033316538822341

In[9]:= 40 - x2b   (* Out: 27S−value *)

Out[9]= 0.0500626567399966683461177659

In[10]:= 2/x2b   (* Out: 30S−value from the improved formula *)

Out[10]= 0.0500626567399966683461177658615

In[11]:= SetPrecision[Sol[[1]], 30]   (* Out: Confirmation of 30S−value *)

Out[11]= 0.0500626567399966683461177658615

Similar Material in AEM: p. 836

EXAMPLE 17.2   FIXED-POINT ITERATION

Find a solution of $x^3 + x - 1 = 0$ by **fixed-point iteration**.

Solution. Transform the given $f(x) = x^3 + x - 1 = 0$ algebraically into the form $x = g(x)$ and solve it by iteration, that is, start from a suitable x_0 and calculate

(A) $$x_1 = g(x_0), \quad x_2 = g(x_1), \quad \ldots, \quad x_n = g(x_{n-1}), \quad \ldots$$

until the desired accuracy is reached. Now $f(x) = 0$ can be cast into the form $x = g(x)$ in several ways. For instance $x = 1 - x^3$ or

(B) $$x = g(x) = 1/(1 + x^2).$$

To find a suitable x_0, sketch or plot $f(x) = x^3 + x - 1$. Type

In[1]:= Plot[x^3 + x - 1, {x, 0, 1.5}, Ticks -> {{0.5, 1, 1.5}, Automatic}]

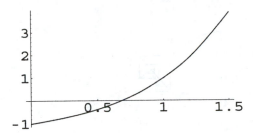

Example 17.2. Given function $f(x)$

Even from a rough sketch you see that there is a zero near 1. Hence you may start the iteration from $x_0 = 1$. You must do the transformation of $f(x) = 0$ into $x = g(x)$ in such a way that you have convergence. If $|g'(x)| \leq K < 1$ in an interval which contains the solution and the starting value and in which $g'(x)$ exists and is continuous, then you have convergence of the iteration (A). This is true if you choose (B) and, say, $x_0 = 1$. Type

In[2]:= `Clear[x]`

In[3]:= `x[0] = 1`   `(* Out: 1 *)`

In[4]:= `g[x_] = 1/(1 + x^2)`   `(* Out:` $\frac{1}{1+x^2}$ `*)`

The calculation is done by using `Do[expression, n]`, which evaluates expression n times. This is called a **do-loop**.

In[5]:= `Do[x[n] = N[g[x[n-1]]], {n, 1, 30}]`

In[6]:= `Table[x[n], {n, 0, 30}]`

Out[6]= {1, 0.5, 0.8, 0.609756, 0.728968, 0.653, 0.701061, 0.670472, 0.689878, 0.677538,

0.685374, 0.680394, 0.683557, 0.681547, 0.682824, 0.682013, 0.682528, 0.682201,

0.682409, 0.682276, 0.68236, 0.682307, 0.682341, 0.682319, 0.682333, 0.682324,

0.68233, 0.682326, 0.682329, 0.682327, 0.682328}

The last value is accurate to 6S, as you can see by typing

In[7]:= `N[Solve[x^3 + x == 1, x]]`

Out[7]= {{x → 0.682328}, {x → −0.341164 − 1.16154 i}, {x → −0.341164 + 1.16154 i}}

To represent the iteration graphically, type the following, where `ListPlot` plots points and segments joining them in the given order.

In[8]:= `P1 = Plot[x, {x, 0, 1}]`

In[9]:= `P2 = Plot[g[x], {x, 0, 1.5}]`

In[10]:= `P3 = ListPlot[{{x[0], 0}, {x[0], x[1]}, {x[1], x[1]}, {x[1], x[2]},`
                    `{x[2], x[2]}, {x[2], x[3]}, {x[3], x[3]}, {x[3], x[4]},`
                    `{x[4], x[4]}, {x[4], x[5]}, {x[5], x[5]}, {x[5], x[6]}},`
                    `PlotJoined -> True]`

In[11]:= `Show[{P1, P2, P3}, AspectRatio -> Automatic,`
           `AxesLabel -> {x, y}, Ticks -> {{0.5, 1, 1.5}, Automatic}]`

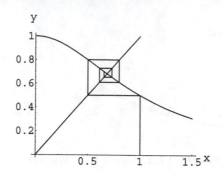

Example 17.2. Graphical representation of the iteration

Similar Material in AEM: pp. 840, 841

EXAMPLE 17.3   **SOLVING EQUATIONS BY NEWTON'S METHOD**

Find the positive solution of $2 \sin x = x$.

Solution. Solve $f(x) = x - 2 \sin x = 0$. To obtain a starting value, sketch $f(x)$. You see that your solution is near 2. As the starting value take $x_0 = 2$. You also need the derivative $f'(x)$. Then set up a do-loop involving the formula $x_{n+1} = x_n - f(x_n)/f'(x_n)$ of Newton's method, whose idea is the approximation of the curve of $f(x)$ by the tangent at x_0, x_1, etc.

In[1]:= `f = x - 2 Sin[x]`   (* Out: $x - 2\,\mathrm{Sin}\,[x]$ *)

In[2]:= `Plot[f, {x, 0, 2.5}]`

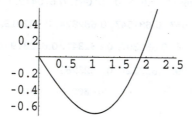

Example 17.3. Curve of the function $f(x)$ whose zero is sought

In[3]:= `g[x_] = f/D[f, x]`   (* Out: $\dfrac{x - 2\,\mathrm{Sin}\,[x]}{1 - 2\,\mathrm{Cos}\,[x]}$ *)

In[4]:= `x[0] = 2`   (* Out: 2 *)

In[5]:= `Do[x[n] = N[x[n-1] - g[x[n-1]]], {n, 1, 6}]`

In[6]:= `S = Table[x[n], {n, 0, 6}]`

Out[6]= $\{2, 1.901, 1.89551, 1.89549, 1.89549, 1.89549, 1.89549\}$

To gain an indication of the accuracy reached, substitute 1.89549 into `f`, obtaining

In[7]:= `f /. x -> S[[4]]`   (* Out: 2.86219×10^{-10} *)

Confirm your result by typing

In[8]:= `Clear[x]`

In[9]:= `FindRoot[f == 0, {x, 2}]`   (* Out: {x → 1.89549} *)

Similar Material in AEM: p. 843

EXAMPLE 17.4   **SOLVING EQUATIONS BY THE SECANT METHOD**

The formula for the secant method is

$$x_{n+1} = x_n - f(x_n)\frac{x_n - x_{n-1}}{f(x_n) - f(x_{n-1})}.$$

Using this method, find the positive solution of $2\sin x = x$. Compare with the previous example.

Solution. Solve $f(x) = x - 2\sin x = 0$. You need two starting values x_0 and x_1 at which f has different signs. From a sketch of f (see the figure in the previous example) you see that you may take, say, $x_0 = 2$ and $x_1 = 1.8$. Accordingly, type

In[1]:= `Clear[x, f]`

In[2]:= `f[x_] = x - 2 Sin[x]`   (* Out: x − 2 Sin [x] *)

In[3]:= `x[0] = 2`   (* Out: 2 *)

In[4]:= `x[1] = 1.8`   (* Out: 1.8 *)

In[5]:= `Do[`

           `x[n+1] = x[n] - N[f[x[n]] (x[n] - x[n-1])/(f[x[n]] - f[x[n-1]])],`

               `{n, 1, 8}]`

`Power::infy :   Infinite expression` $\frac{1}{0.}$ `encountered`

`∞ ::indet :   Indeterminate expression 0. ComplexInfinity encountered`

In[6]:= `Table[x[n], {n, 1, 8}]`

Out[6]= {1.8, 1.88976, 1.89583, 1.89549, 1.89549, 1.89549, 1.89549, 1.89549}

The error message is typical. The denominator of the formula of the method is `f(x[n]) - f(x[n-1])`. It becomes 0 as soon as the two x-values are equal within the number of digits used in the calculation. Here this happens for $n = 8$. Try $n = 7$, and you will see that the error message is gone.

Similar Material in AEM: pp. 845, 846

EXAMPLE 17.5   **SOLVING EQUATIONS BY THE BISECTION METHOD.**
                   MODULE

This explains a module typed as `bisect[f_, a_, b_, acc_] := Module[...]`, which will give you the zero if you simply insert the given function `f`, values `a` and `b` ($>$`a`) at which `f` has different signs, and an error bound `acc` determining the desired accuracy. Since the initial interval has length $b - a$ and is bisected, after n steps you have an interval of length $(b - a)/2^n$, short enough if its length is less than `acc`. Solving this inequality $(b - a)/2^n \le acc$ for n gives $n \ge \ln((b - a)/acc)/\ln 2$. This is used in the module, where `Ceiling[k]` is is the smallest integer greater than equal to k. Bisection of an interval is followed by selecting one of the two resulting subintervals, namely, the one at whose endpoints the function f has different signs. This is the subinterval in which the zero lies. It will be bisected in the next step. Type the module as follows. Apply it to specific equations afterward.

```
In[1]:= bisect[f_, a_, b_, acc_] :=                        (* Line 1 *)
   Module[{left, right, mid, n},
       {
       left = a;                          (* Left end of starting interval *)
     right = b;                          (* Right end of starting interval *)
     n = Ceiling[N[Log[(b - a)/acc]/Log[2]]];

                                       (* n determines the number of steps. *)
   mid = N[(a + b)/2];
       Do[If[N[(f[x] /. x -> left) (f[x] /. x -> mid)] > 0,
           left = mid,
           right = mid];                                   (* Line 10 *)
       mid = N[(left + right)/2], {j, 1, n}]
       };
       Return[mid]]                                        (* Line 13 *)
```

There are several advantages to using a module. The first main advantage is that you merely need to insert your data, that is, the function *f*, the endpoints *a* and *b* of the starting interval, and the required accuracy. The second main advantage is that you can save a module for further use. Do this **before** using it. Proceed as follows.

The choice of a file name is up to you. For instance, in the present case you could call it `bisect.m` (the `.m` is typically used for such Mathematica files). You should make sure that a file of the same name does not already exist. (If it did, Mathematica would *append* your new material – your module – to the material in the existing file.) Accordingly, type,

```
In[2]:= !!bisect.m
```

This will attempt to read the file. If Mathematica informs you that it cannot open `bisect.m`, then you can proceed to save your module. If it displays the contents of a file, you must decide if you wish to keep that file (in which case you should give it a different name). If you do not wish to keep that file, type

```
In[3]:= DeleteFile["bisect.m"]
```

This would delete any previously existing file by that name and thus guarantee that your new file will only contain the desired module `bisect`. Now save the new file by typing

```
In[4]:= Save["bisect.m", bisect]
```

You should then type

```
In[5]:= !!bisect.m
```

to see that the file contains only the module.

If you see from the response to `!!bisect` that the file contains more material than just the module, do the following. Clear all the variables contained in that extra material. Delete the file. Put the cursor in the module and press $\boxed{\texttt{Shift-Enter}}$. Save the file (that will now contain only the module).

The command for later use of the module would be

In[6]:= `<<bisect.m`

(If you get a response to this command, the file contains more than just the module. This might result from your use of the module before your attempt at saving it. In this case repeat the above process of deleting and saving.)

To use the module, you now simply type, for example,

In[7]:= `bisect[Sin, 0.1, 5, 0.00005]`   (* Out: 3.14158 *)

As another example, let

In[8]:= `g[x_] = x^2 - 40 x + 2`   (* Out: $2 - 40\,x + x^2$ *)

This quadratic equation has the solutions (4S-values) 39.95 and 0.05006 (see also Example 17.1 in this Guide). You get the larger root, for instance, from

In[9]:= `bisect[g, 39, 41, 0.00005]`   (* Out: 39.9499 *)

Similarly, you obtain the smaller root, for instance, from

In[10]:= `bisect[g, 0, 10, 0.00005]`   (* Out: 0.0500679 *)

Similar Material in AEM: p. 848

$\boxed{\textbf{EXAMPLE 17.6}}$ **POLYNOMIAL INTERPOLATION**

Polynomial interpolation means the following. Given $n+1$ ordered pairs of numbers $(x_0, y_0), (x_1, y_1), (x_2, y_2), ..., (x_n, y_n)$ (Cartesian coordinates of points in the xy-plane), find the (unique) polynomial $p_n(x)$ of smallest degree that assumes those y-values at the corresponding x-values. In general, p_n will have degree n, but the degree may sometimes be less, depending on those values (on the location of the given points through which the curve of p_n is supposed to pass). p_n is obtained by the command `InterpolatingPolynomial`.

For instance, given (9.0, 2.1972), (9.5, 2.2513), (11.0, 2.3979), you obtain a quadratic polynomial p_2 by typing the vector **u** of the x-values, the vector **v** of the y-values, and then a letter for the independent variable of p_2, say, x. Thus,

In[1]:= `u = {9, 9.5, 11}`   (* Out: {9, 9.5, 11} *)

In[2]:= `v = {2.1972, 2.2513, 2.3979}`   (* Out: {2.1972, 2.2513, 2.3979} *)

In[3]:= `InterpolatingPolynomial[{{u[[1]], v[[1]]}, {u[[2]], v[[2]]},`
`        {u[[3]], v[[3]]}}, x]`

Out[3]= $2.1972 + (0.1082 - 0.00523333(-9.5 + x))(-9 + x)$

If you want a value of p_2 at an x somewhere between the given values, say, at $x = 9.2$, type

In[4]:= `Expand[%]`   (* Out: $0.77595 + 0.205017\,x - 0.00523333\,x^2$ *)

In[5]:= `% /. x -> 9.2`   (* Out: 2.21915 *)

An alternative way that is simpler (in particular, in the case of larger data sets and higher-order polynomials) will be shown below. Rounding to 5S gives 2.2192. The given values are 5S-values of the natural logarithm, and $\ln 9.2 = 2.219203$. Hence your **quadratic interpolation** has given a correct 5S-value. Greater accuracy would require more given data so that an interpolation polynomial of higher degree could be obtained.

As an alternative way, type

In[6]:= `seq = Table[{u[[j]], v[[j]]}, {j, 1, 3}]`

Out[6]= {{9, 2.1972}, {9.5, 2.2513}, {11, 2.3979}}

In[7]:= `InterpolatingPolynomial[seq, x]`

Out[7]= $2.1972 + (0.1082 - 0.00523333(-9.5 + x))(-9 + x)$

In[8]:= `Expand[%]`   (* Out: $0.77595 + 0.205017\,x - 0.00523333\,x^2$ *)

For given data, **Lagrange's** and **Newton's interpolation formulas** (see AEM, pp. 851, 853-856) give the same polynomial (except, perhaps, for small round-off errors), although both formulas look totally different.

Newton's divided difference formula gives p_2 as follows.

In[9]:= `a11 = (v[[2]] - v[[1]])/(u[[2]] - u[[1]])`   (* Out: 0.1082 *)

In[10]:= `a12 = (v[[3]] - v[[2]])/(u[[3]] - u[[2]])`   (* Out: 0.0977333 *)

These are the $n = 2$ first divided differences (there would be more if you had more data). Now comes the $n - 1 = 1$ second divided difference (there is only one; again, there would be more if you had more data).

In[11]:= `a21 = (a12 - a11)/(u[[3]] - u[[1]])`   (* Out: -0.00523333 *)

This gives the interpolation polynomial

In[12]:= `p = v[[1]] + (x - u[[1]]) a11 + (x - u[[1]]) (x - u[[2]]) a21`

Out[12]= $2.1972 + 0.1082\,(-9 + x) - 0.00523333\,(-9.5 + x)\,(-9 + x)$

In[13]:= `Expand[%]`   (* Out: $0.77595 + 0.205017\,x - 0.00523333\,x^2$ *)

This agrees with the above response to `InterpolatingPolynomial` .

Similar Material in AEM: pp. 850, 854

EXAMPLE 17.7   SPLINE INTERPOLATION

Higher order polynomials for interpolating data (x_0, y_0), $(x_1, y_1), \ldots$, (x_n, y_n) tend to oscillate between nodes; see the figure, where $n = 8$ and the nodes are at $x = -4$, $-3, \ldots, 4$, and $y(0) = 1$ and the other y's are zero. Obtain this figure by typing

In[1]:= `u = {-4, -3, -2, -1, 0, 1, 2, 3, 4}` (* Out: $\{-4, -3, -2, -1, 0, 1, 2, 3, 4\}$ *)

In[2]:= `v = {0, 0, 0, 0, 1, 0, 0, 0, 0}`   (* Out: $\{0, 0, 0, 0, 1, 0, 0, 0, 0\}$ *)

In[3]:= `w = Table[{u[[j]], v[[j]]}, {j, 1, 9}]`

Out[3]= $\{\{-4, 0\}, \{-3, 0\}, \{-2, 0\}, \{-1, 0\}, \{0, 1\}, \{1, 0\}, \{2, 0\}, \{3, 0\}, \{4, 0\}\}$

In[4]:= `p = InterpolatingPolynomial[w, x]`

Out[4]= $(1 + x) (2 + x) (3 + x) (4 + x)$

$$\left(\frac{1}{24} + \left(-\frac{1}{24} + \left(\frac{1}{48} + \left(-\frac{1}{144} + \frac{1}{576} (-3 + x) \right) (-2 + x) \right) (-1 + x) \right) x \right)$$

In[5]:= Expand[%]

Out[5]= $1 - \dfrac{205\,x^2}{144} + \dfrac{91\,x^4}{192} - \dfrac{5\,x^6}{96} + \dfrac{x^8}{576}$

In[6]:= N[%]　　　(* Out: $1. - 1.42361\,x^2 + 0.473958\,x^4 - 0.0520833\,x^6 + 0.00173611\,x^8$ *)

P1 plots the points and P2 the curve of the interpolation polynomial. (Points and curve are not shown separately since they appear jointly in the figure.)

In[7]:= P1 = ListPlot[w, Prolog -> AbsolutePointSize[4]]

In[8]:= P2 = Plot[p, {x, -4, 4}]

In[9]:= Show[P1, P2]

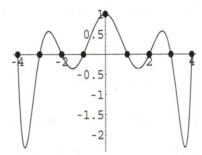

Example 17.7. Oscillations of an interpolation polynomial of 8th degree

Splines. Load the following package.

In[10]:= <<NumericalMath'SplineFit'

Then you can obtain cubic splines for given data (as well as Bezier curves; see AEM, p. 868). Indeed, for the above data set w (pairs of coordinates of points) type

In[11]:= spline = SplineFit[w, Cubic]

Out[11]= SplineFunction[Cubic, {0., 8.}, <>]

This is a natural spline, that is, the second derivative at the two very ends is zero (as it is for the draftman's spline). The data extend from $x = -4$ to $x = 4$, which in the response are labeled 0, 1, ...8 (9 data points). You can now obtain the value of the spline at any point, for instance

In[12]:= spline[0]　　　　　　　　　　(* Out: $\{-4, 0\}$, given *)

In[13]:= spline[0.5]　　　　　　　　　(* Out: $\{-3.5, -0.0115979\}$ *)

So the spline is by no means 0 at intermediate points far away from $x = 0$, but its values are very small, compared to those of the interpolation polynomial of eighth degree. For example,

In[14]:= spline[4]　　　　　　　　　　(* Out: $\{0, 1\}$, given *)

In[15]:= spline[4.4]　　　　　　　　　(* Out: $\{0.4, 0.725196\}$ *)

and so on.

Plotting splines. To plot the spline just discussed, you can type the following, where `Compiled -> False` prevents unnecessary warning messages.

In[16]:= `ParametricPlot[spline[w], {w, 0, 8}, PlotRange -> All,`
        `Compiled -> False]`

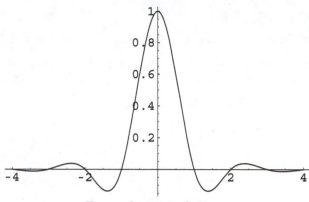

Example 17.7. Spline

Similar Material in AEM: pp. 861-866

| EXAMPLE 17.8 | **NUMERICAL INTEGRATION**

Integrate $f(x) = \exp\left(-x^2\right)$ from 0 to 1 by the usual methods, choosing $n = 10$ steps of size $h = 0.1$.

Solution. Type the integrand and the data $a = 0$, $b = 1$, $M = 10$, $h = (b - a)/M$, which will be the same for all methods.

In[1]:= `Clear[f]`

In[2]:= `f[x_] = Exp[-x^2]`   (* Out: e^{-x^2} *)

In[3]:= `a = 0;    b = 1;    M = 10;    h = (b - a)/M;`

Rectangular rule. This rule is usually not accurate enough. Type

In[4]:= `RectRule = h (Sum[f[x], {x, a + h/2, b - h/2, h}])`

Out[4]= $\dfrac{1}{10}\left(\dfrac{1}{e^{361/400}} + \dfrac{1}{e^{289/400}} + \dfrac{1}{e^{9/16}} + \dfrac{1}{e^{169/400}} + \dfrac{1}{e^{121/400}} + \dfrac{1}{e^{81/400}} + \dfrac{1}{e^{49/400}} + \right.$
$\left. \dfrac{1}{e^{1/16}} + \dfrac{1}{e^{9/400}} + \dfrac{1}{e^{1/400}}\right)$

In[5]:= `N[%]`   (* Out: 0.747131 *)

Trapezoidal rule. Using f, a, b, h as before, type

In[6]:= `TrapezRule = h (f[a]/2 + f[b]/2 + Sum[f[x], {x, a + h, b - h, h}])`

Out[6]= $\dfrac{1}{10}\left(\dfrac{1}{2} + \dfrac{1}{2\,e} + \dfrac{1}{e^{81/100}} + \dfrac{1}{e^{16/25}} + \dfrac{1}{e^{49/100}} + \dfrac{1}{e^{9/25}} + \dfrac{1}{e^{1/4}} + \dfrac{1}{e^{4/25}} + \dfrac{1}{e^{9/100}} + \dfrac{1}{e^{1/25}} + \right.$
$\left. \dfrac{1}{e^{1/100}}\right)$

In[7]:= `N[%]`   (* Out: 0.746211 *)

Simpson's rule. Using f, a, b, h as before, type

In[8]:= `SimpsonRule = h/3 (f[a] + 4 Sum[f[x], {x, a + h, b - h, 2 h}] +`
`2 Sum[f[x], {x, a + 2 h, b - 2 h, 2 h}] + f[1])`

Out[8]= $\frac{1}{30} \left(1 + 2 \left(\frac{1}{e^{16/25}} + \frac{1}{e^{9/25}} + \frac{1}{e^{4/25}} + \frac{1}{e^{1/25}} \right) + \right.$
$\left. 4 \left(\frac{1}{e^{81/100}} + \frac{1}{e^{49/100}} + \frac{1}{e^{1/4}} + \frac{1}{e^{9/100}} + \frac{1}{e^{1/100}} \right) + \frac{1}{e} \right)$

In[9]:= `N[%]`   (* Out: 0.746825 *)

Gauss integration generally requires less work. For nodes and coefficients see AEM, p. 878, and [1], pp. 916-919, listed in Appendix 1 of this Guide. (Note that the nodes are not equally spaced!) Setting $x = (t+1)/2$, you integrate over t from -1 to 1; since $dx = dt/2$, you obtain

(A) $$\int_0^1 \exp(-x^2)\, dx = \frac{1}{2} \int_{-1}^1 \exp\left(-\frac{1}{4}(t+1)^2 \right) dt.$$

Choosing $n = 3$, the nodes (t-values) are $-\sqrt{\frac{3}{5}}$, 0, $\sqrt{\frac{3}{5}}$, and the coefficients are $5/9, 8/9, 5/9$, respectively. You thus obtain from (A)

In[10]:= `1/2 (5/9 Exp[-1/4 (-Sqrt[3/5] + 1)^2] + 8/9 Exp[-1/4] +`
`5/9 Exp[-1/4 (Sqrt[3/5] + 1)^2])`

Out[10]= $\frac{1}{2} \left(\frac{8}{9\,e^{1/4}} + \frac{5}{9} e^{-\frac{1}{4}\left(1-\sqrt{\frac{3}{5}}\right)^2} + \frac{5}{9} e^{-\frac{1}{4}\left(1+\sqrt{\frac{3}{5}}\right)^2} \right)$

In[11]:= `N[%]`   (* Out: 0.746815 *)

The 4D accuracy obtained from this short calculation is astonishing.
Similar Material in AEM: pp. 869-878

Problem Set for Chapter 17

Pr.17.1 **(Quadratic equation)** Solve $x^2 - 30x + 1 = 0$ by the two methods in Example 17.1 in this Guide. (*AEM Ref.* p. 836 (#6))

Pr.17.2 **(Fixed-point iteration)** Solve $x \cosh x = 1$ (appearing in connection with vibrations of beams). (*AEM Ref.* p. 847 (#6))

Pr.17.3 **(Fixed-point iteration)** Find a solution of $x^4 - x - 0.12 = 0$, starting from $x_0 = 1$. (*AEM Ref.* p. 847 (#9))

Pr.17.4 **(Newton's iteration method)** Design a Newton iteration for cube roots and compute the cube root of 7, starting from $x_0 = 2$. (*AEM Ref.* p. 847 (#11))

Pr.17.5 **(Secant method)** Find a zero of $f(x) = 1 - x^2/4 + x^4/64 - x^6/2304$ near $x = 2$ by using $x_0 = 2.0$ and $x_1 = 2.5$. ($f(x)$ is a partial sum of the Maclaurin series of the Bessel function $J_0(x)$, whose zeros are important in connection with vibrations of membranes.) (*AEM Ref.* p. 847 (#23))

Pr.17.6 **(Bisection method)** Solve $x = \cos x$. (See Example 17.5 in this Guide. *AEM Ref.* p. 848 (#25))

Pr.17.7 **(Polynomial interpolation)** Find the values of the Gamma function at 1.01 and 1.03 by quadratic interpolation of the values 1.0000, 0.9888, 0.9784 at 1.00, 1.02, 1.04, respectively, using the command `InterpolatingPolynomial` .
(*AEM Ref.* pp. 860 (#5), A86)

Pr.17.8 **(Newton's forward difference interpolation formula)** Solve Pr.17.7 by the formula

$$f(x) = f_0 + r\,\Delta\,f_0 + \frac{1}{2!}r\,(r-1)\,\Delta^2\,f_0 + \frac{1}{3!}r\,(r-1)\,(r-2)\,\Delta^3\,f_0 + \ldots$$

where $r = (x - x_0)/h$ and h is the distance between the equally spaced adjacent nodes.
(*AEM Ref.* pp. 856, 860 (#11))

Pr.17.9 **(High-order polynomial)** Find the polynomial that interpolates (j, y_j), where $j = -5, -4, \ldots 0, \ldots 4, \ 5$ and $y_0 = 1$, $y_j = 0$ otherwise. (*AEM Ref.* p. 862)

Pr.17.10 **(Experiment on oscillation of interpolation polynomials)** Explore by experimentation and plotting how the oscillations of interpolation polynomials for equally spaced nodes and function values 0 except for a single 1 (as in Example 17.7 in this Guide and in the previous problem) are growing with increasing degree n.

Pr.17.11 **(Spline interpolation)** Find the natural cubic spline for the data in Pr.17.9. Compute some of its values. Plot it. (*AEM Ref.* pp. 861-866)

Pr.17.12 **(Rectangular rule)** Integrate $\exp(-x^2)$ from 0 to 2 by the rectangular rule with $h = 0.2$ that uses the function value at the midpoint of each of the ten subintervals. (*AEM Ref.* p. 869)

Pr.17.13 **(Experiment on accuracy of integration rules)** Integrate $\cos(x^2)$ (not $(\cos x)^2$) from 0 to $\sqrt{\frac{\pi}{2}}$, first by the command `N[Integrate[...]]` to get the exact value, and then by the rectangular rule, the trapezoidal rule, and Simpson's rules. In each case find the (approximate) number of subintervals that gives 5S accuracy, thereby noting the superiority of Simpson's rule. ($\cos(x^2)$ is the integrand of the **Fresnel integral** $C(x)$. Following [1], p. 300 in Appendix 1, Mathematica uses $\cos(\pi t^2/2)$ as the integrand of the Fresnel integral instead of the more common $\cos(x^2)$. To call Help, click on MasterIndex, and then type Fresnel.) (*AEM Ref.* pp. 869-875, A56)

Pr.17.14 **(Simpson's rule)** Obtain a 10D-value of ln 2 by evaluating a suitable integral by Simpson's rule. (*AEM Ref.* pp. 872-875)

Pr.17.15 **(Gauss integration)** Calculate the sine integral Si(x) (defined as the integral of $(\sin u)/u$ from $u = 0$ to $u = x$) for $x = 1$ by Gauss integration with $n = 3$ nodes and find the accuracy by comparing with the value obtained by the command `Integrate` .
(*AEM Ref.* pp. 877, 878, A57)

Chapter 18

Numerical Linear Algebra

Content. Gauss, Doolittle, Cholesky methods (Exs. 18.1-18.3, Prs. 18.1-18.5)
Gauss-Jordan, Gauss-Seidel methods (Exs. 18.4, 18.5, Prs. 18.6-18.8)
Norms, condition numbers (Ex. 18.6, Prs. 18.9-18.12)
Fit by least squares (Ex. 18.7, Prs. 18.13-18.16)
Approximation of eigenvalues (Exs. 18.8-18.10, Prs. 18.17-18.20)

The MatrixManipulation package is used as indicated in the examples. It is loaded by `<<LinearAlgebra'MatrixManipulation'`. Its content can by seen by selecting "Help" on the menu bar, then going three lines down in the menu and selecting "MasterIndex", and typing "MatrixManipulation"; then pressing ⊡Enter⊡ will display the link to "LinearAlgebra'MatrixManipulation'", which you then select.

Examples for Chapter 18

EXAMPLE 18.1	**GAUSS ELIMINATION. PIVOTING**

To keep this chapter independent of Chaps. 6 and 7 on matrices and eigenvalue problems, we begin by describing the implementation of the Gauss elimination on the computer.

Solve the linear system

$$
\begin{aligned}
8\,x_2 + 2\,x_3 &= -7 \\
6\,x_1 + 2\,x_2 + 8\,x_3 &= 26, \\
3\,x_1 + 5\,x_2 + 2\,x_3 &= 8
\end{aligned}
$$

in matrix form $\mathbf{Ax} = \mathbf{b}$, where

$$
\mathbf{A} = [a_{jk}] = \begin{bmatrix} 0 & 8 & 2 \\ 6 & 2 & 8 \\ 3 & 5 & 2 \end{bmatrix} \quad \text{and} \quad \mathbf{b} = \begin{bmatrix} -7 \\ 26 \\ 8 \end{bmatrix}.
$$

Solution. Type the matrix in three lines (or in one line with commas and braces) as shown, then the vector, and then apply the command `LinearSolve`.

```
In[1]:= A = {{0, 8, 2},
            {6, 2, 8},
            {3, 5, 2}}
```
Out[1]= {{0, 8, 2}, {6, 2, 8}, {3, 5, 2}}

In[2]:= %//MatrixForm
$$\left(* \text{ Out: } \begin{pmatrix} 0 & 8 & 2 \\ 6 & 2 & 8 \\ 3 & 5 & 2 \end{pmatrix} *\right)$$

In[3]:= b = {-7, 26, 8}
$$(* \text{ Out: } \{-7, 26, 8\} *)$$

In[4]:= LinearSolve[A, b]
$$\left(* \text{ Out: } \left\{4, -1, \tfrac{1}{2}\right\} *\right)$$

This gave the solution but no indication on the method by which it was obtained.

The **Gauss-Jordan method** is accomplished by first typing

In[5]:= A1 = Transpose[Join[Transpose[A], {b}]]

Out[5]= {{0, 8, 2, -7}, {6, 2, 8, 26}, {3, 5, 2, 8}}

In[6]:= %//MatrixForm

Out[6]//MatrixForm=

$$\begin{pmatrix} 0 & 8 & 2 & -7 \\ 6 & 2 & 8 & 26 \\ 3 & 5 & 2 & 8 \end{pmatrix}$$

which gives the **augmented matrix**, as you see. Then apply the command

In[7]:= B = RowReduce[A1]
$$\left(* \text{ Out: } \left\{\{1, 0, 0, 4\}, \{0, 1, 0, -1\}, \left\{0, 0, 1, \tfrac{1}{2}\right\}\right\} *\right)$$

In[8]:= %//MatrixForm
$$\left(* \text{ Out: } \begin{pmatrix} 1 & 0 & 0 & 4 \\ 0 & 1 & 0 & -1 \\ 0 & 0 & 1 & \tfrac{1}{2} \end{pmatrix} *\right)$$

From this you see that $x_1 = 4$, $x_2 = -1$, $x_3 = 1/2$, as before.

Gauss elimination and **back substitution**. The augmented matrix A1 shows that because of the zero you have to **pivot**. Interchange Rows 1 and 2. These rows can be accessed by the commands

In[9]:= A1[[1]]
$$(* \text{ Out: } \{0, 8, 2, -7\} *)$$

In[10]:= A1[[2]]
$$(* \text{ Out: } \{6, 2, 8, 26\} *)$$

Take Row 1 of A1 as Row 2 of a new matrix by typing

In[11]:= ReplacePart[A1, A1[[1]], 2]

Out[11]= {{0, 8, 2, -7}, {0, 8, 2, -7}, {3, 5, 2, 8}}

Now take Row 2 of the old matrix A1 as Row 1 of the new matrix by typing

In[12]:= A2 = ReplacePart[%, A1[[2]], 1]

Out[12]= {{6, 2, 8, 26}, {0, 8, 2, -7}, {3, 5, 2, 8}}

Pivoting is complete. A2 is the matrix to which you now apply Gauss elimination. Eliminate x_1 from Equation 3 (in Equation 2 it does not occur) by replacing Row 3 of A2 by Row 3 - 1/2 Row 1 of A2 . This gives A3 .

In[13]:= A3 = ReplacePart[A2, A2[[3]] - (1/2) A2[[1]], 3]

Out[13]= {{6, 2, 8, 26}, {0, 8, 2, -7}, {0, 4, -2, -5}}

Eliminate x_2 from Equation 3 by replacing Row 3 of A3 by Row 3 - 1/2 Row 2 of A3 . This gives the triangularized matrix A4

In[14]:= `A4 = ReplacePart[A3, A3[[3]] - (1/2) A3[[2]], 3]`

Out[14]= $\left\{\{6, 2, 8, 26\}, \{0, 8, 2, -7\}, \{0, 0, -3, -\frac{3}{2}\}\right\}$

Triangularization can be done more quickly by do-loops as follows. Type

In[15]:= `<< LinearAlgebra'MatrixManipulation'`

In[16]:= `Clear[j, k, m]`

In[17]:= `A2`   `(* Augmented matrix after pivoting *)`

Out[17]= $\{\{6, 2, 8, 26\}, \{0, 8, 2, -7\}, \{3, 5, 2, 8\}\}$

In[18]:= `m = Dimensions[A2][[1]]`   `(* Out: 3 Number of rows of A2 *)`

In[19]:= `Do [`
```
       Do [
          Print[A2 = ReplacePart[A2, A2[[j]] -
                A2[[k]] A2[[j,k]]/A2[[k,k]], j]],
          {j, k + 1, m}],
          {k, 1, m - 1}
          ];
```

$\{\{6, 2, 8, 26\}, \{0, 8, 2, -7\}, \{3, 5, 2, 8\}\}$ `(* This is A2 *)`

$\{\{6, 2, 8, 26\}, \{0, 8, 2, -7\}, \{0, 4, -2, -5\}\}$ `(* This is A3 *)`

$\left\{\{6, 2, 8, 26\}, \{0, 8, 2, -7\}, \{0, 0, -3, -\frac{3}{2}\}\right\}$ `(* This is A4 *)`

This agrees with the previous results. `A2[[j,k]]` is the entry in Row j and Column k of `A2`.

Back substitution. Type

In[20]:= `m = Dimensions[A3][[1]]`   `(* Out: 3 *)`

In[21]:= `n = Dimensions[A][[2]]`   `(* Out: 3 *)`

This is the number of columns of the coefficient matrix (not of the augmented!).

In[22]:= `x = {t1, t2, t3}`   `(* Out: {t1, t2, t3} *)`

In[23]:= `Do [`
```
       If[A2[[j,j]] != 0,          (* != 0 means different from zero. *)
          x[[j]] = (A2[[j,n + 1]] -
                Sum[A2[[j,k]] x[[k]], {k, j + 1, n}])/A2[[j,j]]],
          {j, m, 1, -1}
          ];
```

In[24]:= `x` $\left(* \text{ Out: } \left\{4, -1, \frac{1}{2}\right\} *\right)$

EXAMPLE 18.2   DOOLITTLE LU-FACTORIZATION

A linear system $\mathbf{Ax} = \mathbf{b}$ of n equations in n unknowns can be solved by first factor-izing $\mathbf{A} = \mathbf{LU}$, where $\mathbf{L}$ is **lower triangular** and $\mathbf{U}$ is **upper triangular**. Then $\mathbf{Ax} = \mathbf{L}(\mathbf{Ux}) = \mathbf{b}$. Set $\mathbf{Ux} = \mathbf{y}$ and solve $\mathbf{Ly} = \mathbf{b}$ for $\mathbf{y}$. Then solve $\mathbf{Ux} = \mathbf{y}$ for $\mathbf{x}$. It can be proved that if **A** is nonsingular, its rows can be reordered so that the resulting matrix has an **LU**-factorization. This factorization is obtained by the command LUDecomposition, as explained below.

Solve the linear system in Example 18.1 in this Guide by **LU**-factorization.

Solution. Type the matrix A and the vector b. Load the package (used in Example 18.1) <<LinearAlgebra'MatrixManipulation'. Then type the command LUDecomposition.

In[1]:= A = {{0, 8, 2}, {6, 2, 8}, {3, 5, 2}};

In[2]:= %//MatrixForm

Out[2]//MatrixForm=

$$\begin{pmatrix} 0 & 8 & 2 \\ 6 & 2 & 8 \\ 3 & 5 & 2 \end{pmatrix}$$

In[3]:= b = {-7, 26, 8}   (* Out: {−7, 26, 8} *)

In[4]:= << LinearAlgebra'MatrixManipulation'

In[5]:= LUDecomp = LUDecomposition[A]

Out[5]= {{{3, 5, 2}, {2, −8, 4}, {0, −1, 6}}, {3, 2, 1}, 1}

This looks mysterious. Click on "Help", then on "Master Index", and type "LU Decomposition". Then you will see that {3, 5, 2} is the first row of U, {0, -8, 4} the second, and {0, 0, 6} the third. The rows of L are {1, 0, 0}, then {2, 1, 0}, and {0, -1, 1}. Thus the 3×3 matrix in the response contains the "nontrivial" entries of both U and L. The {3, 2, 1} part indicates the pivoting, namely, the interchange of Rows 1 and 3 of **A**. (The 1 is an estimate of the size of the condition number of **A** in the l_∞-norm.) To check your result, type

In[6]:= LUMat = LUMatrices[LUDecomp[[1]]]

Out[6]= {{{1, 0, 0}, {2, 1, 0}, {0, −1, 1}}, {{3, 5, 2}, {0, −8, 4}, {0, 0, 6}}}

which produces a list of both L and U which can then be multiplied together.

In[7]:= LUMat[[1]].LUMat[[2]]   (* Out: {{3, 5, 2}, {6, 2, 8}, {0, 8, 2}} *)

The result is the given matrix with Rows 1 and 3 interchanged.

You can now solve $\mathbf{Ax} = (\mathbf{LU})\mathbf{x} = \mathbf{L}(\mathbf{Ux}) = \mathbf{b}$ in two steps, as explained above. Interchange the first and third components of b, as in the pivoting, obtaining b2 = {8, 26, -7}. Thus type

In[8]:= b2 = {8, 26, -7}   (* Out: {8, 26, −7} *)

In[9]:= y = LinearSolve[LUMat[[1]], b2]   (* Out: {8, 10, 3} *)

In[10]:= x = LinearSolve[LUMat[[2]], y]   (* Out: $\left\{4, -1, \frac{1}{2}\right\}$ *)

More quickly,

In[11]:= x = LUBackSubstitution[LUDecomp, b]   (* Out: $\left\{4, -1, \frac{1}{2}\right\}$ *)

Similar Material in AEM: pp. 894-896

EXAMPLE 18.3 | CHOLESKY FACTORIZATION

Solve the following system by Cholesky's method.

$$\begin{aligned} 4\,x_1 \;+\; 2\,x_2 \;+\; 14\,x_3 &= \;\;\;14 \\ 2\,x_1 \;+\; 17\,x_2 \;-\; 5\,x_3 &= -101 \\ 14\,x_1 \;-\; 5\,x_2 \;+\; 83\,x_3 &= \;\;155 \end{aligned}$$

Solution. This method solves linear systems $\mathbf{Ax} = \mathbf{b}$ with symmetric and positive definite $\mathbf{A}$. The present matrix is symmetric. You need not check for positive definiteness; if $\mathbf{A}$ were not positive definite, the calculations would lead into complex. The method uses the factorization $\mathbf{A} = \mathbf{LL}^T$ and then solves $\mathbf{Ax} = \mathbf{L}(\mathbf{L}^T\mathbf{x}) = \mathbf{b}$ in two steps as in the previous example, namely, $\mathbf{Ly} = \mathbf{b}$ and $\mathbf{L}^T\mathbf{x} = \mathbf{y}$. Type the coefficient matrix $\mathbf{A}$, then $\mathbf{b}$. Then load the Cholesky factorization package.

In[1]:= `A = {{4, 2, 14}, {2, 17, -5},{14, -5, 83}}`

In[2]:= `%//MatrixForm`   (* Out: $\begin{pmatrix} 4 & 2 & 14 \\ 2 & 17 & -5 \\ 14 & -5 & 83 \end{pmatrix}$ *)

In[3]:= `b = {14, -101, 155}`   (* Out: $\{14, -101, 155\}$ *)

In[4]:= `<<LinearAlgebra'Cholesky'`

Now type the command for the Cholesky factorization

In[5]:= `U = CholeskyDecomposition[A];MatrixForm[U]`   (* Out: $\begin{pmatrix} 2 & 1 & 7 \\ 0 & 4 & -3 \\ 0 & 0 & 5 \end{pmatrix}$ *)

Note that Mathematica gives an upper triangular matrix $\mathbf{U}$ rather than the lower triangular $\mathbf{L}$ discussed above (and used in AEM); so $\mathbf{U}$ corresponds to $\mathbf{L}^T$ in AEM. Thus

In[6]:= `L = Transpose[U];MatrixForm[L]`   (* Out: $\begin{pmatrix} 2 & 0 & 0 \\ 1 & 4 & 0 \\ 7 & -3 & 5 \end{pmatrix}$ *)

To solve $\mathbf{Ax} = \mathbf{b}$ in two steps, as explained above, type

In[7]:= `y = LinearSolve[L, b]`   (* Out: $\{7, -27, 5\}$ *)

In[8]:= `x = LinearSolve[U, y]`   (* Out: $\{3, -6, 1\}$ *)

We now discuss the formulas for the Cholesky factorization. Type

In[9]:= `n = 3;`

In[10]:= `L = Table[0, {n}, {n}]; MatrixForm[L];`   (* 3×3 **zero matrix** *)

This zero matrix will now get its entries from the formulas for the Cholesky factorization $\mathbf{A} = \mathbf{LL}^T$, which are as follows.

$$l_{11} = \sqrt{a_{11}} \qquad\qquad l_{j1} = \frac{a_{j1}}{l_{11}}, \qquad j = 2,\dots,n$$

$$l_{jj} = \sqrt{a_{jj} - \sum_{s=1}^{j-1} l_{js}^2}, \qquad j = 2,\dots,n$$

$$l_{pj} = \frac{1}{l_{jj}}\Big(pj - \sum_{s=1}^{j-1} l_{js}l_{ps}\Big), \qquad p = j+1,\dots,n, \; j \geq 2.$$

From the first formula you obtain

In[11]:= `L[[1,1]] = Sqrt[A[[1,1]]]`　　　　　　　　　　　　　　　　(* Out: 2 *)

From the second formula you obtain the other two entries of the first column of **L**,

In[12]:= `Do [{L[[j,1]] = A[[j,1]]/L[[1,1]];`
　　　　　`Print["L", j, ", 1] = ", L[[j,1]]]}, {j, 2, n}]`　　　(* Out: 2 *)

L[2, 1] = 1
L[3, 1] = 7

The **printing command** involves variable parts `j`, `L[[j,1]]` and fixed parts (the others). The latter are included in double quotations; thus, `"L["` (the first fixed part), `", 1] = "` (the second fixed part). The other two formulas are interrelated. This leads to two interrelated do-loops, an inner loop, in which `p` runs from `j + 1` to `n` (do not confuse this with the summation index `s` in the sum, which runs from `1` to `j - 1`), and an outer, in which `j` runs from `2` to `n`. This gives you the remaining three nonzero entries of **L**. The explanation of the strange-looking printing commands is the same as that just given.

In[13]:= `Do [`

　　　`L[[j,j]] = Sqrt[A[[j,j]] - Sum[L[[j,s]]^2, {s, 1, j - 1}]];`
　　　　`Print["L[", j, ",", j, "] = ", L[[j,j]]];`
　　　　　`Do [`
　　　　　　`L[[p,j]] = (A[[p,j]] - Sum[L[[j,s]] L[[p,s]],`
　　　　　　　　　　　　　　　`{s, 1, j - 1}])/L[[j,j]];`
　　　　　　　`Print["L[", p, ",", j, "] =", L[[p,j]]],`
　　　　　`{p, j + 1, n}`
　　　　　`],`
　　　　`{j, 2, n}`
　　　`]`　　　　　　　　　　　　　　　　　　　　　　　　　　(* Out: 2 *)

L[2, 2] = 4
L[3, 2] = −3
L[3, 3] = 5

Now solve **L** **y** = **b** for **y** and then **L**^{T} **x** = **y** for **x**.

In[14]:= `y = LinearSolve[L, b]`　　　　　　　　　　(* Out: {7, −27, 5} *)

In[15]:= `x = LinearSolve[Transpose[L], y]`　　　(* Out: {3, −6, 1} *)

Similar Material in AEM: pp. 896-898

| **EXAMPLE 18.4** | **GAUSS-JORDAN ELIMINATION. MATRIX INVERSION** |

The **Gauss-Jordan method** is *not* practical for solving linear systems **Ax** = **b** (because the Jordan reduction needs more calculations than the back substitution does), but the method is useful for obtaining the inverse of a matrix. Gauss's method gives a triangular matrix, and Jordan reduces it further to a diagonal matrix (actually, to the unit matrix). Using the Gauss-Jordan method, solve the following linear system. As a second, separate task, calculate the inverse of the coefficient matrix **A** of the system.

$$
\begin{aligned}
3\,x_1 + 2\,x_2 \qquad\quad &= \quad\;\; 14 \\
12\,x_1 + 13\,x_2 + 6\,x_3 &= \quad\;\; 40 \\
-3\,x_1 + 8\,x_2 + 9\,x_3 &= -28
\end{aligned}
$$

Solution. The system is of the form $\mathbf{Ax} = \mathbf{b}$. A first application of the Gauss-Jordan method to a given system was shown in Example 17.1 in this Guide. Proceed similarly and type **A** and **b** and obtain the augmented matrix **A1**, which by row reduction will yield the Gauss-Jordan form as well as the solution of the given system.

In[1]:= A = {{3, 2, 0}, {12, 13, 6}, {-3, 8, 9}}

Out[1]= {{3, 2, 0}, {12, 13, 6}, {−3, 8, 9}}

In[2]:= b = {14, 40, -28}   (* Out: {14, 40, −28} *)

In[3]:= Join[Transpose[A], {b}]   (* Transpose of the augmented matrix *)

Out[3]= {{3, 12, −3}, {2, 13, 8}, {0, 6, 9}, {14, 40, −28}}

In[4]:= A1 = Transpose[%]   (* Augmented matrix *)

Out[4]= {{3, 2, 0, 14}, {12, 13, 6, 40}, {−3, 8, 9, −28}}

In[5]:= A2 = RowReduce[A1]; MatrixForm[A2]   (* Out:
$\begin{pmatrix} 1 & 0 & 0 & 2 \\ 0 & 1 & 0 & 4 \\ 0 & 0 & 1 & -6 \end{pmatrix}$ *)

From this you can read the answer $x_1 = 2$, $x_2 = 4$, $x_3 = -6$ or by typing the command

In[6]:= x = Table[A2[[j,4]], {j, 1, 3}]   (* Out: {2, 4 −6} *)

Inverse of a matrix A by Gauss-Jordan. You need $\mathbf{B1} = [\mathbf{A}\ \mathbf{I}]$, where **I** is the 3×3 unit matrix. Now the command Join joins rows, not columns. Hence start from $\mathbf{A}^T$ and join **I** to it. Call the result **B0**. Take its transpose **B1**.

In[7]:= B0 = Join[Transpose[A], IdentityMatrix[3]]; MatrixForm[B0]

Out[7]//MatrixForm=

$$
\begin{pmatrix}
3 & 12 & -3 \\
2 & 13 & 8 \\
0 & 6 & 9 \\
1 & 0 & 0 \\
0 & 1 & 0 \\
0 & 0 & 1
\end{pmatrix}
$$

In[8]:= B1 = Transpose[%]; MatrixForm[B1]   (* Out:
$\begin{pmatrix} 3 & 2 & 0 & 1 & 0 & 0 \\ 12 & 13 & 6 & 0 & 1 & 0 \\ -3 & 8 & 9 & 0 & 0 & 1 \end{pmatrix}$ *)

Row reduce B1 . Call the result B2 .

In[9]:= B2 = RowReduce[B1]; MatrixForm[B2]

Out[9]//MatrixForm=

$$
\begin{pmatrix}
1 & 0 & 0 & -\dfrac{23}{15} & \dfrac{2}{5} & -\dfrac{4}{15} \\[2mm]
0 & 1 & 0 & \dfrac{14}{5} & -\dfrac{3}{5} & \dfrac{2}{5} \\[2mm]
0 & 0 & 1 & -3 & \dfrac{2}{3} & -\dfrac{1}{3}
\end{pmatrix}
$$

This reduction amounts to multiplying $\mathbf{B1} = [\mathbf{A} \ \mathbf{I}]$ by $\mathbf{A}^{-1}$, obtaining $\mathbf{A}^{-1}\mathbf{B1} = [\mathbf{A}^{-1}\mathbf{A} \ \ \mathbf{A}^{-1}\mathbf{I}] = [\mathbf{I} \ \mathbf{A}^{-1}]$. Hence $\mathbf{A}^{-1}$ is the submatrix of $\mathbf{B2}$ consisting of the last three columns, that is, the entries in Rows 1 to 3 which are in Columns 4 to 6.

In[10]:= `Inv = Take[B2, {1, 3}, {4, 6}]; MatrixForm[Inv]`

Out[10]//MatrixForm=

$$
\begin{pmatrix}
-\dfrac{23}{15} & \dfrac{2}{5} & -\dfrac{4}{15} \\[2mm]
\dfrac{14}{5} & -\dfrac{3}{5} & \dfrac{2}{5} \\[2mm]
-3 & \dfrac{2}{3} & -\dfrac{1}{3}
\end{pmatrix}
$$

Similar Material in AEM: p. 898

EXAMPLE 18.5   **GAUSS-SEIDEL ITERATION FOR LINEAR SYSTEMS**

Solve the following linear system by **Gauss-Seidel iteration**, starting from the vector $[100 \ \ 100 \ \ 100 \ \ 100]^T$. (Systems of this kind appear in connection with numerical methods for partial differential equations.)

$$
\begin{array}{rcrcrcrcl}
x_1 & - & 0.25\,x_2 & - & 0.25\,x_3 & & & = & 50 \\
-0.25\,x_1 & + & x_2 & & & - & 0.25\,x_4 & = & 50 \\
-0.25\,x_1 & & & + & x_3 & - & 0.25\,x_4 & = & 25 \\
& - & 0.25\,x_2 & - & 0.25\,x_3 & + & x_4 & = & 25
\end{array}
$$

Solution. In $\mathbf{Ax} = \mathbf{b}$ assume that $\mathbf{A}$ has the main diagonal entries 1, ..., 1 (which can be accomplished by division, provided all the main diagonal entries are originally different from 0). Write $\mathbf{A} = \mathbf{I} + \mathbf{L} + \mathbf{U}$, where $\mathbf{I}$ is the unit matrix, $\mathbf{L}$ is lower triangular, and $\mathbf{U}$ is upper triangular. Then $(\mathbf{I} + \mathbf{L} + \mathbf{U})\mathbf{x} = \mathbf{b}$, hence $\mathbf{x} = \mathbf{b} - \mathbf{Lx} - \mathbf{Ux}$. This gives the iteration $\mathbf{x}^{(m+1)} = \mathbf{b} - \mathbf{L}\mathbf{x}^{(m+1)} - \mathbf{U}\mathbf{x}^{(m)}$ because you always use the latest available approximation, which is the $(m + 1)$st below the main diagonal and the mth above it. In terms of components,

$$
x_j^{(m+1)} = b_j - \sum_{k=1}^{j-1} a_{jk}\,x_k^{(m+1)} - \sum_{K=j+1}^{n} a_{jK}\,x_K^{(m)}.
$$

Hence type $\mathbf{A}, \mathbf{b}, \mathbf{x}_0$ and then the program (which includes divisions to make the main diagonal entries 1). n is the number of unknowns and M is the number of steps. Note that you need no superscripts $(m + 1)$ and (m) because if a component has been computed, it is used at once, so that you can overwrite its previous value, which is no longer needed. In the program, `m` counts the steps, and `j` numbers to components of the approximate solution vectors.

```
In[1]:= A = {{1,      -0.25, -0.25,   0},
             {-0.25 ,  1,      0,     -0.25},
             {-0.25,   0,      1,     -0.25},
             { 0,     -0.25, -0.25,   1}}; MatrixForm[A]
```

Out[1]//MatrixForm=

$$\begin{pmatrix} 1 & -0.25 & -0.25 & 0 \\ -0.25 & 1 & 0 & -0.25 \\ -0.25 & 0 & 1 & -0.25 \\ 0 & -0.25 & -0.25 & 1 \end{pmatrix}$$

```
In[2]:= b = {50, 50, 25, 25}                    (* Out: {50, 50, 25, 25} *)

In[3]:= x = {100, 100, 100, 100}                (* Out: {100, 100, 100, 100} *)

In[4]:= n = 4;    M = 7;

In[5]:= Do [
          Do [
            S = Sum[A[[j,k]] x[[k]], {k, 1, j - 1}] +
                Sum[A[[j,K]] x[[K]],  {K, j + 1, n}];
            x[[j]] = -N[(S - b[[j]])/A[[j,j]]],
                {j, 1, n}
          ]
          Print[x], {m, 0, M - 1}
        ]
```

{100., 100., 75., 68.75}

{93.75, 90.625, 65.625, 64.0625}

{89.0625, 88.2813, 63.2813, 62.8906}

{87.8906, 87.6953, 62.6953, 62.5977}

{87.5977, 87.5488, 62.5488, 62.5244}

{87.5244, 87.5122, 62.5122, 62.5061}

{87.5061, 87.5031, 62.5031, 62.5015}

The exact solution is

```
In[6]:= LinearSolve[A, b]                       (* Out: {87.5, 87.5, 62.5, 62.5} *)
```

Similar Material in AEM: pp. 900-903

EXAMPLE 18.6	**VECTOR AND MATRIX NORMS.**
	CONDITION NUMBERS

For information, click on "Help" on the menu bar. A table of 11 lines will appear. Click on Line 3 ("Master Index"). Type "VectorNorm". You will see that you must load the package `LinearAlgebra`MatrixManipulation`. Then you can obtain vector norms, as follows.

In[1]:= <<LinearAlgebra`MatrixManipulation`

Vector norms. There are essentially three practically important vector norms, defined as follows.

In[2]:= x = {x1, x2, x3}   (* Out: {x1, x2, x3} *)

In[3]:= VectorNorm[x, 1]   (* The l_1-norm *)

Out[3]= VectorNorm[{x1, x2, x3}, 1]

Hence you do not get the general formula $|x_1| + |x_2| + |x_3|$. Try a vector whose components are numbers.

In[4]:= y = {1, 4, 7}   (* Out: {1, 4, 7} *)

In[5]:= VectorNorm[y, 1]

VectorNorm::prec : VectorNorm has received a vector with infinite precision.

Out[5]= VectorNorm[{1, 4, 7}, 1]

Here you got a message. It does not seem to like integers. Pretend they are decimal fractions by typing 1 as 1.0. Then it will work.

In[6]:= y = {1.0, 4, 7}   (* Out: {1., 4, 7} *)

In[7]:= VectorNorm[y, 1]   (* Out: 12. The l_1-norm $1 + 4 + 7 = 12$ *)

In[8]:= VectorNorm[y, 2] (* Out: 8.12404 The l_2-norm $\sqrt{66}$; **Euclidean norm** *)

In[9]:= VectorNorm[y]   (* Out: 7. The l_∞-norm *)

You can also obtain any l_p-norm, although these norms will hardly be used in numerical work. For instance, the l_4-norm is the fourth root of the sum of the fourth powers of the components of **y**,

In[10]:= VectorNorm[y, 4]   (* Out: 7.18024 *)

Confirm this by

In[11]:= N[(1^4 + 4^4 + 7^4)^(1/4)]   (* Out: 7.18024 *)

Matrix norms. Some matrix norms result from vector norms. The l_1-vector norm leads to the column "sum" matrix norm. The l_∞-vector norm leads to the row "sum" matrix norm. The definitions are as follows. The column "sum" (row "sum") matrix norm is the largest sum of the n sums of the absolute values of the entries in a column (in a row), respectively.

In[12]:= A = {{a11, a12}, {a21, a22}}; MatrixForm[A]   (* Out: $\begin{pmatrix} a11 & a12 \\ a21 & a22 \end{pmatrix}$ *)

In[13]:= MatrixNorm[A, 1]   (* Out: MatrixNorm[{{a11, a12}, {a21, a22}}, 1] *)

What happened for vectors now happens for matrices. Try integers as entries.

In[14]:= A = {{2, 0, -1}, {4, 3, -2}, {0, 5, 1}}; MatrixForm[A]

Out[14]//MatrixForm=

$$\begin{pmatrix} 2 & 0 & -1 \\ 4 & 3 & -2 \\ 0 & 5 & 1 \end{pmatrix}$$

In[15]:= MatrixNorm[A, 1]

    MatrixNorm::prec : MatrixNorm has received a matrix with infinite precision.

Out[15]= MatrixNorm[{{2, 0, −1}, {4, 3, −2}, {0, 5, 1}}, 1]

Disguise one entry as a decimal fraction.

In[16]:= A = {{2.0, 0, -1}, {4, 3, -2}, {0, 5, 1}}; MatrixForm[A]

Out[16]//MatrixForm=

$$\begin{pmatrix} 2. & 0 & -1 \\ 4 & 3 & -2 \\ 0 & 5 & 1 \end{pmatrix}$$

In[17]:= Norm1 = MatrixNorm[A, 1]   (* Out: 8. **Column "sum" norm** *)

In[18]:= NormInf = RowNorm = MatrixNorm[A]   (* Out: 9. **Row "sum" norm** *)

    The **Frobenius norm** of **A** is the square root of the sum of the squares of all the entries of **A**; thus,

In[19]:= NormFrob = Sqrt[Sum[Sum[A[[j,k]]^2, {j, 1, 3}], {k, 1, 3}]]

Out[19]= 7.74597

    Condition numbers. The definition of the condition number of a matrix **A** is $\kappa(\mathbf{A}) = ||\mathbf{A}|| \, ||\mathbf{A}^{-1}||$ and depends on the choice of a norm, but if $\kappa(\mathbf{A})$ is large in one norm, it tends to be large in others, indicating that **A** is **ill-conditioned**. This is illustrated by the following values for which you first have to calculate the inverse of A, call it B, and its norms.

In[20]:= B = N[Inverse[A]]

Out[20]= {{2.16667, −0.833333, 0.5}, {−0.666667, 0.333333, −7.39968 × 10^{-18}},

       {3.33333, −1.66667, 1.}}

    then

In[21]:= NormB1 = MatrixNorm[B, 1]   (* Out: 6.16667 *)

In[22]:= NormBInf = MatrixNorm[B]   (* Out: 6. *)

In[23]:= NormBFrob = Sqrt[Sum[Sum[B[[j, k]]^2, {j, 1, 3}], {k, 1, 3}]]

Out[23]= 4.59166

In[24]:= kappa1 = Norm1   NormB1   (* Out: 49.3333 *)

In[25]:= kappa2 = NormInf   NormBInf   (* Out: 54. *)

In[26]:= kappa3 = NormFrob   NormBFrob   (* Out: 35.5668 *)

    You can obtain two of these condition numbers by typing

In[27]:= MatrixConditionNumber[A, 1]   (* Out: 49.3333 *)

In[28]:= MatrixConditionNumber[A]   (* Out: 54. *)

    Similar Material in AEM: pp. 909-912

EXAMPLE 18.7   FITTING DATA BY LEAST SQUARES

Fit a straight line and a quadratic polynomial through the four points $(-1.3, 0.103)$, $(-0.1, 1.099)$, $(0.2, 0.808)$, $(1.3, 1.897)$ by the **principle of least squares**. Also find the interpolation polynomial for these data. Plot the results on common axes. For information click on "Help", then on "Master Index", and then type "Least squares fitting". You will see that there are several packages that you can use.

Solution. Load the package for polynomial fit by typing

In[1]:= <<NumericalMath`PolynomialFit`

Then type your data and give the command for fitting a straight line, in which 1 is the degree of the polynomial.

In[2]:= data = {{-1.3, 0.103}, {-0.1, 1.099}, {0.2, 0.808}, {1.3, 1.897}}
Out[2]= {{-1.3, 0.103}, {-0.1, 1.099}, {0.2, 0.808}, {1.3, 1.897}}

In[3]:= f1 = PolynomialFit[data, 1]
Out[3]= FittingPolynomial[<>, 1]

To obtain the value at a given point, say, 0.1, type

In[4]:= f1[0.1]   (* Out: 1.02678   Value at a given point, 0.1. *)

An explicit expression of **f1** is obtained by typing

In[5]:= Clear[x]
In[6]:= p1 = Expand[f1[x]]   (* Out: 0.960074 + 0.667024 x *)

A quadratic polynomial is obtained by typing

In[7]:= f2 = PolynomialFit[data, 2]   (* Out: FittingPolynomial[<>, 2] *)
In[8]:= p2 = Expand[f2[x]]   (* Out: $0.921183 + 0.668065 x + 0.0453236 x^2$ *)

If you request a cubic polynomial, since you have four points (pairs of values), you get the unique interpolation polynomial passing precisely through these points.

In[9]:= f3 = PolynomialFit[data,3]
Out[9]= FittingPolynomial[<>, 3]
In[10]:= p3 = Expand[f3[x]]
Out[10]= $1. - 1. x - 1.97758 \times 10^{-16} x^2 + 1. x^3$

Plot the three polynomials and the points by

In[11]:= P1 = Plot[{p1, p2, p3}, {x, -2, 2}]
In[12]:= P2 = ListPlot[data]
In[13]:= Show[{P1, P2}, Prolog -> AbsolutePointSize[6],
       AspectRatio -> Automatic, PlotRange -> {-0.5, 2.5}]

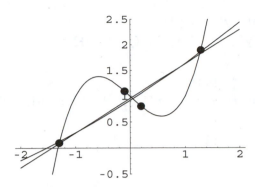

Example 18.7. Least squares straight line and quadratic polynomial.
Cubic interpolation polynomial

Similar Material in AEM: pp. 914-916

EXAMPLE 18.8	**APPROXIMATION OF EIGENVALUES:**
	COLLATZ METHOD

Collatz proved the following. Let $\mathbf{A} = [a_{jk}]$ be a real $n \times n$ matrix with all entries a_{jk} positive. Choose any $\mathbf{x} = [x_j]$ with all n components positive. Compute $\mathbf{y} = [y_j] = \mathbf{A}\mathbf{x}$ and the n quotients $p_j = y_j/x_j$. Then the smallest and largest of these quotients are the endpoints of a closed interval J which contains an eigenvalue λ of $\mathbf{A}$. Hence the midpoint of J is an approximation of λ, and half the length of J is a corresponding error bound.

Write a program for this method and, starting from $[1\ 1\ 1]^T$, apply it to the matrix (20 steps)

$$\mathbf{A} = \begin{bmatrix} 0.49 & 0.02 & 0.22 \\ 0.02 & 0.28 & 0.20 \\ 0.22 & 0.20 & 0.40 \end{bmatrix}.$$

Solution. Calculate $\mathbf{y} = \mathbf{A}\mathbf{x}$ and those quotients. Then sort the latter in ascending order, so that p_1 and p_n are the endpoints of that interval. Note that both quotients are positive because all entries of $\mathbf{A}$ and all components of $\mathbf{x}$ are positive. $n = \text{Length[A]}$ in one of the commands gives the number of rows of $\mathbf{A}$, and $x = y$ copies y to make it the vector $\mathbf{x}$ in the next step. (The response shows Steps 1, 2, 10, and 20.)

In[1]:= A = {{0.49, 0.02, 0.22}, {0.02, 0.28, 0.20},
        {0.22, 0.20, 0.40}}; MatrixForm[A]

Out[1]//MatrixForm=

$$\begin{pmatrix} 0.49 & 0.02 & 0.22 \\ 0.02 & 0.28 & 0.2 \\ 0.22 & 0.2 & 0.4 \end{pmatrix}$$

In[2]:= x = {1, 1, 1};   n = Length[A]; N0 = 20;   (* N is protected. *)

In[3]:= p = Table[0, {i, 1, n}]   (* Zero vector with n components *)

```
In[4]:= Do  [
           y = A.x;   Print["y = ", y];
           Do[p[[j]] = N[y[[j]]/x[[j]]], {j, 1, n}];
           p = Sort[p];
               Print["p = ", p];
               Print["Interval = {", p[[1]], ", ", p[[n]], "}"];
           x = y;
               Print["x = ", x],
           {m, 1, NO}
           ]
```

$y = \{0.73, 0.5, 0.82\}$
$p = \{0.5, 0.73, 0.82\}$
Interval $= 0.5, 0.82\}$
$x = \{0.73, 0.5, 0.82\}$
$y = \{0.5481, 0.3186, 0.5886\}$
$p = \{0.6372, 0.717805, 0.750822\}$
Interval $= 0.6372, 0.750822\}$
$\vdots$
$x = \{0.0577538, 0.0289107, 0.0577876\}$
$y = \{0.0415908, 0.0208076, 0.041603\}$
$p = \{0.719719, 0.71993, 0.720141\}$
Interval $= 0.719719, 0.720141\}$
$\vdots$
$x = \{0.00216309, 0.00108155, 0.00216309\}$
$y = \{0.00155743, 0.000778713, 0.00155743\}$
$p = \{0.72, 0.72, 0.72\}$
Interval $= 0.72, 0.72\}$
$x = \{0.00155743, 0.000778713, 0.00155743\}$

Similar Material in AEM: pp. 923, 924

EXAMPLE 18.9   **APPROXIMATION OF EIGENVALUES:**
                POWER METHOD

Given a real *symmetric* square matrix $\mathbf{A}$, compute $\mathbf{y} = \mathbf{A}\mathbf{x}$, where $\mathbf{x}$ is any nonzero vector. Compute the dot products $m_0 = \mathbf{x} \bullet \mathbf{x}$, $m_1 = \mathbf{x} \bullet \mathbf{y}$, $m_2 = \mathbf{y} \bullet \mathbf{y}$. Then the **Rayleigh quotient** $q = m_1/m_0$ is an approximation of an eigenvalue of $\mathbf{A}$. A corresponding error bound is $\delta = \sqrt{\frac{m_2}{m_0} - q^2}$.

This was the *first step* of the **power method**. In the *second step*, use $\mathbf{y}$ divided by a component that is largest in absolute value as your new $\mathbf{x}$ and proceed as before. (This **"scaling"** will make one component of the new $\mathbf{x}$ equal to 1 and the others smaller than or equal to 1 in absolute value. It has the effect that the sequence of vectors obtained will converge to an eigenvector of that λ approximated by the Rayleigh quotients.) In the further steps you proceed in the same fashion.

Apply this power method to the matrix in the previous example, starting from $\mathbf{x} = [1\ 1\ 1]^T$.

Solution. The matrix **A** is symmetric. Hence the error bounds apply. `Sort[y]` arranges the components of `y` in ascending order. `Sign[z[[1]] + z[[3]]]` equals -1 if the absolutely largest component of `y` is negative, and equals $+1$ if that component is positive. Hence in both cases the next `x` will have a component $+1$. For reasons of space we show only the responses to Steps 1, 2, and 10.

    (sign(0) $= +1$ (or -1) in most computer algebraic systems. `Sign[0] = 0` in Mathematica, so that the scaling step does not work in the (unlikely) case that `Sort[y][[1]] = - Sort[y][[3]]`).

```
In[1]:= A = {{0.49, 0.02, 0.22},
             {0.02, 0.28, 0.2},
             {0.22, 0.2, 0.4}};
In[2]:= x = {1, 1, 1};   NO = 10;                    (* N is protected. *)
In[3]:= Do [
             y = A.x;
                 Print["y = ", y];
             q = N[x.y/x.x];
                 Print["q = ", q];
             delta = N[Sqrt[y.y/x.x - q^2]];
                 Print["delta = ", delta];
             z = Sort[y];
             x = N[y Sign[z[[1]] + z[[3]]]/Max[Abs[y[[1]]], Abs[y[[3]]]]];
                 Print["x = ", x],
             {j, 1, NO}
             ]
y = {0.73, 0.5, 0.82}
q = 0.683333
delta = 0.134743
x = {0.890244, 0.609756, 1.}
y = {0.668415, 0.388537, 0.717805}
q = 0.716048
delta = 0.0388872
  ⋮
x = {0.999414, 0.500293, 1.}
y = {0.719719, 0.36007, 0.71993}
q = 0.72
delta = 0.000140625
x = {0.999707, 0.500146, 1.}
```

    Similar Material in AEM: pp. 925-928

EXAMPLE 18.10	**APPROXIMATION OF EIGENVALUES:**
	QR-FACTORIZATION

This is another iteration method. We explain it for a real symmetric tridiagonal matrix $\mathbf{A} = \mathbf{A}_0$. In the ***first step,*** factor $\mathbf{A}_0 = \mathbf{Q}_0\mathbf{R}_0$, where $\mathbf{Q}_0$ is orthogonal and $\mathbf{R}_0$ is upper triangular. Then compute $\mathbf{A}_1 = \mathbf{R}_0 \mathbf{Q}_0$. In the **second step**, factor $\mathbf{A}_1 = \mathbf{Q}_1 \mathbf{R}_1$. Then compute $\mathbf{A}_2 = \mathbf{R}_1 \mathbf{Q}_1$. And so on. These are similarity transformations, so that the spectrum of $\mathbf{A}$ is preserved. If the eigenvalues of $\mathbf{A}$ are different in absolute value, the process converges to a diagonal matrix whose diagonal entries are the eigenvalues of $\mathbf{A}$. (For a proof, see Wilkinson [4] in Appendix 1. For the transformation of a symmetric matrix to tridiagonal form, see AEM, pp. 929-933.)

Obtain approximations of the eigenvalues of the following matrix by the method of QR-factorization.

$$\mathbf{A} = \begin{bmatrix} 6 & -\sqrt{18} & 0 \\ -\sqrt{18} & 7 & \sqrt{2} \\ 0 & \sqrt{2} & 6 \end{bmatrix}$$

Solution. The response shown corresponds to Steps 1, 2, and 5 of the iteration.

```
In[1]:= N0 = 5;
In[2]:= A = N[{{6, -Sqrt[18], 0}, {-Sqrt[18], 7, Sqrt[2]}, {0, Sqrt[2], 6}}];
```

```
In[3]:= MatrixForm[A]
```
$$(* \; \mathtt{Out:} \; \begin{pmatrix} 6. & -4.24264 & 0. \\ -4.24264 & 7. & 1.41421 \\ 0. & 1.41421 & 6. \end{pmatrix} \; *)$$

```
In[4]:= C2[t_] := {{ Cos[t],  Sin[t], 0},
                   {-Sin[t],  Cos[t], 0},
                   {  0,        0,     1}};

In[5]:= C3[t_] := {{  1,        0,         0},
                   {  0,      Cos[t],    Sin[t]},
                   {  0,     -Sin[t],    Cos[t]}};

In[6]:= Do [
           V2 = C2[t].A;
           t2 = FindRoot[V2[[2,1]] == 0, {t, 0}][[1, 2]];
               Print["t2 =", t2];
           V2  = C2[t2].A;
           R = C3[t].V2;
           t3 = FindRoot[R[[3,2]] == 0, {t, 1}][[1, 2]];
               Print["t3 =", t3];
           R = C3[t3].V2;
               Print["R = ", MatrixForm[R]];
           Q = Transpose[C2[t2]].Transpose[C3[t3]];
               Print["Q = ", MatrixForm[Q]];
           A = R.Q;
               Print["A = ", MatrixForm[A]],
           {m, 1, N0}
        ]
```

t2 $= -0.61548$

t3 $= 0.408638$

$$R = \begin{pmatrix} 7.34847 & -7.50555 & -0.816497 \\ -3.82814 \times 10^{-11} & 3.55903 & 3.44378 \\ -1.65763 \times 10^{-11} & 6.05665 \times 10^{-12} & 5.04715 \end{pmatrix}$$

$$Q = \begin{pmatrix} 0.816497 & 0.529813 & -0.229416 \\ -0.57735 & 0.749269 & -0.324443 \\ 0. & 0.39736 & 0.917663 \end{pmatrix}$$

$$A = \begin{pmatrix} 10.3333 & -2.0548 & -1.70314 \times 10^{-11} \\ -2.0548 & 4.03509 & 2.00553 \\ -1.70313 \times 10^{-11} & 2.00553 & 4.63158 \end{pmatrix}$$

t2 $= -0.196292$

t3 $= 0.513416$

$$R = \begin{pmatrix} 10.5357 & -2.80232 & -0.391146 \\ -5.13638 \times 10^{-8} & 4.0833 & 3.98824 \\ 2.8942 \times 10^{-8} & 2.23918 \times 10^{-14} & 3.06833 \end{pmatrix}$$

$$Q = \begin{pmatrix} 0.980797 & 0.169888 & -0.0957917 \\ -0.195033 & 0.854344 & -0.481723 \\ 0. & 0.491155 & 0.871072 \end{pmatrix}$$

$$A = \begin{pmatrix} 10.8799 & -0.796379 & 2.83862 \times 10^{-8} \\ -0.796379 & 5.44739 & 1.50703 \\ 2.83862 \times 10^{-8} & 1.50703 & 2.67273 \end{pmatrix}$$

t2 $= -0.0730671$

t3 $= 0.273372$

$$R = \begin{pmatrix} 10.909 & -1.19193 & -0.110016 \\ 7.65595 \times 10^{-9} & 5.582 & 2.16877 \\ 2.73343 \times 10^{-8} & 5.53867 \times 10^{-9} & 2.1677 \end{pmatrix}$$

$$Q = \begin{pmatrix} 0.997332 & 0.0702913 & -0.0197091 \\ -0.0730021 & 0.960297 & -0.269259 \\ 0. & 0.26998 & 0.962866 \end{pmatrix}$$

$$A = \begin{pmatrix} 10.9669 & -0.407498 & 2.6857 \times 10^{-8} \\ -0.407498 & 5.9459 & 0.585236 \\ 2.6857 \times 10^{-8} & 0.585236 & 2.08721 \end{pmatrix}$$

t2 $= -0.03714$

t3 $= 0.0984271$

$$R = \begin{pmatrix} 10.9745 & -0.627996 & -0.0217306 \\ 2.63917 \times 10^{-9} & 5.95549 & 0.787108 \\ 2.6727 \times 10^{-8} & 4.54281 \times 10^{-17} & 2.01964 \end{pmatrix}$$

$$Q = \begin{pmatrix} 0.99931 & 0.0369517 & -0.00364884 \\ -0.0371314 & 0.994474 & -0.0982005 \\ 0. & 0.0982683 & 0.99516 \end{pmatrix}$$

$$A = \begin{pmatrix} 10.9902 & -0.221136 & 2.67086 \times 10^{-8} \\ -0.221136 & 5.99993 & 0.198466 \\ 2.67086 \times 10^{-8} & 0.198466 & 2.00986 \end{pmatrix}$$

t2 = −0.0201185

t3 = 0.0330972

$$R = \begin{pmatrix} 10.9924 & -0.341792 & -0.00399254 \\ 8.83819 \times 10^{-10} & 5.99755 & 0.264826 \\ 2.6694 \times 10^{-8} & -2.6131 \times 10^{-14} & 2.00219 \end{pmatrix}$$

$$Q = \begin{pmatrix} 0.999798 & 0.0201061 & -0.000665699 \\ -0.0201171 & 0.99925 & -0.0330845 \\ 0. & 0.0330912 & 0.999452 \end{pmatrix}$$

$$A = \begin{pmatrix} 10.9971 & -0.120653 & 2.66886 \times 10^{-8} \\ -0.120653 & 6.00182 & 0.066255 \\ 2.66886 \times 10^{-8} & 0.066255 & 2.0011 \end{pmatrix}$$

In[7]:= Eigenvalues[A]   (* Out: {11., 6., 2.} *)

You see that the accuracy of the diagonal entries of the last matrix **A** as approximations of the eigenvalues is greater than you would expect from the size of the off-diagonal entries of **A**.

Mathematica gives a QR-factorization of **A** in the form

In[8]:= A = N[{{6, -Sqrt[18], 0}, {-Sqrt[18], 7, Sqrt[2]}, {0, Sqrt[2], 6}}];

In[9]:= MatrixForm[%]   (* Out: $\begin{pmatrix} 6. & -4.24264 & 0. \\ -4.24264 & 7. & 1.41421 \\ 0. & 1.41421 & 6. \end{pmatrix}$ *)

In[10]:= A2 = QRDecomposition[A]; MatrixForm[A2]

Out[10]//MatrixForm=

$$\begin{pmatrix} \begin{pmatrix} -0.816497 \\ 0.57735 \\ 0. \end{pmatrix} & \begin{pmatrix} -0.529813 \\ -0.749269 \\ -0.39736 \end{pmatrix} & \begin{pmatrix} 0.229416 \\ 0.324443 \\ -0.917663 \end{pmatrix} \\ \begin{pmatrix} -7.34847 \\ 7.50555 \\ 0.816497 \end{pmatrix} & \begin{pmatrix} 0. \\ -3.55903 \\ -3.44378 \end{pmatrix} & \begin{pmatrix} 0. \\ 0. \\ -5.04715 \end{pmatrix} \end{pmatrix}$$

Comparing with the previous calculation in this example, you see that the response is $-\mathbf{Q}$ (the first three rows) and $-\mathbf{R}^T$ (the last three rows). This is the first step of the iteration.

You can obtain $-\mathbf{Q}^T$ by the command

In[11]:= `A2[[1]];`

In[12]:= `MatrixForm[%]`

Out[12]//MatrixForm=

$$\begin{pmatrix} -0.816497 & 0.57735 & 0. \\ -0.529813 & -0.749269 & -0.39736 \\ 0.229416 & 0.324443 & -0.917663 \end{pmatrix}$$

Similarly, $-\mathbf{R}$ is obtained by the command

In[13]:= `A2[[2]];`

In[14]:= `MatrixForm[%]`

Out[14]//MatrixForm=

$$\begin{pmatrix} -7.34847 & 7.50555 & 0.816497 \\ 0. & -3.55903 & -3.44378 \\ 0. & 0. & -5.04715 \end{pmatrix}$$

To check that this QR-factorization is correct, type

In[15]:= `Transpose[A2[[1]]].A2[[2]]`

Out[15]= $\{\{6., -4.24264, 5.44269 \times 10^{-16}\}, \{-4.24264, 7., 1.41421\}, \{0., 1.41421, 6.\}\}$

This is the given matrix represented as $(-\mathbf{Q})(-\mathbf{R}) = \mathbf{QR}$.

Interchanging the two factors (and dropping the minuses), you obtain the new matrix $\mathbf{RQ}$ after the first step, call it `M`.

In[16]:= `M = A2[[2]].Transpose[A2[[1]]];`

In[17]:= `MatrixForm[%]`

Out[17]//MatrixForm=

$$\begin{pmatrix} 10.3333 & -2.0548 & 4.52329 \times 10^{-16} \\ -2.0548 & 4.03509 & 2.00553 \\ 0. & 2.00553 & 4.63158 \end{pmatrix}$$

The previous calculation can now be done by the following program.

In[18]:= `Clear[B]`

In[19]:= `NO = 5;    B[1] = A;`

In[20]:= `Do [`
```
    B[j + 1] =
    QRDecomposition[B[j]][[2]].Transpose[QRDecomposition[B[j]][[1]]],
    {j, 1, NO}
    ]
```

In[21]:= `Table[B[j]; MatrixForm[B[j]], {j, 1, 5}]`

$$\text{Out[21]} = \left\{ \begin{pmatrix} 6. & -4.24264 & 0. \\ -4.24264 & 7. & 1.41421 \\ 0. & 1.41421 & 6. \end{pmatrix}, \begin{pmatrix} 10.3333 & -2.0548 & -3.33067 \times 10^{-16} \\ -2.0548 & 4.03509 & 2.005533 \\ 0. & 2.00553 & 4.63158 \end{pmatrix}, \right.$$

$$\begin{pmatrix} 10.8799 & -0.796379 & 0. \\ -0.796379 & 5.44739 & 1.507023 \\ 0. & 1.50703 & 2.67273 \end{pmatrix}, \begin{pmatrix} 10.9669 & -0.407498 & 1.52656 \times 10^{-16} \\ -0.407498 & 5.9459 & 0.585236 \\ 0. & 0.585236 & 2.08721 \end{pmatrix},$$

$$\left. \begin{pmatrix} 10.9902 & -0.2211356 & 1.9082 \times 10^{-16} \\ -0.221136 & 5.99993 & 0.198466 \\ 0. & 0.198466 & 2.00986 \end{pmatrix} \right\}$$

(The form of the small numbers in the upper right-hand corner may differ.)

Similar Material in AEM: pp. 933-937

Problem Set for Chapter 18

Pr.18.1 (Gauss elimination) Solve the following linear system by each of the three methods in Example 18.1 in this Guide. (*AEM Ref.* p. 893 (#7))

$$\begin{aligned} 6\,x_2 + 13\,x_3 &= 61 \\ 6\,x_1 \qquad\quad - 8\,x_3 &= -38 \\ 13\,x_1 - 8\,x_2 \qquad\quad &= 79 \end{aligned}$$

Pr.18.2 (Gauss elimination) Solve the following linear system. (*AEM Ref.* p. 893 (#10))

$$\begin{aligned} 4\,x_1 + 4\,x_2 + 2\,x_3 &= 0 \\ 3\,x_1 - \quad x_2 + 2\,x_3 &= 0 \\ 3\,x_1 + 7\,x_2 + \quad x_3 &= 0 \end{aligned}$$

Pr.18.3 (Doolittle factorization) Solve the following linear system by Doolittle's method. (*AEM Ref.* p. 899 (#5))

$$\begin{aligned} 5\,x_1 + \quad 4\,x_2 + \quad x_3 &= 3.4 \\ 10\,x_1 + \quad 9\,x_2 + \quad 4\,x_3 &= 8.8 \\ 10\,x_1 + 13\,x_2 + 15\,x_3 &= 19.2 \end{aligned}$$

Pr.18.4 (Doolittle factorization) Solve Pr.18.1 by Doolittle's method. (*AEM Ref.* pp. 894-896)

Pr.18.5 (Cholesky factorization) Solve the following linear system by Cholesky's method. (*AEM Ref.* p. 899 (#7))

$$\begin{aligned} 9\,x_1 + \quad 6\,x_2 + 12\,x_3 &= 17.4 \\ 6\,x_1 + 13\,x_2 + 11\,x_3 &= 23.6 \\ 12\,x_1 + 11\,x_2 + 26\,x_3 &= 30.8 \end{aligned}$$

Pr.18.6 (Gauss-Jordan elimination) Solve Pr.18.5 by Gauss-Jordan elimination. (*AEM Ref.* p. 898)

Pr.18.7 (Inverse) Find the inverse of the coefficient matrix **A** in Pr.18.5 by the Gauss-Jordan method. (*AEM Ref.* p. 898)

Pr.18.8 **(Gauss-Seidel iteration)** Solve the following linear system by Gauss-Seidel iteration, starting from $[1\ 1\ 1]^T$. Do 5 steps. (*AEM Ref.* p. 905 (#1))

$$\begin{aligned} 5x_1 +\ x_2 + 2x_3 &= 19 \\ x_1 + 4x_2 - 2x_3 &= -2 \\ 2x_1 + 3x_2 + 8x_3 &= 39 \end{aligned}$$

Pr.18.9 **(Vector norms)** Find the l_1, l_2, l_∞ norms of the vectors $\mathbf{x} = [7\ \ -12\ \ 5\ \ 0]^T$ and $y = 10x$ on the computer. (*AEM Ref.* p. 912 (#1))

Pr.18.10 **(Matrix norms)** Find the row "sum", column "sum", and Frobenius norms of the coefficient matrix in Pr.18.8 and of its square. (*AEM Ref.* pp. 903, 909).

Pr.18.11 **(Condition number)** Find the condition numbers κ of the coefficient matrix in Pr.18.5 for the three matrix norms defined in Example 18.6 in this Guide, by the command `MatrixConditionNumber`. Check the results by using the definition of κ. (*AEM Ref.* pp. 910-912)

Pr.18.12 **(Experiment on Hilbert matrix)** The $n \times n$ Hilbert matrix is $\mathbf{H} = [h_{jk}]$, where $h_{jk} = 1/(j+k-1)$. Find the condition numbers for the three matrix norms defined in Example 18.6 in this Guide when $n = 3$. Conclude that $\mathbf{H}$ is ill-conditioned. Go on to $n = 4$ and larger n. Study the decrease of the sequence of the values of the determinants of these matrices. (*AEM Ref.* p. 913 (#19))

Pr.18.13 **(Least squares, straight line)** Fit a straight line to the data $(x, y) = (0, 2)$, $(1, 2)$, $(2, 1)$, $(3, 3)$, $(5, 2)$, $(7, 3)$, $(9, 3)$, $(10, 4)$, $(11, 3)$ by least squares. Plot the points and the straight line. (See Example 18.7 in this Guide.)

Pr.18.14 **(Least squares, straight line)** Fit a straight line through the points $(x,y) = (400, 580)$, $(500, 1030)$, $(600, 1420)$, $(700, 1880)$, $(750, 2100)$ by least squares. (*AEM Ref.* p. 916 (#6))

Pr.18.15 **(Least squares, quadratic parabola)** Fit a quadratic parabola to the points $(x, y) = (2, 0)$, $(3, 3)$, $(5, 4)$, $(6, 3)$, $(7, 1)$ by least squares. Plot the points and the parabola. (*AEM Ref.* p. 916 (#11))

Pr.18.16 **(Least squares, cubic parabola)** Fit a cubic parabola to the points $(x, y) = (-2, -8)$, $(-1, 0)$, $(0, 1)$, $(1, 2)$, $(2, 12)$, $(4, 80)$ by least squares. Plot the points and the parabola. (*AEM Ref.* p. 917 (#17))

Pr.18.17 **(Eigenvalues, eigenvectors)** Find the eigenvalues and eigenvectors of the matrix in Example 18.8 in this Guide by the commands `Eigenvalues`, `Eigenvectors`, `Eigensystem`. (*AEM Ref.* pp. 917-919)

Pr.18.18 **(Collatz method)** Find an approximation of an eigenvalue of the following matrix by Collatz's method, starting from $[1\ 1\ 1]^T$ and doing 10 steps. (See Example 18.8 in this Guide. *AEM Ref.* pp. 923, 924)

$$\mathbf{A} = \begin{bmatrix} 3 & 2 & 3 \\ 2 & 6 & 6 \\ 3 & 6 & 3 \end{bmatrix}$$

Pr.18.19 **(Power method for eigenvalues)** Find an approximation and error bounds for an eigenvalue of the matrix in Pr.18.18 by the power method with scaling, starting from $[1\ 1\ 1]^T$ and doing 10 steps. (*AEM Ref.* pp. 925-927)

Pr.18.20 (QR-factorization) Using the QR-method (5 steps), find approximations of the eigenvalues of the following matrix. (*AEM Ref.* p. 938 (#8))

$$A = \begin{bmatrix} 14.2 & -0.1 & 0 \\ -0.1 & -6.3 & 0.2 \\ 0 & 0.2 & 2.1 \end{bmatrix}$$

Chapter 19

Numerical Methods for Differential Equations

Content. Euler methods (Exs. 19.1, 19.2, Prs. 19.1-19.5)
Runge-Kutta methods (Exs. 19.3, 19.5, 19.6, Prs. 19.6, 19.7, 19.10-19.12)
Multistep methods (Ex. 19.4, Prs. 19.8, 19.9)
Laplace equation (Ex. 19.7, Pr. 19.13)
Heat equation (Ex. 19.8, Prs. 19.14, 19.15)

The commands in this chapter are those used in Chaps. 1 and 2 and (for systems of ODE's) in Chaps. 3 and 7.

Examples for Chapter 19

EXAMPLE 19.1 | EULER METHOD

The Euler method solves **initial value problems** $y' = f(x, y)$, $y(x_0) = y_0$ numerically, according to the formula

(A) $$y_{n+1} = y_n + h f(x_n, y_n) \qquad (n = 0, 1, 2, ...).$$

It is a **step-by-step method** which calculates approximate values of the solution at $x = x_0$, $x_1 = x_0 + h$, $x_2 = x_1 + h = x_0 + 2h$, etc. You choose the **stepsize** h, e.g., $h = 0.1$. Geometrically, the idea of the method is the approximation of the unknown solution curve by a polygon whose first side is tangent to the curve at x_0. The method is too inaccurate in practice, but is a nice illustration of the idea of step-by-step methods.

For instance, solve $y' + 5x^4 y^2 = 0$, $y(0) = 1$. Choose $h = 0.2$. Do $M = 10$ steps. Plot the exact solution curve, obtained by separating variables or by DSolve, and the approximate values obtained.

Solution. Type f in the differential equation $y' = f(x, y) = -5x^4 y^2$ as shown below. Type h, M, and the initial condition $y(0) = 1$ in the form x[0] = 0; y[0] = 1. Then calculate the approximate y-values by a do-loop involving (A). Then use Table to make up a table showing $x(n)$, $y(n)$, and the error. The latter is $1/(1 + x(n)^5) - y(n)$, with the exact solution $y = 1/(1 + x^5)$ obtained by separating variables or by DSolve.

```
In[1]:= Clear[f, x, y]
In[2]:= f[x_, y_] = -5 x^4 y^2                    (* Out: -5 x^4 y^2 *)
In[3]:= h = 0.2;   M = 10;   x[0] = 0; y[0] = 1;   (* N is protected! *)
In[4]:= Do [y[n+1] = N[y[n] + h f[x[n], y[n]]];
            x[n+1] = N[x[n] + h], {n, 0, M}]
In[5]:= Table[{x[n], y[n], 1/(1 + x[n]^5) - y[n]}, {n, 0, M}]
```

Out[5]= {{0, 1, 0}, {0.2, 1., −0.000319898}, {0.4, 0.9984, −0.00853621},

{0.6, 0.972882, −0.0450315}, {0.8, 0.850216, −0.097022},

{1., 0.554129, −0.0541294}, {1.2, 0.24707, 0.0396009},

{1.4, 0.12049, 0.036293}, {1.6, 0.0647183, 0.0223461},

{1.8, 0.0372688, 0.0129934}, {2., 0.022688, 0.00761501}}

Now plot the approximate values as points as well as the (exact) solution curve on common axes.

In[6]:= `P1 = ListPlot[Table[{x[n], y[n]}, {n, 0, M}],`

`     Prolog -> AbsolutePointSize[4]]`

In[7]:= `P2 = Plot[1/(1 + x^5), {x, 0, 2}]`

In[8]:= `Show[P1, P2]`

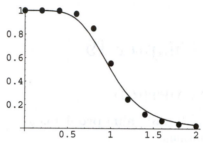

Example 19.1. Solution $y = 1/(1 + x^5)$ and approximation by Euler's method

Similar Material in AEM: pp. 942-944, 951 (#3)

EXAMPLE 19.2	**IMPROVED EULER METHOD**

In the **improved Euler method** (or **Euler-Cauchy method**) you compute first an auxiliary value $y_{n+1}^* = y_n + h\,f(x_n, y_n)$ and then the new value $y_{n+1} = y_n + \frac{1}{2}h[f(x_n, y_n) + f(x_{n+1}, y_{n+1}^*)]$.

For instance, apply this method to the initial value problem $y' = f(x, y) = 1 + y^2$, $y(0) = 0$. Do $M = 10$ steps with $h = 0.1$.

Solution. Type the right-hand side of the differential equation,

In[1]:= `f[x_, y_] = 1 + y^2`   (* Out: $1 + y^2$ *)

Type a do-loop similar to that in the previous example. Solve the initial value problem exactly by separating variables, obtaining $y = \tan x$, so that you can calculate the error to find out about the accuracy of the method. You will see that the values obtained lie almost on the curve of the exact solution. The printout gives the x-values, y-values, and errors.

In[2]:= `h = 0.1;    M = 10;    x[0] = 0;    y[0] = 0;`

In[3]:= `Do [`

`      ystar = y[n] + h f[x[n], y[n]];`   (* Auxiliary value *)

```
y[n+1] = y[n] + h/2 (f[x[n], y[n]] + f[x[n]+ h, ystar]);
   x[n+1] = x[n] + h,
   {n, 0, M}
   ]
```

In[4]:= `Table[{x[n], y[n], Tan [x[n]] - y[n]}, {n, 0, M}]`

Out[4]= {{0, 0, 0}, {0.1, 0.1005, −0.000165328}, {0.2, 0.203035, −0.000325292},

         {0.3, 0.309814, −0.000477536}, {0.4, 0.423408, −0.000615128},

         {0.5, 0.547024, −0.000721811}, {0.6, 0.684899, −0.000762192},

         {0.7, 0.842949, −0.000660151}, {0.8, 1.02989, −0.000248368},

         {0.9, 1.2593, 0.00085887}, {1., 1.55379, 0.00361822}}

For plotting the solution curve and the approximate values as points, type

In[5]:= `P1 = ListPlot[Table[{x[n], y[n]}, {n, 0, M}],`
       `Prolog ->  AbsolutePointSize[4]]`

In[6]:= `P2 = Plot[Tan[x], {x, 0, 1}]`

In[7]:= `Show[P1, P2]`

Example 19.2. Solution $y = \tan x$ and values
obtained by the improved Euler method

Similar Material in AEM: pp. 945, 946

| EXAMPLE 19.3 | CLASSICAL RUNGE-KUTTA METHOD (RK).
MODULE |

The (classical) Runge-Kutta method is an important and very accurate step-by-step method for solving initial value problems $y' = f(x,y)$, $y(x_0) = y_0$. Using this method, solve $y' = 1 + y^2$, $y(0) = 0$, doing $N = 10$ steps of size $h = 0.1$. Determine the error by using the exact solution $y(x) = \tan x$, obtained by separating variables.

Explanation of the method and solution. In each step of the method you first calculate four auxiliary values k1 , k2 , k3 , k4 by the four formulas in Lines 4-7 of the subsequent program. The new y[n+1] is then obtained from y[n] by adding the linear combination of these k 's in Line 8. In Line 9 you set x to the new value x[n+1] . This is a do-loop which begins in Line 3 and ends in Line 10.

Line 18 contains the given initial values, where x[0] and y[0] mean x and y for $n = 0$, as in the previous examples.

Lines 3-10 is the **actual program.** Lines 1, 2 and 11 make it into a module. The opening Line 1 shows a name for it (whose choice is up to you) and the list of quantities which you must specify later.

```
In[1]:= RK[f_, x_, y_, h_, M_] :=                        (* Line 1 *)

         Module[{n},
            Do [
                k1 = h f[x[n], y[n]];                    (* Line 4 *)

                k2 = h f[x[n] + h/2, y[n] + k1/2];       (* Line 5 *)

                k3 = h f[x[n] + h/2, y[n] + k2/2];       (* Line 6 *)

                k4 = h f[x[n] + h, y[n]+ k3];            (* Line 7 *)

                y[n+1] = y[n] + 1/6 (k1 + 2 k2 + 2 k3 + k4);  (* Line 8 *)

                x[n+1] = x[n] + h,                       (* Line 9 *)
                {n, 0, M}
                   ]

              ]                                          (* Line 11 *)
```

Future use of a module. As discussed in Example 17.5, you must **save** this module for future use under some name, say, RK, *before using any further commands*, in particular, before applying the module to your problem. To do this, type

```
In[2]:= DeleteFile["RK.m"]                               (* Line 12 *)
```

```
In[3]:= Clear[f, x, y, h, M]
```

```
In[4]:= Save["RK.m", RK]                                 (* Line 14 *)
```

```
In[5]:= !!RK.m            (* To make sure the file is correct *)
```

where RK is the name of the module (see Line 1) to be saved and RK.m is your choice for the file name under which it is saved. !! shows the contents of the file.

To use the module again (after having filed it as just shown), type << RK.m. This will load it, ready for your new use. (See Example 19.4 in this Guide.)

Now apply the module to your present problem. Type the given data

```
In[6]:= f[x_, y_] = 1 + y^2;                             (* Line 17 *)
```

```
In[7]:= x[0] = 0;      y[0] = 0;      h = 0.1;      M = 10;   (* Line 18 *)
```

Then call the module by typing

```
In[8]:= RK[f, x, y, h, M]                                (* Line 19 *)
```

Now use the command Table to obtain the desired values. And since you know the exact solution (which will *not* be the case in practice), you can obtain the errors of the approximate values computed and see that the Runge-Kutta method is much more accurate than, say, the method in the previous example.

```
In[9]:= Table[{x[n], y[n],  Tan[x[n]] - y[n]}, {n, 0, 10}]    (* Line 20 *)
```

Out[9]= {{0, 0, 0}, {0.1, 0.100335, 8.30073×10^{-8}}, {0.2, 0.20271, 1.57277×10^{-7}},

{0.3, 0.309336, 2.10265×10^{-7}}, {0.4, 0.422793, 2.25884×10^{-7}},

{0.5, 0.546302, 1.8226×10^{-7}}, {0.6, 0.684137, 5.16886×10^{-8}},

{0.7, 0.842289, -1.88932×10^{-7}}, {0.8, 1.02964, -5.04065×10^{-7}},

{0.9, 1.26016, -5.65241×10^{-7}}, {1., 1.55741, 1.28181×10^{-6}}}

What is the **advantage** of a module? Well, if you now want to solve another differential equation for other initial values, all you have to do is to change your data accordingly, instead of retyping the whole program.

Similar Material in AEM: pp. 947-949

EXAMPLE 19.4   **ADAMS-MOULTON MULTISTEP METHOD**

The Adams-Moulton method is a step-by-step method for solving initial value problems $y' = f(x, y)$, $y(x_0) = y_0$. It is obtained as follows. Integrate the differential equation on both sides from x_n to x_{n+1}, obtaining

$$y(x_{n+1}) - y(x_n) = \int_{x_n}^{x_{n+1}} f(x, y(x))\ dx.$$

On the right replace f by an interpolation polynomial p. Then on the left you have an approximation $y_{n+1} - y_n$. For p take the cubic polynomial that at x_n, x_{n-1}, x_{n-2}, x_{n-3} has the values $f(x_n, y_n), ..., f(x_{n-3}, y_{n-3})$, respectively. Then you get the **predictor formula** in Line (A) below. Next take for p the cubic polynomial that at x_{n+1}, x_n, x_{n-1}, x_{n-2} has the values $f(x_{n+1}, y_{n+1}^*)$, $f(x_n, y_n)$, $f(x_{n-1}, y_{n-1})$, $f(x_{n-2}, y_{n-2})$. Then you get the **corrector formula** in Line (B).

The method is called a **multistep method** because in each step it uses the results of several previous steps. Since at the beginning no values are available, the method needs **starting values** at x_0 (there you can take the initial value y_0), x_1, x_2, x_3 which you must obtain by some other method, say, by Runge-Kutta, so that they are as accurate as possible. You can load the RK module saved in the previous example, where it was saved by `Save["RK.m", RK]` . There it was explained that you can load it for further use by the command `<< RK.m` . Hence you can now design a module for Adams-Moulton with built-in Runge-Kutta, as follows.

Begin by loading the Runge-Kutta module `RK` by typing

```
In[1]:= << RK.m

In[2]:= AdamsMoultonRK[f_, x_, y_, h_, M_] := Module[{n, ystar},
        Do [
            k1 = h f[x[n], y[n]];
            k2 = h f[x[n] + h/2, y[n] + k1/2];
            k3 = h f[x[n] + h/2, y[n] + k2/2];
            k4 = h f[x[n] + h, y[n] + k3];
            y[n + 1] = y[n] + 1/6 (k1 + 2 k2 + 2 k3 + k4);
            x[n + 1] = x[n] + h,
            {n, 0, 3}
            ];
```

```
Do [
    x[n + 1] = x[n] + h;
    ystar = y[n] + h/24 (55 f[x[n], y[n]] -
            59 f[x[n - 1], y[n - 1]] +  37 f[x[n - 2], y[n - 2]] -
            9 f[x[n - 3], y[n - 3]]);              (* Line A *)
    y[n + 1] = y[n] + h/24 (9 f[x[n + 1], ystar] +
            19 f[x[n], y[n]] - 5 f[x[n - 1], y[n - 1]] +
            f[x[n - 2], y[n - 2]]),                (* Line B *)
    {n, 3, M}
    ]

    ]
```

You can save the module for later use by typing

In[3]:= DeleteFile["AMRK.m"]

In[4]:= Clear[f, x, y, m, ystar, k1, k2, k3, k4]

In[5]:= Save["AMRK.m", AdamsMoultonRK]

In[6]:= !!AMRK.m

Solve the initial value problem

$$y' = (y - x - 1)^2 + 2, \qquad y(0) = 1$$

by Adams-Moulton, doing 7 steps of size 0.1.

Solution. Type the right-hand side of the differential equation, then the initial values and $h = 0.1$ and $M = 7$, and then call the Adam-Moulton module with built-in Runge Kutta. Finally, use Table to print the values that will be of interest, also showing the accuracy of the method.

In[7]:= f[x_, y_] = (y - x - 1)^2 + 2   (* Out: $2 + (-1 - x + y)^2$ *)

In[8]:= x[0] = 0; y[0] = 1; h = 0.1; M = 7;

In[9]:= AdamsMoultonRK[f, x, y, h, M]

In[10]:= Table[{x[n], y[n], Tan [x[n]] + x[n] + 1 - y[n]}, {n, 0, 7}]

Out[10]= {{0, 1, 0}, {0.1, 1.20033, 8.30073×10^{-8}},

     {0.2, 1.40271, 1.57277×10^{-7}}, {0.3, 1.60934, 2.10265×10^{-7}},

     {0.4, 1.8228, -4.86249×10^{-6}}, {0.5, 2.04631, -0.0000124163},

     {0.6, 2.28416, -0.0000242425}, {0.7, 2.54233, -0.0000435021}}

This table gives x, the approximate solution y, and the error, which you obtain by setting $u = y - x - 1$ and separating variables in the differential equation, or by DSolve, using letters not used above; say, t instead of x and z instead of y,

In[11]:= ode = D[z[t], t] == (z[t] - t - 1)^2 + 2;

In[12]:= DSolve[{ode, z[0] == 1}, z[t], t]

Out[12]= {{z[t] $\rightarrow 1 + t + $ Tan [t]}}

Similar Material in AEM: pp. 952-956

EXAMPLE 19.5	CLASSICAL RUNGE-KUTTA METHOD FOR SYSTEMS (RKS)

RKS solves initial value problems for first-order systems $\mathbf{y}' = \mathbf{f}(x, \mathbf{y})$, $\mathbf{y}(x_0) = \mathbf{y}_0$, where $\mathbf{y} = [y_1, y_2, ..., y_m]$ and $\mathbf{f} = [f_1, f_2, ..., f_m]$. (We write m since n will denote the steps of the iteration.)

```
In[1]:= Clear[f, x, y, h, M]                    (* Use M since N is protected! *)
In[2]:= RKS[f_, x_, y_, h_, M_] := Module[{n, k1, k2, k3, k4},
         Do [
             k1 = h f[x[n], y[n]];
             k2 = h f[x[n] + h/2, y[n] + k1/2];
             k3 = h f[x[n] + h/2, y[n] + k2/2];
             k4 = h f[x[n] + h, y[n] + k3];
             y[n + 1] = y[n] + 1/6 (k1 + 2 k2 + 2 k3 + k4);
             x[n + 1] = x[n] + h,
             {n, 0, M - 1}
             ];
            ]
```

For later use you can save this module by typing

```
In[3]:= DeleteFile["RKS.m"]
In[4]:= Save["RKS.m", RKS]
In[5]:= !!RKS.m                    (* !! shows the content of the file. *)
```

To reload it for further use, you will have to type `<< RKS.m`, then the data, then `RKS[f, x, y, h, M]` and then `Table[...]`.

Using this module, obtain the solution (the **Airy function**) Ai(x) of the **Airy equation** $y'' = xy$. Do 5 steps with $h = 0.2$.

Solution. Setting $y_1 = y$, $y_2 = y_1' = y'$ (the usual conversion of a second-order equation to a system), you obtain $y_1' = y_2$, $y_2' = xy_1$. Hence the vector function $\mathbf{f}$ on the right-hand side of $\mathbf{y}' = \mathbf{f}(x, \mathbf{y})$ is $\mathbf{f} = [y_2, \ xy_1]$.

```
In[6]:= f[x_, y_] := {y[[2]], x y[[1]]}  (* Note the colon : No response. *)
```

Now standard linearly independent solutions of Airy's ODE are Ai(x) and Bi(x) (Click on "Help", then on "Master Index", then type "Airy functions".) See also Ref. [1], pp. 446 and 475, in Appendix 1). Hence you obtain Ai(x) by choosing the initial conditions $y_1(0) = $ Ai(0), $y_2(0) = $ Ai$'(0)$. Thus type, in vector form,

```
In[7]:= x[0] = 0;    y[0] = {0.35502805, -0.25881940};   h = 0.2;   M = 5;
In[8]:= RKS[f, x, y, h, M]
In[9]:= Table[{x[n], y[n]}, {n, 0, M}]
Out[9]= {{0, {0.355028, -0.258819}}, {0.2, {0.303703, -0.252405}},
         {0.4, {0.254742, -0.235831}}, {0.6, {0.2098, -0.212792}},
         {0.8, {0.169846, -0.186412}}, {1., {0.135292, -0.159147}}}
```

This shows x and approximations of y and y'.

Similar Material in AEM: pp. 958, 959

EXAMPLE 19.6 | CLASSICAL RUNGE-KUTTA-NYSTROEM METHOD (RKN)

The RKN method extends the classical Runge-Kutta method to initial value problems for second-order differential equations $y'' = f(x, y, y')$, $y(x_0) = K_0$, $y'(x_0) = K_1$. The program below shows the following. In each step you have to compute four quantities k_1, k_2, k_3, k_4, in turn involving two quantities K and L (introduced merely for simplifying the notations of the k's). From these you calculate the new y-value y[n+1] and − this is the new feature of RKN over RK − a new value yp[n+1] for the derivative needed in the next step, where p suggests 'prime'. Finally, you type Table to display the values computed.

Using RKN (5 steps of size 0.2), solve **Airy's equation** $y'' = xy$ for the initial values $y(0) = 0.35502805$, $y'(0) = -0.25881940$. The exact solution is Ai(x). Obtain the exact values by Mathematica and calculate the error of the RKN values.

Solution. Type $f = xy$, then the initial values, and then a do-loop that will calculate the auxiliary values as well as the values for y and y'. Finally, use Table to get the results as well as the errors. You may also use DSolve to see a general solution involving the **Airy functions** Ai and Bi.

In[1]:= Clear[y]

```
In[2]:= RKN[f_, x_, y_, yp_, h_, M_] := Module[{n},
        Do [
            k1 = h/2 f[x[n], y[n], yp[n]];
            K = h/2 (yp[n] + k1/2);
            k2 = h/2 f[x[n] + h/2, y[n] + K, yp[n] + k1];
            k3 = h/2 f[x[n] + h/2, y[n] + K, yp[n] + k2];
            L = h (yp[n] + k3);
            k4 = h/2 f[x[n] + h, y[n] + L, yp[n] + 2 k3];
            x[n + 1] = x[n] + h;
            y[n + 1] = y[n] + h (yp[n] + 1/3 (k1 + k2 + k3));
            yp[n + 1] = yp[n] + 1/3 (k1 + 2 k2 + 2 k3 + k4),
            {n, 0, M}
            ]
        ]
```

First save this module for further use by typing

In[3]:= DeleteFile["RKN.m"]

In[4]:= Clear[f, x, y, yp, h, M]

In[5]:= Save["RKN.m", RKN]

In[6]:= !!RKN.m

and then get the approximate solution by

In[7]:= `f[x_, y_, yp_] = x y`   (* Out: xy *)

In[8]:= `x[0] = 0;    y[0] = 0.355028;    yp[0] = -0.258819;`
        `h = 0.2;    M = 5;`

In[9]:= `RKN[f, x, y, yp, h, M]`

In[10]:= `Table[{x[n], y[n], yp[n]}, {n, 0, M}]`

Out[10]= {{0, 0.355028, −0.258819}, {0.2, 0.303703, −0.252405},

          {0.4, 0.254742, −0.235831}, {0.6, 0.2098, −0.212792},

          {0.8, 0.169846, −0.186411}, {1., 0.135292, −0.159146}}

Similar Material in AEM: p. 960

| EXAMPLE 19.7 | **LAPLACE EQUATION. BOUNDARY VALUE PROBLEM**

Solve the **Dirichlet problem** for the **Laplace equation** $u_{xx} + u_{yy} = 0$ numerically in the square in the figure with the grid shown, when the boundary potential equals 0 on the upper edge and 100 volts on the three other edges (equivalently: the temperature on the upper edge is 0 and on the other edges it is 100° C).

Solution. The Laplace equation is replaced by a difference equation. This gives a linear system of 4 equations in the 4 unknown potentials $u_{jk} = u(x, y) = u(4j, 4k) = u(P_{jk})$, $j = 1, 2$ and $k = 1, 2$, at the 4 inner points of the grid in the figure. At each such point, -4 times the potential plus the sum of the potentials at the 4 closest neighbors equals 0. In these equations you take the 4 inner points in the order $P_{11}, P_{21}, P_{12}, P_{22}$. Thus for P_{11} the equation is $-4u_{11} + u_{10} + u_{21} + u_{12} + u_{01} = -4u_{11} + 100 + u_{21} + u_{12} + 100 = 0$. Hence the coefficient matrix of the linear system $\mathbf{Ax} = \mathbf{b}$ has $[-4 \ 1 \ 1 \ 0]$ as the first row and -200 as the first component of $\mathbf{b}$. Similarly for the other equations. You thus obtain $\mathbf{Ax} = \mathbf{b}$ and its solution (the potential at the 4 inner points) as follows. The matrix $\mathbf{A}$ is

In[1]:= `A = {{-4, 1, 1, 0}, {1, -4, 0, 1}, {1, 0, -4, 1}, {0, 1, 1, -4}}`

Out[1]= {{−4, 1, 1, 0}, {1, −4, 0, 1}, {1, 0, −4, 1}, {0, 1, 1, −4}}

In[2]:= `%//MatrixForm`

Out[2]//MatrixForm=

$$\begin{pmatrix} -4 & 1 & 1 & 0 \\ 1 & -4 & 0 & 1 \\ 1 & 0 & -4 & 1 \\ 0 & 1 & 1 & -4 \end{pmatrix}$$

The grid for the difference equation is obtained by the commands

In[3]:= `P1 = ContourPlot[x, {x, 0, 12}, {y, 0, 12}, Contours -> 2,`
        `ContourShading -> False]`

In[4]:= `P2 = ContourPlot[y, {x, 0, 12}, {y, 0, 12}, Contours -> 2,`
        `ContourShading -> False]`

In[5]:= Show[P1, P2]

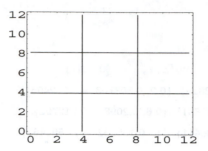

Example 19.7. Grid in the xy-plane

Now construct **b** as follows. On each of the 4 sides of the grid there are 2 points whose potentials contribute to an equation. Going counterclockwise around the boundary, you have 100 (at $(x, y) = (4, 0)$), 100 (at $(8, 0)$), 100 (at $(12, 4)$), 100 (at $(12, 8)$, 0 (at $(8, 12)$, 0 (at $(4, 12)$, 100 (at $(0, 8)$), 100 (at $(0, 4)$). This gives the vector

In[6]:= c = {100, 100, 100, 100, 0, 0, 100, 100};

Obtain from $\mathbf{c} = [c_j]$ the vector $\mathbf{b} = -\mathbf{Bc}$ with the 4×8 matrix $\mathbf{B} = [b_{jk}]$ constructed as follows. (It is practical to introduce the minus resulting from taking the boundary terms from the left to the right in all equations.)

In[7]:= <<LinearAlgebra'MatrixManipulation'

In[8]:= B = ZeroMatrix[4, 8]

In[9]:= B[[1,1]] = 1; B[[1,8]] = 1; B[[2,2]] = 1; B[[2,3]] = 1;
        B[[3,6]] = 1; B[[3,7]] = 1; B[[4,4]] = 1; B[[4,5]] = 1;

In[10]:= B; MatrixForm[B]

Out[10]//MatrixForm=

$$\begin{pmatrix} 1 & 0 & 0 & 0 & 0 & 0 & 0 & 1 \\ 0 & 1 & 1 & 0 & 0 & 0 & 0 & 0 \\ 0 & 0 & 0 & 0 & 0 & 1 & 1 & 0 \\ 0 & 0 & 0 & 1 & 1 & 0 & 0 & 0 \end{pmatrix}$$

$b_{11} = 1$ and $b_{18} = 1$ because c_1 and c_8 contribute to the first equation; hence minus their sum - 200 (minus because you took the terms to the right!) is the first component b_1. Similarly for the other components of **b**. You thus obtain

In[11]:= b = -B.c   (* Out: {−200, −200, −100, −100} *)

Now solve the system $\mathbf{Ax} = \mathbf{b}$ obtained.

In[12]:= x = N[LinearSolve[A, b]]   (* Out: {87.5, 87.5, 62.5, 62.5} *)

Finer grid with 3×3 **inner points.** Obtain the coefficient matrix **A** from a **band matrix** by changing four entries from 1 to 0, as follows.

In[13]:= p = 3;

In[14]:= `A = Table[Switch[j - k, 0, -4, 1, 1, -1, 1, p, 1, -p, 1, _, 0],`
           `{j, p^2}, {k, p^2}]; MatrixForm[A]`

Out[14]//MatrixForm=

$$\begin{pmatrix} -4 & 1 & 0 & 1 & 0 & 0 & 0 & 0 & 0 \\ 1 & -4 & 1 & 0 & 1 & 0 & 0 & 0 & 0 \\ 0 & 1 & -4 & 1 & 0 & 1 & 0 & 0 & 0 \\ 1 & 0 & 1 & -4 & 1 & 0 & 1 & 0 & 0 \\ 0 & 1 & 0 & 1 & -4 & 1 & 0 & 1 & 0 \\ 0 & 0 & 1 & 0 & 1 & -4 & 1 & 0 & 1 \\ 0 & 0 & 0 & 1 & 0 & 1 & -4 & 1 & 0 \\ 0 & 0 & 0 & 0 & 1 & 0 & 1 & -4 & 1 \\ 0 & 0 & 0 & 0 & 0 & 1 & 0 & 1 & -4 \end{pmatrix}$$

Explanation. For $j - k = 0$ (the main diagonal) take -4 as entries. This explains `j - k, 0, -4`. For $j - k = 1$ (the sloping line above the main diagonal) take 1 as entries. This explains `1, 1`. For $j - k = -1$ (the line below the diagonal) take 1 as entries. For $j - k = p \, (= 3)$ take 1 as entries. For $j - k = -p \, (= -3)$ take 1 as entries. For all other values of $j - k$ (represented by `_`) take 0 as entries.

To produce the coefficient matrix you now change four 1's into 0's as follows

In[15]:= `A[[p, p + 1]] = 0; A[[p + 1, p]] = 0; A[[2 p, 2 p + 1]] = 0;`
           `A[[2 p + 1, 2 p]] = 0; MatrixForm[A]`

Out[15]//MatrixForm=

$$\begin{pmatrix} -4 & 1 & 0 & 1 & 0 & 0 & 0 & 0 & 0 \\ 1 & -4 & 1 & 0 & 1 & 0 & 0 & 0 & 0 \\ 0 & 1 & -4 & 0 & 0 & 1 & 0 & 0 & 0 \\ 1 & 0 & 0 & -4 & 1 & 0 & 1 & 0 & 0 \\ 0 & 1 & 0 & 1 & -4 & 1 & 0 & 1 & 0 \\ 0 & 0 & 1 & 0 & 1 & -4 & 0 & 0 & 1 \\ 0 & 0 & 0 & 1 & 0 & 0 & -4 & 1 & 0 \\ 0 & 0 & 0 & 0 & 1 & 0 & 1 & -4 & 1 \\ 0 & 0 & 0 & 0 & 0 & 1 & 0 & 1 & -4 \end{pmatrix}$$

Construct **b** as before. The vector **c** now has $4 \times 3 = 12$ components, 3 resulting from each side,

In[16]:= `c = {100, 100, 100, 100, 100, 100, 0, 0, 0, 100, 100, 100}`

The matrix **B** is 9×12 because you have 9 equations, one from each inner point, and 12 contributing points around the boundary. Thus create a zero matrix

In[17]:= `B = ZeroMatrix[9, 12]`

Now 12 entries are the 1's of the 12 boundary values, each contributing to one of the 9 equations, the first and the last values to the first equation, the second value to the second equation, the third to the third, the fourth to the sixth equation, etc. Type these 12 nonzero entries as

```
In[18]:= B[[1,1]]   = 1;  B[[1,12]] = 1;  B[[2,2]] = 1;  B[[3,3]] = 1;
         B[[3,4]]   = 1;  B[[4,11]] = 1;  B[[6,5]] = 1;  B[[7,9]] = 1;
         B[[7,10]]  = 1;  B[[8,8]]  = 1;  B[[9,6]] = 1;  B[[9,7]] = 1;

In[19]:= B; MatrixForm[B]
```

Out[19]//MatrixForm=

$$
\begin{pmatrix}
1 & 0 & 0 & 0 & 0 & 0 & 0 & 0 & 0 & 0 & 0 & 1 \\
0 & 1 & 0 & 0 & 0 & 0 & 0 & 0 & 0 & 0 & 0 & 0 \\
0 & 0 & 1 & 1 & 0 & 0 & 0 & 0 & 0 & 0 & 0 & 0 \\
0 & 0 & 0 & 0 & 0 & 0 & 0 & 0 & 0 & 0 & 1 & 0 \\
0 & 0 & 0 & 0 & 0 & 0 & 0 & 0 & 0 & 0 & 0 & 0 \\
0 & 0 & 0 & 0 & 1 & 0 & 0 & 0 & 0 & 0 & 0 & 0 \\
0 & 0 & 0 & 0 & 0 & 0 & 0 & 0 & 1 & 1 & 0 & 0 \\
0 & 0 & 0 & 0 & 0 & 0 & 0 & 1 & 0 & 0 & 0 & 0 \\
0 & 0 & 0 & 0 & 0 & 1 & 1 & 0 & 0 & 0 & 0 & 0
\end{pmatrix}
$$

Partition this matrix into 3×3 matrices; then you can see that the pattern of the distribution of ones is quite simple. Obtain **b** by typing

```
In[20]:= b = -B.c          (* Out: {-200, -100, -200, -100, 0, -100, -100, 0, -100} *)
```

The solution of $\mathbf{Ax} = \mathbf{b}$, that is, the potential at the 9 inner points, is

```
In[21]:= N[LinearSolve[A, b]]
```

Out[21]= {92.8571, 90.1786, 92.8571, 81.25, 75., 81.25, 57.1429, , 47.3214, 57.1429}

```
In[22]:= X = Partition[%, p]; //MatrixForm[X]
```

Out[22]//MatrixForm=

$$
\begin{pmatrix}
92.8571 & 90.1786 & 92.8571 \\
81.25 & 75. & 81.25 \\
57.1429 & 47.3214 & 57.1429
\end{pmatrix}
$$

To get these values in the same positions as those of the corresponding points, type

```
In[23]:= Answer = {X[[3]], X[[2]], X[[1]]}; MatrixForm[Answer]
```

Out[23]//MatrixForm=

$$
\begin{pmatrix}
57.1429 & 47.3214 & 57.1429 \\
81.25 & 75. & 81.25 \\
92.8571 & 90.1786 & 92.8571
\end{pmatrix}
$$

The grid in the figure is obtained similarly to the previous one by typing

```
In[24]:= P3 = ContourPlot[x, {x, 0, 12}, {y, 0, 12}, Contours -> 3,
         ContourShading ->  False]
In[25]:= P4 = ContourPlot[y, {x, 0, 12}, {y, 0, 12}, Contours -> 3,
         ContourShading -> False]
In[26]:= Show[P3, P4]
```

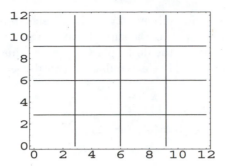

Example 19.7. Refined grid

Similar Material in AEM: pp. 962-966

| EXAMPLE 19.8 | **HEAT EQUATION. CRANK-NICOLSON METHOD** |

The **one-dimensional heat equation** $u_t = c^2 u_{xx}$ is a **parabolic equation** that governs, for instance, the heat flow in a bar, where $u(x,t)$ is the temperature at a point x and time t. Solve the corresponding difference equation with $c^2 = 1$ on the interval $0 \leq x \leq 1$ (the bar extending from $x = 0$ to $x = 1$ along the x-axis) subject to the initial temperature $u(x,0) = \sin \pi x$ by the **Crank-Nicolson method** with x-step $h = 0.2$ and time step $k = 0.04$, doing 5 time steps.

Solution. The method proceeds by time steps. For each $t = 0$, 0.04, etc. you have 4 points of the grid at which the 4 values of the temperature are to be determined from a linear system of 4 equations in these unknown values. In this method the discretization of the heat equation is done in such a way that the resulting linear system $\mathbf{Av} = \mathbf{b}$ has the following coefficient matrix, where $r = i/h^2$ ($= 1$ in the present case) and i is the time step (in AEM denoted by k, which we need as a summation index). Switch used here is explained on p. 223.

In[1]:= `n = 4;    r = 1;    h = 0.2;`

In[2]:= `A = Table[Switch[j - k,  0, 4, 1, -1, -1, -1, _, 0], {j, n}, {k, n}];`

In[3]:= `MatrixForm[A]`

$$\text{Out[3]=} \begin{pmatrix} 4 & -1 & 0 & 0 \\ -1 & 4 & -1 & 0 \\ 0 & -1 & 4 & -1 \\ 0 & 0 & -1 & 4 \end{pmatrix}$$

Obtain the initial temperature distribution and the preparation for a plot by typing

In[4]:= `n = 4;`

In[5]:= `Clear[u]`

In[6]:= `Do[u[k] = N[Sin[Pi k h]], {k, 0, n + 1}]`

In[7]:= `Table[u[k], {k, 0, n + 1}]`

Out[7]= $\{0., 0.587785, 0.951057, 0.951057, 0.587785, 1.22461 \times 10^{-16}\}$

These are the initial values (the initial temperatures).

Now obtain the values of the temperature $u(x,t)$ for $t = 0.04, 0.08, ..., 0.20$ by iteration with respect to time, solving $\mathbf{Av} = \mathbf{b}$ in each step. In this program you obtain the mth component `b[m]` as the sum of two previous values (see Line 3 of the do-loop). The command before the do-loop sets up an auxiliary table into which the program will feed values that will be accessible afterwards; see the further explanation below.

In[8]:= `Table[Temp[k], {k, 1, n + 1}];`

Out[8]= `{Temp[1], Temp[2], Temp[3], Temp[4], Temp[5]}`

In[9]:= `M = 5;`

In[10]:= `Do [`

```
          b = Table[0, {i, 1, n}];
          Do[b[[k]] = u[k - 1] + u[k + 1], {k, 1, n}];        (* Line 3 *)
          v = N[LinearSolve[A, b]];
              Print[v]; Temp[j] = v;
          Do[u[k] = v[[k]], {k, 1, n}],
          {j, 1, M}
          ];
```

`{0.399274, 0.646039, 0.646039, 0.399274}`
`{0.271221, 0.438844, 0.438844, 0.271221}`
`{0.184236, 0.2981, 0.2981, 0.184236}`
`{0.125149, 0.202495, 0.202495, 0.125149}`
`{0.0850118, 0.137552, 0.137552, 0.0850118}`

In[11]:= `v = Table[Table[Temp[k][[j]], {j, 1, M - 1}], {k, 1, n + 1}]`

Out[11]= `{{0.399274, 0.646039, 0.646039, 0.399274},`
`{0.271221, 0.438844, 0.438844, 0.271221},`
`{0.184236, 0.2981, 0.2981, 0.184236},`
`{0.125149, 0.202495, 0.202495, 0.125149},`
`{0.0850118, 0.137552, 0.137552, 0.0850118}}`

The point of this second table is this. Whereas in the first table you could access only the last value because the others are overwritten in the process of the do-loop, in the second table you can access all the values, as you will need them for plotting. For instance,

In[12]:= `v[[3]]`   `(* Out: {0.184236, 0.2981, 0.2981, 0.184236} *)`

In[13]:= `v[[3, 2]]`   `(* Out: 0.2981 *)`

Extend the table by including the boundary values 0.

In[14]:= `w = Table[Join[{0}, v[[j]], {0}], {j, 1, M}]`

Out[14]= `{{0, 0.399274, 0.646039, 0.646039, 0.399274, 0},`
`{0, 0.271221, 0.438844, 0.438844, 0.271221, 0},`
`{0, 0.184236, 0.2981, 0.2981, 0.184236, 0},`
`{0, 0.125149, 0.202495, 0.202495, 0.125149, 0},`
`{0, 0.0850118, 0.137552, 0.137552, 0.0850118, 0}}`

In[15]:= `w[[1]]`   `(* Out: {0, 0.399274, 0.646039, 0.646039, 0.399274, 0} *)`

Now begin a plot of temperatures at constant times by polygons, as follows.

In[16]:= `T0 = Table[{0.2 k, u[k]}, {k, 0, n + 1}]`

Out[16]= {{0, 0.}, {0.2, 0.587785}, {0.4, 0.951057},

       {0.6, 0.951057}, {0.8, 0.587785}, {1., 1.22461 × 10^{-16}}}

In[17]:= `T = Table[Table[{0.2 i, w[[p]][[i + 1]]}, {i, 0, n + 1}], {p, 1, M}]`

Out[17]= {{{0, 0}, {0.2, 0.399274}, {0.4, 0.646039},

       {0.6, 0.646039}, {0.8, 0.399274}, {1., 0}},

       {{0, 0}, {0.2, 0.271221}, {0.4, 0.438844}, {0.6, 0.438844},

       {0.8, 0.271221}, {1., 0}}, {{0, 0}, {0.2, 0.184236},

       {0.4, 0.2981}, {0.6, 0.2981}, {0.8, 0.184236}, {1., 0}},

       {{0, 0}, {0.2, 0.125149}, {0.4, 0.202495}, {0.6, 0.202495},

       {0.8, 0.125149}, {1., 0}}, {{0, 0}, {0.2, 0.0850118},

       {0.4, 0.137552}, {0.6, 0.137552}, {0.8, 0.0850118}, {1., 0}}}

In[18]:= `T[[1]]`

Out[18]= {{0, 0}, {0.2, 0.399274}, {0.4, 0.646039},

       {0.6, 0.646039}, {0.8, 0.399274}, {1., 0}}

In[19]:= `P0 = ListPlot[T0, PlotJoined -> True]`

In[20]:= `P1 = ListPlot[T[[1]], PlotJoined -> True]`

In[21]:= `P2 = ListPlot[T[[2]], PlotJoined -> True]`

In[22]:= `P3 = ListPlot[T[[3]], PlotJoined -> True]`

In[23]:= `P4 = ListPlot[T[[4]], PlotJoined -> True]`

In[24]:= `P5 = ListPlot[T[[5]], PlotJoined -> True]`

In[25]:= `Show[P0, P1, P2, P3, P4, P5]`

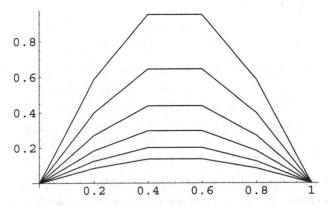

Example 19.8. Temperatures for time $t = 0$, 0.04, 0.08, ...

Each polygon corresponds to one of the times t_j considered, and is obtained by joining the temperatures $v(x, t_j)$ computed for $x = 0.2,\ 0.4,\ 0.6,\ 0.8$ Now use splines for obtaining a similar graph that gives a better impression of the actual temperatures.

In[26]:= <<Graphics'Spline'

In[27]:= S0 = Show[Graphics[Spline[T0, Cubic]]]

In[28]:= S1 = Show[Graphics[Spline[T[[1]], Cubic]]]

In[29]:= S2 = Show[Graphics[Spline[T[[2]], Cubic]]]

In[30]:= S3 = Show[Graphics[Spline[T[[3]], Cubic]]]

In[31]:= S4 = Show[Graphics[Spline[T[[4]], Cubic]]]

In[32]:= S5 = Show[Graphics[Spline[T[[5]], Cubic]]]

In[33]:= Show[S0, S1, S2, S3, S4, S5, Axes -> True]

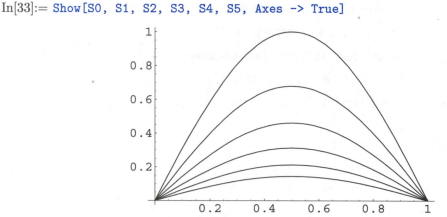

Example 19.8. Temperatures for time $t = 0,\ 0.04,\ 0.08,\ ...$

Similar Material in AEM: pp. 978-980

Problem Set for Chapter 19

Pr.19.1 (**Euler's method**) Solve $y' + 0.1y = 0$, $y(0) = 2$, doing 10 steps of size 0.1. Find the error. Plot the solution curve and the approximate values obtained.
(*AEM Ref.* p. 951 (#1))

Pr.19.2 (**Euler's method, Riccati equation**) Solve $y' = (x+y)^2$, $y(0) = 0$, 10 steps of size 0.1. Find the exact solution and determine the error. (*AEM Ref.* p. 951 (#4))

Pr.19.3 (**Experiment on stepsize for Euler's method**) In Pr.19.1 find out experimentally for what stepsize the error of $y(1)$ will decrease to 0.0001 (approximately).
(*AEM Ref.* p. 951 (#1))

Pr.19.4 (**Improved Euler method. Logistic population**) Solve the **Verhulst equation** $y' = y - y^2$ subject to the initial condition $y(0) = 0.5$. Do 10 steps of size 0.1. Find the exact solution and the error. Plot the exact solution. (*AEM Ref.* p. 951 (#6))

Pr.19.5 (**Experiment on improved Euler method. Unbounded solution**) Try to extend the calculation in Example 19.2 in this Guide to x-values near $\pi/2$ (where the solution becomes infinite). Experiment with $h = 0.1,\ 0.05,\ 0.025$ and study the increase of the error with x for constant h and the decrease of the error under halving h for constant x. (*AEM Ref.* pp. 945-947)

Pr.19.6 **(Classical Runge-Kutta method RK)** Solve $y' = (y - x - 1)^2 + 2$, $y(0) = 1$ by RK. Do 10 steps of size 0.1. Solve the problem exactly by setting $y - x - 1 = u$ and separating variables, or by `DSolve`. Calculate the error. (*AEM Ref.* pp. 948-950)

Pr.19.7 **(RK, step halving)** Solve $y' = -0.2xy$, $y(0) = 1$ by RK. Do 50 steps of size 0.2. Do 100 steps of size 0.1. Solve the problem exactly. Compare the errors of the two values for $x = 10$. (*AEM Ref.* p. 952 (#17))

Pr.19.8 **(Adams-Moulton method)** Solve the initial value problem $y' = x/y$, $y(1) = 3$, by the module `AdamsMoultonRK`. Do 10 steps of size 0.2. Find the exact solution. Determine the errors. (*AEM Ref.* p. 955 (#6))

Pr.19.9 **(Multistep method)** Using the module in Example 19.4 in this Guide, solve $y' = x + y$, $y(0) = 0$. Do 10 steps of size 0.1. (*AEM Ref.* pp. 953, 955)

Pr.19.10 **(Mass-spring system)** Solve $y'' + 2y' + 0.75y = 0$, $y(0) = 3$, $y'(0) = -2.5$ by RKS (Runge-Kutta for systems). Do 5 steps of size 0.2. Find the exact solution and compute the error. (*AEM Ref.* pp. 958, 959)

Pr.19.11 **(RKS Runge-Kutta for systems)** Solve $y_1' = 2y_1 - 4y_2$, $y_2' = y_1 - 3y_2$, $y_1(0) = 3$, $y_2(0) = 0$ by RKS for x from 0 to 5 with $h = 0.1$. Find the exact solution and the errors of y_1 and y_2. Plot y_1 and y_2. (*AEM Ref.* p. 961 (#5))

Pr.19.12 **(Runge-Kutta-Nystroem method)** Solve the initial value problem in Pr.19.10 by RKN (see Example 19.7 in this Guide) and compare the accuracy with that in Pr.19.10. (*AEM Ref.* p. 960)

Pr.19.13 **(Laplace equation. Dirichlet problem)** Solve the Laplace equation numerically for the refined grid in Example 19.7 in this Guide and boundary values 0 on the bottom, 110 V on the right edge, 220 V on the top, 110 V on the left edge. (*AEM Ref.* pp. 964-966)

Pr.19.14 **(Crank-Nicolson. "Triangular" initial temperature)** Solve the heat equation in Example 19.8 in this Guide when the initial temperature is x for $0 \le x \le 1/2$ and $1 - x$ for $1/2 \le x \le 1$, using the same grid and r, h, and i as in that example. (*AEM Ref.* p. 978)

Pr.19.15 **(Crank-Nicolson method, grid refinement)** In Example 19.8 in this Guide choose $n = 24$ (instead of $n = 4$), thus $h = 1/25 = 0.04$, $r = i/h^2 = 1$, thus $i = 1/625 = 0.0016$, and do enough steps so that you obtain values corresponding to those given in Example 19.8 and compare. (*AEM Ref.* pp. 978, 979)

PART F. OPTIMIZATION, GRAPHS

Content. Steepest descent, linear programming (Chap. 20)
  Graphs, digraphs, shortest spanning trees, network flows (Chap. 21)

Chapter 20

Unconstrained Optimization. Linear Programming

Content. Steepest descent (Ex. 20.1, Prs. 20.1-20.3)
  Linear programming (Ex. 20.2, Prs 20.4, 20.5)

This concerns finding maxima or minima using calculus methods under no constraints (steepest descent) or matrix methods under constraints (restriction of the region in which the maximum or minimum is to be found) (linear programming).

Examples for Chapter 20

EXAMPLE 20.1 | METHOD OF STEEPEST DESCENT

This is a method of unconstrained optimization for determining a minimum of a function $f(x, y)$. **"Unconstrained"** means that the variables x, y are not restricted to certain parts (regions) in the xy-plane. The method uses the fact that at each point of the surface S $(x, y, f(x, y))$ of f over the xy-plane the direction of **steepest descent** is given by $-\operatorname{grad} f$, the **gradient** being $\operatorname{grad} f = [\partial f/\partial x, \ \partial f/\partial y]$. (Exceptional points may occur.) You begin at some point $\mathbf{x}_0 = (X_0, Y_0)$ and follow the straight line in the direction of $-\operatorname{grad} f(\mathbf{x}_0)$. This line is given by

(A)
$$\mathbf{z}(t) = \mathbf{x}_0 - t \operatorname{grad} f(\mathbf{x}_0).$$

It gives a curve $C : [z_1, z_2, f(\mathbf{z})]$ on S, where $\mathbf{z} = [z_1, z_2]$. Follow the straight line until you reach the lowest point on C, the point at which

(B)
$$g(t) = f(\mathbf{z}(t)) \qquad \mathbf{z}(t) = [z_1(t), z_2(t)]$$

has a minimum. Determine the latter by finding $t = t_1$ where the derivative is zero,

(C)
$$g'(t) = 0.$$

Take $\mathbf{z}(t_1)$ as your new $\mathbf{x} = \mathbf{x_1}$ and begin the next step. The new gradient will be perpendicular to the old. Why? (*Answer.* You reach the level curve of f at $\mathbf{x}_1$ tangentially, whereas the new gradient is perpendicular to that level curve.)

For instance, find a minimum of the function $f(\mathbf{x}) = x^2 + 3y^2$, $\mathbf{x} = [x, y]$, starting from $\mathbf{x}_0 = [6, 3]$.

Solution. f has a minimum at $\mathbf{x} = \mathbf{0}$, as inspection shows. Hence you can see how the method approaches the solution. The level curves are ellipses. A module for the method is as follows. Explanations are given afterwards.

```
In[1]:= Clear[x, y, z, t, S, T, f]

In[2]:= <<Calculus`VectorAnalysis`

In[3]:= SetCoordinates[Cartesian[x,y,u]]          (* Out: Cartesian[x, y, u] *)

In[4]:= SD[f_, X_, Y_, M_] := Module[{j},          (* Line 1 *)
           z = {x, y, 0} - t Grad[f[x, y]];         (* Line 2 *)
           g = f[X, Y] /. {X -> z[[1]], Y -> z[[2]]};   (* Line 3 *)
           gprime = D[g, t];                        (* Line 4 *)
           Do[SOL = Solve[gprime == 0, t];          (* Line 5 *)
              T[j] = N[SOL[[1, 1, 2]] /. {x -> X[j], y -> Y[j]}];
              X[j + 1] =  z[[1]] /. {x -> X[j], y -> Y[j], t -> T[j]};
              Y[j + 1] = z[[2]] /. {x -> X[j], y -> Y[j], t -> T[j]},
           {j, 0, M}];  ]                           (* Line 9 *)
```

Future use of a module

Save the module by typing the following, with `SD` (suggesting 'steepest descent') being your choice for the file name.

```
In[5]:= DeleteFile["SD.m"]

In[6]:= Save["SD.m", SD]

In[7]:= !!SD.m
```

To **use the module again**, later, you will have to type

```
In[8]:= << SD.m
```

Explanations. In Line 1 you see the data needed, where `M` is the number of steps to be done. You also see the formulas (A), (B), and the derivative in (C) (Lines 2, 3, 4), a do-loop (Lines 5-9) that gives the solution of (C) (Line 5, denoted by `SOL`), the current `x` and `y` substituted into `SOL` (Line 6), and the new `x` and `y` (Lines 7 and 8).

To apply the module, type `f`, the initial data, and `M`. Then call the module and use `Table` to obtain the values $(x, y, f(x,y))$.

```
In[9]:= f[x_, y_] = x^2 + 3 y^2                     (* Out: x^2 + 3y^2 *)

In[10]:= X[0] = 6;  Y[0] = 3;  T[0] = 0;  M = 6;

In[11]:= SD[f, X, Y, M]

In[12]:= Z = Table[N[{X[j], Y[j], f[X[j], Y[j]]}], {j, 0, M}]

Out[12]= {{6, 3, 63}, {3.48387, -0.774194, 13.9355}, {1.32719, 0.663594, 3.0825},
          {0.770626, -0.17125, 0.681844}, {0.293572, 0.146786, 0.150823},
          {0.170461, -0.0378802, 0.0333617}, {0.0649375, 0.0324688, 0.00737954}}
```

Commands for plotting the surface of f and a broken line of segments joining the points $(x_j, y_j, f(x_j, y_j))$ on the surface (exhibiting the change of search direction at those points) are as follows.

In[13]:= << Graphics`Graphics3D`

In[14]:= P1 = ScatterPlot3D[Z, PlotJoined -> True,
     PlotStyle -> Thickness[0.01], AspectRatio -> 0.5]

In[15]:= P2 = Plot3D[f[x, y], {x, -7, 7}, {y, -6, 6}]

In[16]:= Show[{P1, P2}, PlotRange -> {{-7, 7}, {-6, 6}, {0, 150}},
     ViewPoint -> {-5, -3, 200}]

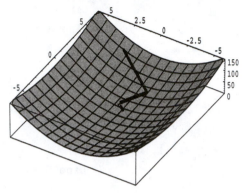

Example 20.1. Surface of $f(x, y)$ and polygonal line marking the descent on it

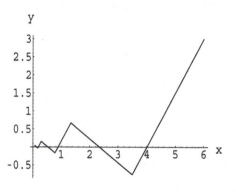

Example 20.1. Search path of steepest descent in the xy-plane

The second figure shows the search path in the xy-plane. Type

In[17]:= S = Table[N[{X[j], Y[j]}], {j, 0, 6}]

Out[17]= {{6., 3.}, {3.48387, −0.774194}, {1.32719, 0.663594}, {0.770626, −0.17125},

   {0.293572, 0.146786}, {0.170461, −0.0378802}, {0.0649375, 0.0324688}}

In[18]:= L = Line[S];

Out[18]= Line[{{6., 3.}, {3.48387, −0.774194}, {1.32719, 0.663594}, {0.770626, −0.17125},

   {0.293572, 0.146786}, {0.170461, −0.0378802}, {0.0649375, 0.0324688}}]

In[19]:= Show[Graphics[L], Axes -> True, AxesLabel -> {x, y}]

Similar Material in AEM: pp. 991-993

| EXAMPLE 20.2 | **SIMPLEX METHOD OF CONSTRAINED OPTIMIZATION** |

The function $z = f(x_1, x_2)$ to be maximized (profit, e.g.) is linear, say, for instance, $z = 40x_1 + 88x_2$, subject to linear inequalities (and linear equations) that determine a region ("**feasibility region**") in the x_1x_2-plane (bounded by a polygon) to which (x_1, x_2) is restricted and in which the maximum is to be found. For instance, $x_1 \geq 0$, $x_2 \geq 0$ (the first quadrant), and

(A)
$$2\,x_1 + 8\,x_2 \leq 60$$
$$5\,x_1 + 2\,x_2 \leq 60$$

Find the maximum of z under these **constraints**.

Solution. Convert the inequalities (A) to equations by introducing **slack variables** x_3, x_4. Including the function z to be maximized, this gives the **normal form** of the problem

$$\begin{aligned} z - 40\,x_1 - 88\,x_2 & & = 0 \\ 2\,x_1 + 8\,x_2 + x_3 & & = 60 \\ 5\,x_1 + 2\,x_2 & + x_4 & = 60 \end{aligned}$$

with x_1, x_2, x_3, x_4 being nonnegative. This is a linear system of equations. Find the maximal z, that is, find (x_1, x_2) satisfying the constraints and such that $z = f(x_1, x_2)$ is maximum. You can do this by suitably transforming the corresponding **augmented matrix** $\mathbf{T}_0$ (also called **simplex table**).

In[1]:= `T0 = {{1, -40, -88, 0, 0, 0},`

`{0, 2, 8, 1, 0, 60},`

`{0, 5, 2, 0, 1, 60}}; MatrixForm[T0]`

Out[1]//MatrixForm=

$$\begin{pmatrix} 1 & -40 & -88 & 0 & 0 & 0 \\ 0 & 2 & 8 & 1 & 0 & 60 \\ 0 & 5 & 2 & 0 & 1 & 60 \end{pmatrix}$$

Proceed by elimination of variables. This is similar to Gauss's method in Chaps. 6 and 18, but the choice of pivots is quite different, as follows. The first *negative* entry in Row 1 is -40 in Column 2, the column of x_1. Eliminate x_1 from 2 of the 3 equations. Choose 5 as the pivot because it gives the smallest of the quotients $60/2 = 30$ (in Row 2) and $60/5 = 12$ (in Row 3). To eliminate x_1 from Row 1 and then from Row 2, add $40/5$ times Row 3 (the pivot row) to Row 1 and then add $-2/5$ of Row 3 to Row 2.

In[2]:= `T1 = {T0[[1]] + (40/5) T0[[3]], T0[[2]] - 2/5 T0[[3]], T0[[3]]};`

`MatrixForm[T1]`

Out[2]//MatrixForm=

$$\begin{pmatrix} 1 & 0 & -72 & 0 & 8 & 480 \\ 0 & 0 & \dfrac{36}{5} & 1 & -\dfrac{2}{5} & 36 \\ 0 & 5 & 2 & 0 & 1 & 60 \end{pmatrix}$$

The next (and only further) negative entry in Row 1 is -72 in Column 3, the column of x_2. Eliminate x_2 from 2 of the 3 equations. Choose Row 2 as the pivot row because $36/(36/5) = 5$ is the smaller of this quotient and $60/2 = 30$ (in Row 3). To eliminate x_2 from Row 1 and then from Row 3, add $72/(36/5) = 10$ times Row 2 to Row 1 and then add $-2/(36/5) = -10/36$ times Row 2 to Row 3.

In[3]:= `T2 = {T1[[1]] + 10 T1[[2]], T1[[2]], T1[[3]] - 10/36 T1[[2]]};`

`MatrixForm[T2]`

Out[3]//MatrixForm=

$$\begin{pmatrix} 1 & 0 & 0 & 10 & 4 & 840 \\ 0 & 0 & \dfrac{36}{5} & 1 & -\dfrac{2}{5} & 36 \\ 0 & 5 & 0 & -\dfrac{5}{18} & \dfrac{10}{9} & 50 \end{pmatrix}$$

Since Row 1 has no more negative entries, you have reached the maximum. It is 840 (see Row 1) and occurs at $x_1 = 50/5 = 10$ (see Row 3) and $x_2 = 36/(36/5) = 5$ (see

Row 2). This is the right upper vertex of the quadrangle in the figure. To plot the figure, type

In[4]:= `ListPlot[{{0, 0},{12, 0}, {10, 5}, {0, 7.5}}, PlotJoined -> True,`
             `AxesLabel -> {x1, x2}]`

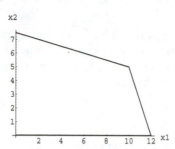

Example 20.2. Feasibility region

Similar Material in AEM: pp. 995, 998-1001

Problem Set for Chapter 20

Pr.20.1 (Steepest descent) Apply the method of steepest descent to $f(x,y) = x^2 + 10\,y^2$, starting from $(3, -2)$ and plotting the search path in the xy-plane. How many steps do you need to reach a point where $|f(x,y)| < 0.0001$? (*AEM Ref.* pp. 991, 992)

Pr.20.2 (Steepest descent) Apply the method of steepest descent to $f(x,y) = xy$, starting from $(1,2)$ and plotting the search path in the xy-plane. First guess what will happen. Then calculate. (*AEM Ref.* pp. 991, 992)

Pr.20.3 (Steepest descent) Apply the method of steepest descent to $f(x,y) = 5\,x^2 - y^2$, starting from $(1/2, 1/2)$ and plotting the search path in the xy-plane. First guess what the path might look like. Then calculate. (*AEM Ref.* pp. 991, 992)

Pr.20.4 (Linear programming) Maximize $z = 4\,x_1 + x_2 + 2\,x_3$ subject to $x_1 + x_2 + x_3 \leq 1$, $x_1 + x_2 - x_3 \leq 0$ and x_1, x_2, x_3 nonnegative. Proceed stepwise, showing the matrices. (*AEM Ref.* p. 1007 (#9))

Pr.20.5 (Linear programming) Maximize the daily output in producing x_1 glass plates by a process P_1 and x_2 glass plates by a process P_2 subject to the following constraints (labor hours, machine hours, raw material supply). Show the steps as in Example 20.2 in this Guide. (*AEM Ref.* p. 1007 (#3))

$$2\,x_1 + 3\,x_2 \leq 130, \qquad 3\,x_1 + 8\,x_2 \leq 300, \qquad 4\,x_1 + 2\,x_2 \leq 140.$$

PART G. PROBABILITY AND STATISTICS

Content. Data analysis, binomial, Poisson, hypergeometric, normal distributions
(Chap. 22)
Random numbers, confidence intervals, tests, regression (Chap. 23)

Packages you will need in these two chapters are loaded by the commands

<<Statistics'DescriptiveStatistics'　　　　See Example 22.1.
<<Graphics'Graphics'　　　　　　　　　　　See Example 22.2.
<<Statistics'DiscreteDistributions'　　　　See Example 22.3.
<<Statistics'ContinuousDistributions'　　　See Example 22.4.

Chapter 22

Data Analysis. Probability Theory

Content. Data analysis of samples (Exs. 22.1, 22.2, Prs. 22.1, 22.2)
Discrete distributions (Ex. 22.3, Prs. 22.3, 22.5, 22.6, 22.9-22.12)
Normal distribution (Ex. 22.4, Prs. 22.13-22.15)

Probability theory provides models (probability distributions) of random experiments
in this chapter and the mathematical justification of the statistical methods in the
next chapter.

Examples for Chapter 22

EXAMPLE 22.1　　**DATA ANALYSIS: MEAN, VARIANCE,**
STANDARD DEVIATION

Given a sample S of **size** n, that is, consisting of n values,

$$S = [x_1, x_2, ..., x_n],$$

you can arrange these values in ascending or descending order by the command Sort.
For instance,

In[1]:= S = {89, 84, 87, 81, 89, 86, 91, 90, 78, 89, 87, 99, 83, 89}
Out[1]= {89, 84, 87, 81, 89, 86, 91, 90, 78, 89, 87, 99, 83, 89}
In[2]:= Sort[%]　　　　(* Out: {78, 81, 83, 84, 86, 87, 87, 89, 89, 89, 89, 90, 91, 99} *)
In[3]:= -Sort[-S]　　　(* Out: {99, 91, 90, 89, 89, 89, 89, 87, 87, 86, 84, 83, 81, 78} *)

The **sample mean** $\bar{x}$ is

(1)
$$\bar{x} = \frac{1}{n} \sum_{j=1}^{n} x_j.$$

This is the arithmetic mean of the sample values. It measures the average size of these values. You obtain it (for the above S, which has size $n = 14$) by loading

In[4]:= `<<Statistics'DescriptiveStatistics'`

and then typing

In[5]:= `N[Mean[S]]`   (* Out: 87.2857 *)

The **sample variance** s^2 is

(2)
$$s^2 = \frac{1}{n-1} \sum_{j=1}^{n} (x_j - x)^2.$$

It measures the spread of the sample values. You obtain it (for the above S) by typing

In[6]:= `var = N[{Variance}[S]]`   (* Out: 25.1429 *)

The **sample standard deviation** is the nonnegative square root of the sample variance. Thus,

In[7]:= `Sqrt[var]`   (* Out: 5.01427 *)

or directly

In[8]:= `N[StandardDeviation[S]]`   (* Out: 5.01427 *)

Similar Material in AEM: pp. 1053, 1054

EXAMPLE 22.2   **DATA ANALYSIS: HISTOGRAMS**

These are graphical representations of data. A **histogram** of a sample shows the **absolute frequencies** $a(x) = $ *Number of times the value x occurs in that sample.* Actually, to obtain a better general impression of the essential features of a sample, **group** it into **classes**. For instance, take the sample in Example 22.1 and order it,

In[1]:= `S = {89, 84, 87, 81, 89, 86, 91, 90, 78, 89, 87, 99, 83, 89}`

In[2]:= `Sort[%]`   (* Out: {78, 81, 83, 84, 86, 87, 87, 89, 89, 89, 89, 90, 91, 99} *)

Look at the data and decide that good **class marks** (midpoints of class intervals) would be 80, 85, 90, 95, 100. That is, take the class intervals 77.5-82.5, 82.5-87.5, 87.5-92.5, 92.5-97.5, 97.5-102.5. The corresponding **absolute class frequencies** are 2 (because 78 and 81 are in the first interval), 5, 6, 0, 1, respectively. Now plot a histogram by typing

In[3]:= `<< Graphics'Graphics'`

In[4]:= `Histogram[S, HistogramScale -> Length[S], Ticks -> IntervalBoundaries,`
`        HistogramCategories -> {77.5, 82.5, 87.5, 92.5, 97.5, 102.5}]`

`HistogramCategories` is optional. It fixes the endpoints of the class intervals of your choice. Try without. `Ticks -> IntervalBoundaries` shows those endpoints in the figure. Without this command you will get the class marks. Try it. `Length[S]` is the size of the sample (14), and `HistogramScale -> Length[S]` makes the total area of the histogram equal to 1. Try without and note how the scale on the vertical axis changes.

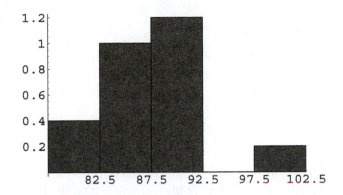

Example 22.2. Histogram of the grouped sample

Similar Material in AEM: pp. 1051-1053

| EXAMPLE 22.3 | **DISCRETE PROBABILITY DISTRIBUTIONS**

In this chapter you will need the binomial, Poisson, and hypergeometric distributions, which are discrete, and the uniform and normal distributions, which are continuous. All these are available in Mathematica. Also the chi-square, t, and F distributions needed in Chap. 23 are available – as well as many other distributions.

Discrete distributions are given by their **probability function** $f(x)$, in Mathematica called PDF (suggesting 'probability density function'), or, equivalently, by their distribution function $F(x)$, in Mathematica CDF (suggesting 'cumulative distribution function'). You call these functions as shown below for the distributions to be discussed.

Binomial distribution. The probability function is

$$f(x) = \binom{n}{x} p^x (1-p)^{n-x}, \qquad x = 0, 1, ..., n.$$

This is the probability of x successes in n independent trials when the probability of success in a single trial is p. For instance, if $p = 0.55$ and $n = 5$, you obtain numerical values of f for $x = 0, ..., 5$ by typing

```
In[1]:= f[x_] := Binomial[5, x]  0.55^x  0.45^(5 - x)
In[2]:= S = Table[{n, f1[n] == N[f[n]], F[n] == N[Sum[f[j], {j, 0, n}]]},
        {n, 0, 5}]
```

Out[2]= {{0, f1[0] == 0.0184528, F[0] == 0.0184528},

     {1, f1[1] == 0.112767, F[1] == 0.13122},

     {2, f1[2] == 0.275653, F[2] == 0.406873},

     {3, f1[3] == 0.336909, F[3] == 0.743783},

     {4, f1[4] == 0.205889, F[4] == 0.949672},

     {5, f1[5] == 0.0503284, F[5] == 1.}}

Individual values can be accessed either by f or by S. For instance,

In[3]:= f[4]   (* Out: 0.205889 *)

In[4]:= S[[5, 2, 2]]   (* Out: 0.205889 *)

In[5]:= S[[5, 3, 2]]   (* Out: 0.949672 *)

More directly, you obtain values of the probability function and of the distribution function of the binomial distribution by loading and typing the following.

In[6]:= << Statistics`DiscreteDistributions`

In[7]:= BinDist = BinomialDistribution[5, 0.55]

Out[7]= BinomialDistribution[5, 0.55]

In[8]:= Table[PDF[BinDist, j], {j, 0, 5}]

Out[8]= {0.0184528, 0.112767, 0.275653, 0.336909, 0.205889, 0.0503284}

In[9]:= Seq = Table[CDF[BinDist, j], {j, 0, 5}]

Out[9]= {0.0184528, 0.13122, 0.406873, 0.743783, 0.949672, 1}

Mean and **variance** are obtained from the defining formulas or from a direct command, as follows.

In[10]:= mu = Sum[x f[x], {x, 0, 5}]   (* Out: 2.75 *)

In[11]:= var = Sum[(x - mu)^2 f[x], {x, 0, 5}]   (* Out: 1.2375 *)

In[12]:= Mean[BinomialDistribution[5, 0.55]]   (* Out: 2.75 *)

In[13]:= Variance[BinomialDistribution[5, 0.55]]   (* Out: 1.2375 *)

Bar graphs of probability functions can be obtained by the following commands, where 0.1 (or any other positive number less than 1) gives the width of the bars.

In[14]:= S = Table[{x, PDF[BinDist, x], 0.1}, {x, 0, 5}]

Out[14]= {{0, 0.0184528, 0.1}, {1, 0.112767, 0.1}, {2, 0.275653, 0.1},

     {3, 0.336909, 0.1}, {4, 0.205889, 0.1}, {5, 0.0503284, 0.1}}

In[15]:= << Graphics`Graphics`

In[16]:= GeneralizedBarChart[S]   (* See the figure. *)

Graphs of distribution functions. In the discrete case these are step functions, with the stepsizes equal to the values of the corresponding probability function. You can obtain the graph of the above distribution function by typing the following, where Seq[[x+1]] accesses terms in Seq.

In[17]:= G = Table[{{x, Seq[[x+1]]}, {x+1, Seq[[x+1]]}}, {x, 0, 5}]

Out[17]= {{{0, 0.0184528}, {1, 0.0184528}}, {{1, 0.13122}, {2, 0.13122}},

{{2, 0.406873}, {3, 0.406873}}, {{3, 0.743783}, {4, 0.743783}},

{{4, 0.949672}, {5, 0.949672}}, {{5, 1}, {6, 1}}}

In[18]:= G1 = Join[Flatten[G, 1], {{10, 1}}]

Out[18]= {{0, 0.0184528}, {1, 0.0184528}, {1, 0.13122}, {2, 0.13122},

{2, 0.406873}, {3, 0.406873}, {3, 0.743783}, {4, 0.743783},

{4, 0.949672}, {5, 0.949672}, {5, 1}, {6, 1}, {10, 1}}

You see that Flatten has the effect that from the nested list (of pairs of pairs) G you get a single list of pairs, to which you add the single pair 10, 1. You now obtain the graph of $F(x)$ in the second figure by typing

In[19]:= L = Line[G1];

Out[19]= Line[{{0, 0.0184528}, {1, 0.0184528}, {1, 0.13122}, {2, 0.13122},

{2, 0.406873}, {3, 0.406873}, {3, 0.743783}, {4, 0.743783},

{4, 0.949672}, {5, 0.949672}, {5, 1}, {6, 1}, {10, 1}}]

In[20]:= Show[Graphics[L], Axes -> True]

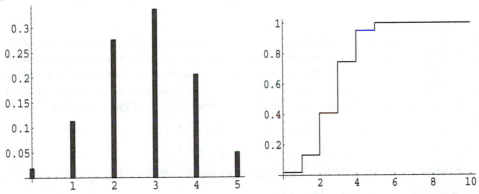

Example 22.3. Probability function of the binomial distribution with $n = 5$ and $p = 0.55$

Example 22.3. Distribution function of the binomial distribution with $n = 5$ and $p = 0.55$

Poisson distribution. The probability function is

$$f(x) = \frac{\mu^x}{x!} e^{-\mu}, \qquad x = 0, 1, \dots .$$

It is the limiting case of the binomial distribution if one lets $n \to \infty$ and $p \to 0$ so that the mean $\mu = np$ approaches a finite value. Its graph has infinitely many bars whose lengths decrease to zero very fast. For instance, to plot f with $\mu = 5$, type

In[21]:= Poi = PoissonDistribution[5]   (* Out: PoissonDistribution[5] *)

In[22]:= probabilities = Table[N[PDF[Poi, x]], {x, 0, 15}]

Out[22]= {0.00673795, 0.0336897, 0.0842243, 0.140374, 0.175467,

　　　　　0.175467, 0.146223, 0.104445, 0.065278, 0.0362656, 0.0181328,

　　　　　0.00824218, 0.00343424, 0.00132086, 0.000471736, 0.000157245}

In[23]:= BarPoi = Table[{x, probabilities[[x + 1]], 0.3}, {x, 0, 15}]

Out[23]= {{0, 0.00673795, 0.3}, {1, 0.0336897, 0.3},

　　　　　{2, 0.0842243, 0.3}, {3, 0.140374, 0.3}, {4, 0.175467, 0.3},

　　　　　{5, 0.175467, 0.3}, {6, 0.146223, 0.3}, {7, 0.104445, 0.3},

　　　　　{8, 0.065278, 0.3}, {9, 0.0362656, 0.3}, {10, 0.0181328, 0.3},

　　　　　{11, 0.00824218, 0.3}, {12, 0.00343424, 0.3}, {13, 0.00132086, 0.3},

　　　　　{14, 0.000471736, 0.3}, {15, 0.000157245, 0.3}}

In[24]:= GeneralizedBarChart[BarPoi]　　　　　　　　　　(* See figure. *)

Hypergeometric distribution. The probability function is

$$f(x) = \frac{\binom{M}{x}\binom{N-M}{n-x}}{\binom{N}{n}}.$$

This is the probability of obtaining x red balls in drawing n balls from a lot of N balls, M of which are red, and the drawing is done one-by-one **without replacement** (that is, balls drawn are not returned to the lot). Instead of red balls you could think of defective screws in a lot of screws produced. For instance, when $N = 100$ and $M = 20$, and you draw 5 balls, you obtain the probability function by typing

In[25]:= HypDist = HypergeometricDistribution[5, 20, 100];

In[26]:= hyp = Table[N[PDF[HypDist, x]], {x, 0, 5}]

Out[26]= {0.319309, 0.420144, 0.207344, 0.0478486, 0.00514826, 0.000205931}

and the distribution function by typing

In[27]:= Table[N[CDF[HypDist, x]], {x, 0, 5}]

Out[27]= {0.319309, 0.739453, 0.946797, 0.994646, 0.999794, 1.}

For plotting, similarly as before, type

In[28]:= BarHyp = Table[{x, hyp[[x+1]], 0.1}, {x, 0, 5}]

Out[28]= {{0, 0.319309, 0.1}, {1, 0.420144, 0.1}, {2, 0.207344, 0.1},

　　　　　{3, 0.0478486, 0.1}, {4, 0.00514826, 0.1}, {5, 0.000205931, 0.1}}

In[29]:= GeneralizedBarChart[BarHyp]

Mean and variance of the hypergeometric distribution, with NO written for the protected N, are obtained by

In[30]:= Clear[M]

In[31]:= Mean[HypergeometricDistribution[n, M, NO]]　　　　(* Out: $\frac{Mn}{NO}$ *)

In[32]:= Variance[HypergeometricDistribution[n, M, NO]]

$$\text{Out[32]}= \frac{\text{Mn}\left(1 - \frac{\text{M}}{\text{NO}}\right)(-\text{n} + \text{NO})}{(-1 + \text{NO})\,\text{NO}}$$

In[33]:= `Simplify[%]` $\left(* \text{ Out: } \frac{\text{Mn}\,(\text{M}-\text{NO})\,(\text{n}-\text{NO})}{(-1+\text{NO})\,\text{NO}^2} \;*\right)$

To obtain specific values, for instance when $n = 2$, $M = 3$, and $N = 10$, type

In[34]:= `PDF[HypergeometricDistribution[2, 3, 10], 0]` $\left(* \text{ Out: } \frac{7}{15} \;*\right)$

In[35]:= `PDF[HypergeometricDistribution[2, 3, 10], 1]` $\left(* \text{ Out: } \frac{7}{15} \;*\right)$

In[36]:= `CDF[HypergeometricDistribution[2, 3, 10], 1]` $\left(* \text{ Out: } \frac{14}{15} \;*\right)$

In[37]:= `CDF[HypergeometricDistribution[2, 3, 10], 2]` $(* \text{ Out: } 1 \;*)$

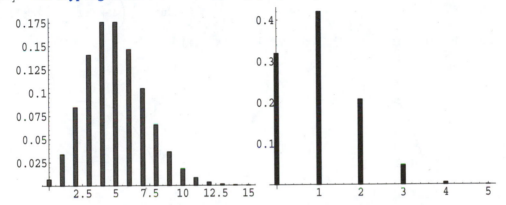

Example 22.3. Probability function
of the Poisson distribution
with mean $\mu = 5$

Example 22.3. Probability function
of the hypergeometric distribution
with $N = 100$, $M = 20$, and $n = 5$

Similar Material in AEM: pp. 1079-1083

EXAMPLE 22.4   NORMAL DISTRIBUTION

The **standardized normal distribution** (that is, the normal distribution with mean
0 and variance 1) is a **continuous distribution** with **density**

(A) $$f(x) = \frac{1}{\sqrt{2\pi}} \exp\left(-\frac{x^2}{2}\right) \qquad (-\infty < x < \infty).$$

Hence its distribution function is

(B) $$\Phi(z) = \frac{1}{\sqrt{2\pi}} \int_{-\infty}^{z} e^{-u^2/2}\,du.$$

Mathematica knows this function as well as its inverse (see below), and various numerical tables of $\Phi(z)$ can be found in the literature.

Setting $u = (v - \mu)/\sigma$, you have $du = dv/\sigma$ and $z = (x - \mu)/\sigma$. You thus obtain
the distribution function of the normal distribution with mean μ and variance σ^2

(C) $$F(x) = \Phi(\frac{x - \mu}{\sigma}) = \frac{1}{\sigma\sqrt{2\pi}} \int_{-\infty}^{x} \exp\left[-\frac{1}{2}\left(\frac{v - \mu}{\sigma}\right)^2\right]\,dv.$$

To plot the density (A), type the following, which shows that PDF denotes the density ('**probability density function**').

In[1]:= <<Statistics`ContinuousDistributions`

In[2]:= ND = NormalDistribution[0, 1]   (* Out: NormalDistribution[0,1] *)

In[3]:= PDF[ND, x]   (* Out: $\dfrac{e^{-\frac{x^2}{2}}}{\sqrt{2\,\pi}}$ *)

In[4]:= P1 = Plot[%, {x, -3, 3}, AxesLabel -> {x,y}]

To plot the distribution function (B), type the following, where $\mathrm{Erf}(x/\sqrt{2})$ is the error function (the integral of $\exp(-t^2)$ from 0 to $x/\sqrt{2}$, times $2/\sqrt{Pi}$).

In[5]:= CDF[ND, x]   (* Out: $\dfrac{1}{2}\left(1 + \mathrm{Erf}\left[\dfrac{x}{\sqrt{2}}\right]\right)$ *)

In[6]:= P2 = Plot[%, {x, -3, 3}, AxesLabel -> {x, y}]

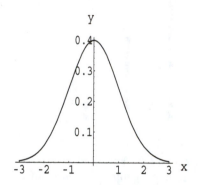 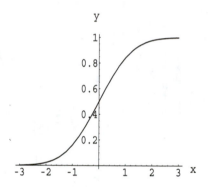

Example 22.4. Density (A) of the standardized normal distribution

Example 22.4. Distribution function (B) of the standardized normal distribution

Two basic tasks in connection with any distribution. Illustration in terms of a random variable X that is normal with mean 5.0 and variance 0.04, hence standard deviation 0.2.

First task. Given $x = 5.3$, find the probability that in a trial (an experiment), X will assume any value not exceeding 5.3.

Solution. You need the **distribution function** CDF (which Mathematica calls 'cumulative distribution function'). Note that the command involves the standard deviation 0.2, not the variance. The answer is 93.3%, approximately.

In[7]:= ND1 = NormalDistribution[5, 0.2];

Out[7]= NormalDistribution[5, 0.2]

In[8]:= CDF[ND1, 5.3]   (* Out: 0.933193 *)

Second task. Given the probability $P = 0.95 = 95\%$, find $x = c$ such that with probability 0.95 the random variable X will assume any value not exceeding c. (This task will arise in Chap. 23 for several important distributions.)

Solution. You need the **inverse** of the distribution function for which you have to type

In[9]:= Quantile[ND1, 0.95]　　　　　　　　　　　(* Out: 5.32897 *)

Hence the answer is $x = c = 5.33$, approximately.

Similar Material in AEM: pp. 1085-1089, A89, A90

Problem Set for Chapter 22

Pr.22.1 **(Mean and variance)** Find the mean filling [grams] and the variance of the sample of fillings 203, 199, 198, 201, 200, 201, 201. (*AEM Ref.* p. 1054 (#10))

Pr.22.2 **(Mean, median, standard deviation)** Find the mean, median, and standard deviation of the sample 17, 18, 17, 16, 17, 16, 18, 16. (*AEM Ref.* p. 1054 (#2))

Pr.22.3 **(Binomial distribution with $p = 0.5$)** Find the probability function and its values of the binomial distribution for 10 trials and probability of success 1/2. From it find the values of the distribution function as well as the mean and the variance. Confirm mean and variance by the appropriate commands that give those values directly. (*AEM Ref.* pp. 1079, 1080)

Pr.22.4 **(Probability)** A circuit contains 10 automatic switches. You want, with a probability of 95%, that during a given time interval all the switches are working. What probability of failure per time interval can you admit for a single switch? (*AEM Ref.* p. 1064 (#12))

Pr.22.5 **(Experiment on binomial distribution)** For what values of p will $f(x)$ with constant n be (a) large for small x, (b) large for large x? Experiment with plots and n's of your choice (10, 20, 100, or whatever). (*AEM Ref.* pp. 1079, 1080)

Pr.22.6 **(Binomial distribution)** Find and plot (as a bar graph) the probabilities of x successes in 40 independent trials with probability of success 1/2 in a single trial. Why is the figure symmetric? Does the figure remind you of something you have seen in connection with the normal distribution? For which x's are these probabilities very small? (*AEM Ref.* pp. 1079, 1080)

Pr.22.7 **(Permutations)** In how many ways can you assign 8 workers to 8 jobs? First guess, then calculate. (*AEM Ref.* p. 1065)

Pr.22.8 **(Experiment on Stirling formula)** A convenient approximation for (inconvenient) large factorials is $n! \approx \sqrt{2 \pi n} \left(\frac{n}{e}\right)^n$. Find conjectures about the absolute error and the relative error for growing n by experimentation. (*AEM Ref.* p. 1067)

Pr.22.9 **(Poisson distribution)** Investigate the Poisson distribution graphically. What happens if you let μ increase? μ is the mean as well as the variance. Do the graphs give you that impression? Does the distribution approach some kind of symmetry? For what μ? (*AEM Ref.* p. 1081)

Pr.22.10 **(Poisson distribution)** Find and plot the first ten values of the probability function of the Poisson distribution with mean 1. (*AEM Ref.* p. 1081)

Pr.22.11 **(Hypergeometric distribution)** Find and plot the values of the probability function $f(x)$ of the hypergeometric distribution corresponding to randomly drawing 10 screws without replacement from a lot of 100 screws, half of which are right-handed and the other half left-handed, where x is the number of right-handed screws drawn. Is the distribution symmetric? If so, why? (*AEM Ref.* p. 1082)

Pr.22.12 **(Hypergeometric distribution)** If a carton of 20 fuses contains 5 defectives and 3 fuses are randomly drawn from it without replacement, what are the probabilities of obtaining 0, 1, 2, 3 defective fuses? (*AEM Ref.* p. 1084 (#10))

Pr.22.13 **(Uniform distribution)** The uniform distribution on an interval $a \leq x \leq b$ has the density $f(x) = 1/(b-a)$ if $a \leq x \leq b$ and 0 otherwise. Find the mean and the variance on the computer by using the definitions. (*AEM Ref.* p. 1075)

Pr.22.14 **(Normal distribution)** Let X be normal with mean 105 and variance 25. Find $P(X < 112.5)$, $P(X > 100)$, $P(110.5 < X < 111.25)$ on the computer. (*AEM Ref.* p. 1090 (#2))

Pr.22.15 **(Normal distribution)** If sick-leave time X used by employees of some company in one month is (very roughly) normal with mean 1000 hours and standard deviation 100 hours, how much time t should be budgeted for sick leave during the next month if t is to be exceeded with a probability of only 20%? (*AEM Ref.* p. 1091 (#13))

Chapter 23

Mathematical Statistics

Content. Random numbers, samples (Ex. 23.1, Pr. 23.1)
Confidence intervals (Exs. 23.2-23.4, Prs. 23.2-23.5)
Tests (Exs. 23.5-23.10, Prs. 23.6-23.13)
Regression (Ex. 23.11, Prs. 23.14, 23.15)

The **binomial**, **Poisson**, **hypergeometric**, and **normal distributions** discussed in Chap. 22 of this Guide will occur again. For their Mathematica commands see Examples 22.3 and 22.4. New distributions in this chapter are **Student's *t*-distribution** (see Example 23.3), the **chi-square distribution** (Example 23.4), and **Fisher's *F*-distribution** (Example 23.9; also known as **variance-ratio distribution**). These will be needed in connection with confidence intervals and tests. Packages needed in this chapter are listed in the part opening before Chap. 22. An additional package needed in Example 23.11 is <<Statistics`MultiDescriptiveStatistics`.

Examples for Chapter 23

EXAMPLE 23.1 **RANDOM NUMBERS**

Random numbers can be used for obtaining samples (this always means *random samples*, by definition) from populations. Suppose you want to draw a sample of 10 items from a population of 80 items (screws, animals, humans or whatever). Number the items and type for a single value

In[1]:= Random[Integer, {1, 80}]   (* Out: 16   Random number generator *)

and for a sample of size 10

In[2]:= Table[Random[Integer, {1, 80}], {j, 1, 10}]

Out[2]= {40, 65, 36, 67, 63, 34, 48, 33, 22, 31}

(Expect to get different numbers.)
If you call the generator again, you will get another sample

In[3]:= Table[Random[Integer, {1, 80}], {j, 1, 10}]

Out[3]= {61, 27, 12, 33, 23, 54, 71, 74, 63, 56}

and so on. If you have reasons to obtain the same sample (or sequence of samples) with your generator, you can type SeedRandom[i], where i is some positive integer, and then a few samples, say,

In[4]:= SeedRandom[14]

In[5]:= Table[Random[Integer, {1, 80}], {j, 1, 10}]

Out[5]= {26, 45, 4, 29, 66, 3, 67, 21, 40, 69}

In[6]:= Table[Random[Integer, {1, 80}], {j, 1, 10}]

Out[6]= {16, 35, 5, 42, 15, 38, 44, 52, 8, 30}

and if you now type SeedRandom[14], you will get the same samples as before; thus,

In[7]:= SeedRandom[14]

In[8]:= Sa1 = Table[Random[Integer, {1, 80}], {j, 1, 10}]

Out[8]= {26, 45, 4, 29, 66, 3, 67, 21, 40, 69}

In[9]:= Sa2 = Table[Random[Integer, {1, 80}], {j, 1, 10}]

Out[9]= {16, 35, 5, 42, 15, 38, 44, 52, 8, 30}

And so on. (What will you get if you type SeedRandom[3] or SeedRandom[0] ?)

Mean, variance, and standard deviation of samples are obtained as explained in Example 22.1 in this Guide. They vary from sample to sample. Type

In[10]:= <<Statistics'DescriptiveStatistics'

In[11]:= Mean[Sa1]   (* Out: 37 *)

In[12]:= N[Mean[Sa2]]   (* Out: 28.5 *)

In[13]:= N[Variance[Sa1]]   (* Out: 613.778 *)

In[14]:= N[Variance[Sa2]]   (* Out: 268.944 *)

In[15]:= N[StandardDeviation[Sa1]]   (* Out: 24.7745 *)

In[16]:= N[StandardDeviation[Sa2]]   (* Out: 16.3995 *)

Similar Material in AEM: pp. 1053, 1105

EXAMPLE 23.2   **CONFIDENCE INTERVAL FOR THE MEAN OF THE NORMAL DISTRIBUTION WITH KNOWN VARIANCE**

Find a confidence interval for the mean μ of the normal distribution with known variance $\sigma^2 = 9$. Use a sample of 100 values with sample mean $\bar{x} = 5$. Choose the confidence level $\gamma = 95\%$.

Solution. Regard $\bar{x}$ as an observed value of a random variable $\bar{X} = (1/n)(X_1 + \ldots + X_n)$, where $X_1, \ldots, X_n$ are independent random variables all having the same distribution of some random variable X. It can be shown that if X is normal with mean μ and variance σ^2, then $\bar{X}$ is normal with mean μ and variance σ^2/n.

Type the given data, denoting the confidence level γ by g.

In[1]:= <<Statistics'ContinuousDistributions'

In[2]:= n = 100;   xbar = 5;   var = 9;   g = 0.95;

Consider the standardized normal distribution NormalDistribution[0, 1] (see Example 22.4). You obtain the shortest interval on the x-axis corresponding to $\gamma = 95\%$ of the area under the density curve (that is, corresponding to the probability 0.95) if you choose that interval symmetrically located with respect to the mean 0. Then its endpoints $-c$ and $+c$ correspond to the probabilities 2.5% and 97.5%. You get the latter by typing

In[3]:= c = Quantile[NormalDistribution[0, 1], 0.975]   (* Out: 1.95996 *)

For $\bar{X}$ this corresponds to $\mu - k$ and $\mu + k$, where $k = c\sigma/\sqrt{n}$. Accordingly, type

In[4]:= k = c Sqrt[var]/Sqrt[n]   (* Out: 0.587989 *)

You now get the confidence interval by replacing $\mu - k$ with $\bar{x} - k$ and $\mu + k$ with $\bar{x} + k$

In[5]:= conf1 = xbar - k   (* Out: 4.41201 *)

In[6]:= conf2 = xbar + k   (* Out: 5.58799 *)

The confidence interval is $CONF_{0.95}(4.41 \leq \mu \leq 5.59)$.

In this approach you used the standardized normal distribution, as it has been tabulated. On the computer you can proceed more directly by noting that the midpoint of the confidence interval is $\bar{x}$ and the endpoints are the 2.5%-point and the 97.5%-point of the distribution of $\bar{X}$, which is normal with variance $\sigma^2/n = 9/100$, hence with standard deviation $\sigma/\sqrt{n} = 0.3$. You thus obtain directly

In[7]:= Quantile[NormalDistribution[xbar, 0.3], 0.025]   (* Out: 4.41201 *)

In[8]:= Quantile[NormalDistribution[xbar, 0.3], 0.975]   (* Out: 5.58799 *)

Similar Material in AEM: p. 1110

EXAMPLE 23.3	**CONFIDENCE INTERVAL FOR THE MEAN OF THE NORMAL DISTRIBUTION WITH UNKNOWN VARIANCE. *t*-DISTRIBUTION**

Find a confidence interval for the mean of the normal distribution with unknown variance σ^2. Use the sample 144, 147, 146, 142, 144 (flashpoint in degrees Fahrenheit of a certain kind of Diesel oil). Choose the confidence level $\gamma = 99\%$.

Solution. Regard $\bar{x}$ as an observed value of a random variable $\bar{X}$, as in the previous example. $k = c\sigma/\sqrt{n}$ can no longer be used because σ is no longer known. The idea is to replace σ by the sample standard deviation s as defined in Example 22.1 in this Guide, and to regard s as an observed value of a random variable S. To use S, one must know its probability distribution. Student (pseudonym for W. S. Gosset) has shown that if the population random variable X is normal with mean μ and variance σ^2, then $t = \dfrac{\bar{x} - \mu}{s/\sqrt{n}}$ is an observed value of a random variable T which has a *t-distribution* with $n - 1$ **degrees of freedom (df)**, and he gave the formula for the probability density of T, which is symmetric with respect to 0. Accordingly, type the given data, and calculate $\bar{x}$ and s.

In[1]:= sample = {144, 147, 146, 142, 144};

In[2]:= <<Statistics`DescriptiveStatistics`

In[3]:= <<Statistics`ContinuousDistributions`

In[4]:= xbar = N[Mean[sample]]   (* Out: 144.6 *)

In[5]:= s = N[StandardDeviation[sample]]   (* Out: 1.94936 *)

Now obtain the 99.5%-point c for the t-distribution with $n - 1 = 4$ degrees of freedom and calculate $K = Cs/\sqrt{5}$, the counterpart of k in the previous example.

In[6]:= c = Quantile[StudentTDistribution[4], 0.995]   (* Out: 4.60409 *)

In[7]:= K = c s/Sqrt[5]   (* Out: 4.01376   (n = 5 is the sample size.) *)

You now get the confidence interval by replacing $\mu - K$ with $\bar{x} - K$ and $\mu + K$ with $\bar{x} + K$.

In[8]:= conf1 = xbar - K   (* Out: 140.586 *)

In[9]:= conf2 = xbar + K   (* Out: 148.614 *)

This gives the confidence interval $CONF_{0.99}(140 \leq \mu \leq 149)$. This is rather large, but keep in mind that your sample was small. σ was unknown. If it were known and equal to s, you should get a shorter interval because you use more information. Can you calculate this?

Similar Material in AEM: p. 1113

EXAMPLE 23.4   CONFIDENCE INTERVAL FOR THE VARIANCE OF THE NORMAL DISTRIBUTION. χ^2-DISTRIBUTION

Find a confidence interval for the unknown variance σ^2 of the normal distribution, using the following sample of size $n = 14$ and choosing the confidence level $\gamma = 95\%$. (The population mean μ need not be known.)

$$89 \; 84 \; 87 \; 81 \; 89 \; 86 \; 91 \; 90 \; 78 \; 89 \; 87 \; 99 \; 83 \; 89$$

Solution. It can be shown that under the normality assumption the quantity

$$y = (n-1)s^2/\sigma^2$$

is an observed value of a random variable Y that has a **chi-square distribution** with $n - 1 = 13$ degrees of freedom. Here, s^2 is the sample variance, as before. Type the sample and then $nva = (n - 1) \; s^2$.

In[1]:= <<Statistics`DescriptiveStatistics`

In[2]:= <<Statistics`ContinuousDistributions`

In[3]:= sample = {89, 84, 87, 81, 89, 86, 91, 90, 78, 89, 87, 99, 83, 89}

Out[3]= {89, 84, 87, 81, 89, 86, 91, 90, 78, 89, 87, 99, 83, 89}

In[4]:= N[Variance[sample]]   (* Out: 25.1429 *)

In[5]:= nva = % 13   (* Out: 326.857 *)

Determine the 2.5%-point and the 97.5%-point of the chi-square distribution (which is not symmetric!) with 13 degrees of freedom,

In[6]:= c1 = Quantile[ChiSquareDistribution[13], 0.025]   (* Out: 5.00875 *)

In[7]:= c2 = Quantile[ChiSquareDistribution[13], 0.975]   (* Out: 24.7356 *)

From this you obtain the endpoints of the confidence interval

In[8]:= conf1 = nva/c1   (* Out: 65.2572 *)

In[9]:= conf2 = nva/c2   (* Out: 13.214 *)

This gives the confidence interval $CONF_{0.95}(13.21 \leq \sigma^2 \leq 65.26)$.

Similar Material in AEM: pp. 1114-1116

EXAMPLE 23.5	TEST FOR THE MEAN OF THE NORMAL DISTRIBUTION

You want to buy 500 coils of wire. Test the manufacturer's claim that the breaking limit X of the wire is $\mu = \mu_0 = 200$ lb (or more). Assume that X has a normal distribution.

Solution. Test the **hypothesis** $\mu = \mu_0 = 200$ against the **alternative** $\mu = \mu_1 < 200$, an undesirable weakness. Hence this test is **left-sided**, the **rejection region** extends from a **critical point** c to the left.

To obtain a sample, select some of the coils, say, 25, at random. Cut a piece from each coil and determine the breaking limit experimentally. Suppose that this sample of $n = 25$ values has the mean $\bar{x} = 197$ lb (somewhat less than the claim!) and the standard deviation $s = 6$ lb. Then (as in Example 23.3 in this Guide)

In[1]:= `t = N[(197 - 200)/(6/Sqrt[25])]`   (* Out: −2.5 *)

is an observed value of a random variable T that has a t–distribution with $n - 1 = 24$ degrees of freedom. Choose a **significance level** of the test, say $\alpha = 5\%$. Find the critical c by typing (similarly as in Example 23.3 in this Guide)

In[2]:= `<<Statistics'ContinuousDistributions'`

In[3]:= `c = Quantile[StudentTDistribution[24], 0.05]`   (* Out: −1.71088 *)

Reason as follows. If the hypothesis is true, the probability of obtaining a $t < c$ is very small, namely, equal to $\alpha = 5\%$, so that it would happen only about once in 20 tests. Hence if it happens, as in the present case, where $-2.50 < -1.71$, cast doubt on the truth of the hypothesis. Hence **reject the hypothesis** and assert that $\mu < 200$ and the manufacturer had promised too much.

Similar Material in AEM: pp. 1118, 1119

EXAMPLE 23.6	TEST FOR THE MEAN: POWER FUNCTION

You make an **error of the first kind** if you reject a hypothesis although it is true. You do this with probability α, the significance level of the test. You make an **error of the second kind** if you accept a hypothesis although the alternative is true. The corresponding probability is denoted by β. The quantity $\eta = 1 - \beta$ (thus the probability of avoiding an error of the second kind) is called the **power** of the test. β depends on the alternative μ. One calls $\beta(\mu)$ the **operating characteristic (OC)** and $\eta(\mu) = 1 - \beta(\mu)$ the **power function** of the test.

Let X be normal with mean μ and known variance $\sigma^2 = 9$. Then $\bar{X}$ is normal with mean μ and variance σ^2/n, where n is the size of the sample used in the test (see Example 23.2 in this Guide). Let the hypothesis be $\mu_0 = 24$. Choose $\alpha = 5\%$. Let $n = 10$. Then the standard deviation of $\bar{X}$ is $sd = 3/\sqrt{10}$.

1. Left-sided test (as in the previous example). The **critical region** (rejection region) extends from the critical $c = c_1$ to the left. Obtain the critical c_1 by typing

In[1]:= `<<Statistics'ContinuousDistributions'`

In[2]:= `sd = 3/Sqrt[10]`   (* Out: $\frac{3}{\sqrt{10}}$ *)

In[3]:= `c1 = Quantile[NormalDistribution[24, sd], 0.05]`   (* Out: 22.4396 *)

Power of the left-sided test

The power is the area under the density curve of $\bar{X}$ with the alternative μ being true, from $-\infty$ to c_1. This curve, and hence the area, depends on μ. The power is practically 1 at $\mu = 20$, decreases monotone to 0.05 at $\mu = 24$ and practically to 0 at $\mu = 26$; see the figure. **Erf** is the **error function** (see AEM, p. A56), in terms of which Mathematica expresses the distribution function of the normal distribution.

In[4]:= Clear[mu]

In[5]:= powerleft = CDF[NormalDistribution[mu, sd], c1]

Out[5]= $\frac{1}{2}\left(1 + \text{Erf}\left[\frac{1}{3}\sqrt{5}\,(22.4396 - \text{mu})\right]\right)$

In[6]:= Simplify[%]

Out[6]= $\frac{1}{2}\left(1 + \text{Erf}[16.7255 - 0.745356\,\text{mu}]\right)$

2. Right-sided test. The critical region now extends from the critical $c = c_2$ to the right. Obtain the critical $c = c_2$ by typing

In[7]:= c2 = Quantile[NormalDistribution[24, sd], 0.95]   (* Out: 25.5604 *)

Power of the right-sided test

The power is the area under the density curve of $\bar{X}$ with the alternative μ being true, from c_2 to ∞, and again depends on μ.

In[8]:= powerright = 1 - CDF[NormalDistribution[mu, sd], c2]

Out[8]= $1 + \frac{1}{2}\left(-1 - \text{Erf}\left[\frac{1}{3}\sqrt{5}\,(25.5604 - \mu)\right]\right)$

In[9]:= Simplify[%]   (* Out: $\frac{1}{2}\,(1 - \text{Erf}[19.0516 - 0.745356\,\text{mu}])$ *)

3. Two-sided test. The critical region now consists of two parts, from $-\infty$ to a lower critical point $c = c_{3a}$ and from an upper critical point $c = c_{3b}$ to ∞. These are the 2.5%- and 97.5%-points of the distribution of $\bar{X}$ with the hypothesis $\mu_0 = 24$ being true. Obtain these points by typing

In[10]:= c3a = Quantile[NormalDistribution[24, sd], 0.025]   (* Out: 22.1406 *)

In[11]:= c3b= Quantile[NormalDistribution[24, sd], 0.975]   (* Out: 25.8594 *)

Power of the two-sided test

The power is the area under the density curve of $\bar{X}$ with the alternative being true, from $-\infty$ to c_{3a} and from c_{3b} to ∞. This is the U-shaped curve in the figure.

In[12]:= powertwosided = CDF[NormalDistribution[mu, sd], c3a] +
            1 - CDF[NormalDistribution[mu, sd], c3b]

Out[12]= $1 + \frac{1}{2}\left(1 + \text{Erf}\left[\frac{1}{3}\sqrt{5}\,(22.1406 - \mu)\right]\right) + \frac{1}{2}\left(-1 - \text{Erf}\left[\frac{1}{3}\sqrt{5}\,(25.8594 - \mu)\right]\right)$

In[13]:= Simplify[%]

Out[13]= $\frac{1}{2}\left(2 + \text{Erf}[16.5026 - 0.745356\,\text{mu}] - \text{Erf}[19.2744 - 0.745356\,\mu]\right)$

Plotting all three curves on common axes is now quite simple. (P1 needs PlotRange .
Try without.)

In[14]:= P1 = Plot[powerleft, {mu, 20, 30}, PlotRange -> {0, 1}]

In[15]:= P2 = Plot[powerright, {mu, 20, 30}]

In[16]:= P3 = Plot[powertwosided, {mu, 20, 30}]

In[17]:= Show[P1, P2, P3]

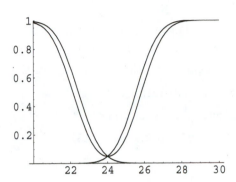

Example 23.6. Power functions of tests of $\mu_0 = 24$
against $\mu < 24$, $\mu > 24$, and $\mu \neq 24$

Similar Material in AEM: pp. 1122-1124

| EXAMPLE 23.7 | **TEST FOR THE VARIANCE OF THE NORMAL DISTRIBUTION** |

Using a sample of size $n = 15$ and sample variance $s^2 = 13$ from a normal population,
test the hypothesis $\sigma^2 = \sigma_0{}^2 = 10$ against the alternative $\sigma^2 = \sigma_1{}^2 = 20$.

Solution. It can be shown that under the normality assumption and under the hypothesis the quantity

$$y = (n-1)s^2/\sigma_0{}^2 = 14 \times 13/10 = 18.2$$

is an observed value of a random variable $Y = 14\,S^2/10$ that has a **chi-square distribution** with $n - 1 = 14$ degrees of freedom. (See also Example 23.4 in this Guide.)
Since the alternative is greater than the hypothesis, the test is right-sided. Choose a
significance level, say, $\alpha = 5\%$. Determine the 95%-point c of that distribution from

In[1]:= <<Statistics`ContinuousDistributions`

In[2]:= c = Quantile[ChiSquareDistribution[14], 0.95]   (* Out: 23.6848 *)

Since $y < c$ (and the test is right-sided), accept the hypothesis.

If the alternative is true, $Y_1 = (n-1)S^2/\sigma_1{}^2 = 14S^2/20 = Y/2$ has a chi-square
distribution with 14 degrees of freedom. Hence the **power** is the area under the
density curve of Y_1 from $c/2$ to ∞. That is,

In[3]:= power = 1 - CDF[ChiSquareDistribution[14], c/2]   (* Out: 0.618958 *)

This leaves a probability of 38% for committing an error of the second kind. This is
too large, and you should repeat the test with a larger sample (if available!).

Similar Material in AEM: pp. 1124, 1125

EXAMPLE 23.8   COMPARISON OF MEANS

Test the hypothesis that two normal distributions with the same variance have the same mean, $\mu_1 = \mu_2$, against the alternative that they have different means. Choose the significance level $\alpha = 5\%$. Use two samples $x_1, x_2, ..., x_{n_1}$ and $y_1, y_2, ..., y_{n_2}$ which are independent. (Dependence would mean that some x-values are related to some y-values, for instance, if they came from the two front tires of the same car, from test scores of the same student, etc. Equality of variances will be tested in the next example.)

$$111 \quad 114 \quad 91 \quad 109 \quad 108 \quad 107 \quad 113$$
$$89 \quad 92 \quad 84 \quad 97 \quad 103 \quad 107 \quad 111 \quad 97$$

Solution. Calculate the mean $\bar{x}$ and the variance $s_x{}^2$ of the first sample by typing

In[1]:= <<Statistics`DescriptiveStatistics`

In[2]:= <<Statistics`ContinuousDistributions`

In[3]:= Sa1 = {111, 114, 91, 109, 108, 107, 113}

In[4]:= xbar = N[Mean[Sa1]]   (* Out: 107.571 *)

In[5]:= xvar = N[Variance[Sa1]]   (* Out: 59.9524 *)

Calculate the mean $\bar{y}$ and the variance $s_y{}^2$ of the second sample by typing

In[6]:= Sa2 = {89, 92, 84, 97, 103, 107, 111, 97}

In[7]:= ybar = N[Mean[Sa2]]   (* Out: 97.5 *)

In[8]:= yvar = N[Variance[Sa2]]   (* Out: 84. *)

It can be shown that if the hypothesis is true, the quantity

$$t_0 = \sqrt{\frac{n_1 \, n_2 \, (n_1 + n_2 - 2)}{n_1 + n_2}} \; \frac{\bar{x} - \bar{y}}{\sqrt{(n_1 - 1) \, s_x{}^2 + (n_2 - 1) \, s_y{}^2}}$$

is an observed value of a random variable T that has a t-distribution with $n_1 + n_2 - 2$ degrees of freedom. The test is two-sided. Since $\alpha = 5\%$, determine the 2.5%-point c_1 and the 97.5%-point c_2 of the t-distribution with $n_1 + n_2 - 2 = 8 + 7 - 2 = 13$ degrees of freedom by typing

In[9]:= Quantile[StudentTDistribution[13], 0.025]   (* Out: −2.16037 *)

In[10]:= Quantile[StudentTDistribution[13], 0.975]   (* Out: 2.16037 *)

If t_0 lies between these values (inclusively), accept the hypothesis. Otherwise reject it. Calculate

In[11]:= t0 = Sqrt[8 7 (8 + 7 - 2)/(8 + 7)] (xbar - ybar)/Sqrt[6 xvar + 7 yvar]

Out[11]= 2.27915

Reject the hypothesis and assert that the populations from which the samples were drawn have different means.

Similar Material in AEM: pp. 1125, 1126

EXAMPLE 23.9   COMPARISON OF VARIANCES. *F*-DISTRIBUTION

Test the hypothesis that the variances of the two normal distributions in the previous example are equal against the alternative that they are different. Choose the significance level 5%.

Solution. Type the samples and calculate their variances.

In[1]:= <<Statistics`DescriptiveStatistics`

In[2]:= <<Statistics`ContinuousDistributions`

In[3]:= Sa1 = {111, 114, 91, 109, 108, 107, 113}

In[4]:= xvar = N[Variance[Sa1]]   (* Out: 59.9524 *)

In[5]:= Sa2 = {89, 92, 84, 97, 103, 107, 111, 97}

In[6]:= yvar = N[Variance[Sa2]]   (* Out: 84. *)

It can be shown that if the hypothesis is true, the ratio $v_0 = s_x^2/s_y^2$ is an observed value of a random variable V which has an **F-distribution** with $(n_1 - 1, \ n_2 - 1) = (6, \ 7)$ degrees of freedom. Since $\alpha = 5\%$ and the test is two-sided, determine the 2.5%-point c_1 and the 97.5%-point c_2 of this distribution by typing

In[7]:= c1 = Quantile[FRatioDistribution[6, 7], 0.025]   (* Out: 0.175578 *)

In[8]:= c2 = Quantile[FRatioDistribution[6, 7], 0.975]   (* Out: 5.1186 *)

Accept the hypothesis because v_0 lies between c_1 and c_2. Indeed,

In[9]:= v0 = xvar/yvar   (* Out: 0.713719 *)

Similar Material in AEM: p. 1126

EXAMPLE 23.10   CHI-SQUARE TEST FOR GOODNESS OF FIT

With this test you find out how well the distribution of a sample fits the hypothetical distribution of the population. For instance, can you claim on the 5%-level that a die is fair if in 20,000 trials you obtain $x = 1, 2, ..., 6$ with the following absolute frequencies (actual classical data obtained by R. Wolf in Switzerland).

$$3407 \quad 3631 \quad 3176 \quad 2916 \quad 3448 \quad 3422$$

Solution. If the die is fair, each of the 6 values is equally likely, hence the expected absolute frequency is $e = 3333.33$. For each x calculate the observed value minus e, square it and divide the result by e. The sum of the 6 numbers thus obtained is an observed value of a random variable χ which is (asymptotically) chi-square distributed with $6 - 1 = 5$ degrees of freedom. Call this sum χ_0 . Clearly, it measures the discrepancy between observations and expectation. Type

In[1]:= <<Statistics`DescriptiveStatistics`

In[2]:= <<Statistics`ContinuousDistributions`

In[3]:= data = {3407, 3631, 3176, 2916, 3448, 3422}

In[4]:= e = 20000/6   (* Out: $\frac{10000}{3}$ *)

In[5]:= chi0 = N[Sum[(data[[j]] - e)^2/e, {j, 1, 6}]]   (* Out: 94.189 *)

The test is right-sided. The rejection region (critical region) extends from the critical

c to the right and corresponds to a probability of 5%. Hence c is the 95%-point of the chi-square distribution with 5 degrees of freedom,

In[6]:= c = Quantile[ChiSquareDistribution[5], 0.95]   (* Out: 11.0705 *)

You see that χ_0 is much larger than c. Reject the hypothesis and assert that Wolf's die was not fair or there were serious flaws in throwing and/or counting.
      Similar Material in AEM: p. 1140 (#4)

EXAMPLE 23.11   REGRESSION

In regression analysis you choose values $x_1, x_2, ..., x_n$ of an ordinary variable x (for instance, x may be time) and observe corresponding values $y_1, y_2, ..., y_n$ of a random variable Y (for instance, temperature at some place). This gives a sample of n pairs $(x_1, y_1), ..., (x_n, y_n)$. You assume that the mean of Y depends linearly on x, say, $\mu(x) = \kappa_0 + \kappa_1 x$. This is the **regression line of the population.** Let the sample be (x = pressure in atmospheres, y = decrease of volume of leather in %)

x	4000	6000	8000	10000
y	2.3	4.1	5.7	6.9

From the sample you obtain the **sample regression line** $y = k_0 + k_1 x$. You fit this line through the given points (given pairs of coordinate values in the xy-plane) by the **least squares principle**, as follows.

In[1]:= <<Statistics`DescriptiveStatistics`

In[2]:= <<Statistics`ContinuousDistributions`

In[3]:= xSa = {4000, 6000, 8000, 10000}   (* (the x-values) *)

In[4]:= ySa = {2.3, 4.1, 5.7, 6.9}   (* (the y-values) *)

In[5]:= data = Table[{xSa[[j]], ySa[[j]]}, {j, 1, 4}]

Out[5]= {{4000, 2.3}, {6000, 4.1}, {8000, 5.7}, {10000, 6.9}}

In[6]:= f = Fit[data, {1, x}, x]   (* Out: $-0.64 + 0.00077\,x$ *)

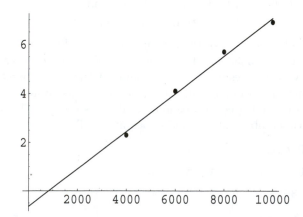

Example 23.11. Sample regression line and given data (4 points)

This is the sample regression line. Obtain the figure by the subsequent commands.

From the figure you see that for $x = 4000 \ldots 10000$ the line fits the data reasonably well (but would be useless near $x = 0$ because there is no decrease of volume when the pressure is 0).

In[7]:= `Clear[P1, P2]`

In[8]:= `P1 = Plot[f, {x, 0, 10000}]`

In[9]:= `P2 = ListPlot[data]`

In[10]:= `Show[{P1, P2}, Prolog -> AbsolutePointSize[3]]`

The least squares principle is purely geometric. Probability enters through random variables. To obtain confidence intervals or tests for the **regression coefficient** κ_1, you must make assumptions about the probability distribution of Y. To apply the theory on the normal distribution (as in the previous examples), make the reasonable assumption that Y is normal and its variance σ^2 is the same for all x.

For obtaining a confidence interval for κ_1 you will need the variance of the x-values, the variance of the y-values, and the covariance. Accordingly, type

In[11]:= `n = 4;`   `(* Sample size (= number of pairs = number of points) *)`

In[12]:= `xbar = Mean[xSa]`   `(* Out: 7000 *)`

In[13]:= `ybar = Mean[ySa]`   `(* Out: 4.75 *)`

In[14]:= `xvar = N[Variance[xSa]]`   `(* Out: 6.66667 × 10^6 *)`

In[15]:= `yvar = Variance[ySa]`   `(* Out: 3.98333 *)`

In[16]:= `<<Statistics'MultiDescriptiveStatistics'`

In[17]:= `xycovar = Covariance[xSa, ySa]`   `(* Out: 5133.33 *)`

Then type the formula for the sample regression coefficient k_1 and for an auxiliary quantity q_0.

In[18]:= `k1 = xycovar/xvar`   `(* Out: 0.00077 *)`

In[19]:= `q0 = (n - 1) (yvar - k1^2 xvar)`   `(* Out: 0.092 *)`

It can be shown that under your assumptions the random variable Y has a Student's t-distribution with $n - 2 = 2$ degrees of freedom. Choose a confidence level, say, 95%. Then type the commands for the 2.5%-point and the 97.5%-point of that distribution, actually, only the latter, c, because the former is $-c$, by the symmetry of the distribution.

In[20]:= `c = Quantile[StudentTDistribution[n - 2], 0.975]`

Out[20]= `4.30265`

Half the length of the confidence interval is

In[21]:= `k = c Sqrt[q0/((n - 2)(n - 1) xvar)]`   `(* Out: 0.000206348 *)`

With this you obtain the confidence interval $CONF_{0.95}(0.00056 \le \kappa_1 \le 0.00098)$ because its endpoints are

In[22]:= `conf1 = k1 - k`   `(* Out: 0.000563652 *)`

In[23]:= `conf2 = k1 + k`   `(* Out: 0.000976348 *)`

Similar Material in AEM: pp. 1145-1149

Problem Set for Chapter 23

Pr.23.1 (**Experiment on sample mean**) Find out experimentally how sample means vary from sample to sample. Plot a histogram of their frequency function. *Suggestion:* 100 samples of size 5 obtained by the random generator from a population of size 50. Use `Seed[6];`, so that you can reproduce your samples if needed.
(*AEM Ref.* p. 1105)

Pr.23.2 (**Confidence interval for the mean**) Find a 99% confidence interval for the mean of a normal population with standard deviation 2.5, using the sample 30.8, 30.0, 29.9, 30.1, 31.7, 34.0. (*AEM Ref.* p. 1117 (#3))

Pr.23.3 (**Length of confidence interval**) Plot the length of a 95% confidence interval as a function of sample size n and measured in multiples of σ, for the mean of the normal distribution with known variance (*AEM Ref.* p. 1112)

Pr.23.4 (**Confidence interval for the mean**) What confidence interval would you obtain in Example 23.3 in this Guide if σ were known and equal to $s = 1.9494$ (the value in that example), the other data being as before?

Pr.23.5 (**Confidence interval for the variance**) Find a 95% confidence interval for the variance of the normal distribution, using the sample of carbon monoxide emission (grams/mile) of a passenger car cruising at a speed of 55 mph. 17.3, 17.8, 18.0, 17.7 18.2, 17.4, 17.6, 18.1. (*AEM Ref.* p. 1117 (#16))

Pr.23.6 (**Test for the mean**) Test the hypothesis $\mu_0 = 24$ against the alternative $\mu_1 = 27$, choosing $\alpha = 5\%$ and using a sample of size 10 with mean 25.8 from a normal population with variance 9. Is the power of the test sufficiently large?
(*AEM Ref.* pp. 1122-1124)

Pr.23.7 (**Test for the mean**) If a standard treatment cures about 75% of patients suffering from a certain disease, and a new treatment cured 310 of the first 400 patients on whom it was tried, can you conclude that the new treatment is better? First guess. Then calculate, choosing $\alpha = 5\%$ and using the fact that $X = $ *Number of cases cured in 400 cases* is about normal with mean np and variance $np(1-p)$.
(*AEM Ref.* p. 1127 (#12))

Pr.23.8 (**Dependence of power on sample size**) How does the figure in Example 23.6 in this Guide change if you take a larger sample (of size $n = 100$, for instance)? Give the reason. Plot a new figure for $n = 100$. (*AEM Ref.* pp. 1123, 1124)

Pr.23.9 (**Test for the variance**) Suppose that in the past the standard deviation of weights of certain 100.0-oz packages filled by a machine was 0.8 oz. Test the hypothesis $H_0 : \sigma = 0.8$ against the alternative $H_1 : \sigma > 0.8$ (an undesirable increase), using a sample of 20 packages with standard deviation 1.0 oz, assuming normality and choosing $\alpha = 5\%$. (*AEM Ref.* p. 1127 (#15))

Pr.23.10 (**Comparison of means**) Will an increase of temperature increase the yield (measured in grams/min) of some chemical process? Test this, using the following independent samples, assuming normality and choosing $\alpha = 5\%$. (*AEM Ref.* p. 1125)

Yield x at 40°C   116 123 121 105 138 135 119 111 115 125

Yield y at 65°C   130 125 134 112 145 137 122 139

Pr.23.11 **(Paired comparison of means)** Measure the electric voltage in a circuit at the same instants simultaneously by two kinds of voltmeters. Test the hypothesis that there is no difference in the calibration of the two kinds of instruments against the alternative that there is a difference. In this case you merely need a sample of differences of corresponding measurements ("**paired comparison**"), say, 0.4, −0.6, 0.2, 0.0, 1.0, 1.4, 0.4, 1.6. Assume normality and choose $\alpha = 5\%$. (*AEM Ref.* p. 1127 (#13))

Pr.23.12 **(Comparison of variances)** Test that the variances of the populations in Pr.23.10 are equal against the alternative that they are different. Choose $\alpha = 5\%$. (*AEM Ref.* p. 1126)

Pr.23.13 **(Goodness of fit)** Can you assert that the traffic on the three lanes of an expressway (in one direction) is about the same on each lane if a count gives 920, 870, 750 cars on the right, middle, and left lanes, respectively, during the same interval of time? (*AEM Ref.* p. 1141 (#13))

Pr.23.14 **(Linear regression)** If a sample of 9 pairs of values $[x_j, y_j]$ has the variance of the x-values 118.000, the variance of the y-values 215.125, and the covariance −155.750, what can you say about the sample regression line? What about a 95% confidence interval for the regression coefficient (the slope) κ_1 if you assume Y to be normal with variance independent of x? (*AEM Ref.* pp. 1145-1149)

Pr.23.15 **(Quadratic regression parabola)** Fit a quadratic parabola through the following data. Plot the curve and the given data (as points) on common axes. Use the command `Fit` (see Example 23.11 in this Guide).

x	1	2	4	5	7	8
y	7	5	2	1	2	4

Appendix 1

References

[1] Abramowitz, M. and I. A. Stegun (eds.), *Handbook of Mathematical Functions.* 10th printing, with corrections. Washington, DC: National Bureau of Standards, 1972. (Also New York: Dover, 1965.)

[2] Kreyszig, E., *Advanced Engineering Mathematics.* 8th ed. New York: Wiley, 1999.

[3] Kreyszig, H., and E. Kreyszig, *Student Solutions Manual.to Accompany Advanced Engineering Mathematics*, Eighth Edition. New York: Wiley, 2000.

[4] Wilkinson, J. H., The Algebraic Eigenvalue Problem. Oxford: Clarendon, 1988.

[5] Wolfram, S., The MATHEMATICA Book, 4th ed., Champaign, Il: Wolfram Media/Cambridge University Press, 1999.

Appendix 2

Answers to Odd-Numbered Problems

CHAPTER 1,   page 14

Pr.1.1.

```
In[1]:= ode = y'[x] == y[x]^2                          (* Out: y'[x] = y[x]^2 *)
In[2]:= << Graphics'PlotField'
In[3]:= p1 = PlotVectorField[{1, y^2}, {x, -3, 3}, {y, 0, 3}]
In[4]:= sol1 = DSolve[{ode, y[0] == 1}, y[x], x]
In[5]:= sol2 = DSolve[{ode, y[0] == 2}, y[x], x]
In[6]:= sol3 = DSolve[{ode, y[0] == 3}, y[x], x]
In[7]:= p2 = Plot[{sol1[[1, 1, 2]], sol2[[1, 1, 2]], sol3[[1, 1, 2]]},
        {x, -3, 3}]
In[8]:= Show[p1, p2, PlotRange -> {0, 3}]
```

Pr.1.3.

```
In[1]:= Clear[y]
In[2]:= sol = DSolve[{y'[t] == Exp[0.2 t], y[0] == 2}, y[t], t]
In[3]:= Plot[sol[[1, 1, 2]], {t, 0, 10}]
```

Pr.1.5.

```
In[1]:= ode = y'[t] == k y[t]
In[2]:= yp = DSolve[{ode, y[0] == 4}, y[t], t]      (* Out: {{y[t] → 4 e^{k t}}} *)
In[3]:= eq = yp /. {t -> 1, y[1] -> 2}               (* Out: {{y[1] → 4 e^k}} *)
In[4]:= k0 = Solve[eq[[1, 1, 2]] == 2, k]            (* Out: {{k → -Log[2]}} *)
In[5]:= yp2 = yp /. k -> k0[[1, 1, 2]]               (* Out: {{y[t] → 2^{2-t}}} *)
```

Pr.1.7.

```
In[1]:= y1 = Tan[x]
In[2]:= D[y1, x] == 1 + y1^2
In[3]:= Simplify[%]                                   (* Out: True *)
```

Pr.1.9.

```
In[1]:= M = x^3 + 3 x y^2
In[2]:= N1 = 3 x^2 y + y^3                             (* N is protected! *)
```

A2

In[3]:= D[M, y] - D[N1, x]   (* Out: 0   The equation is exact. *)

In[4]:= u1 = Integrate[M, x] $\left(* \text{ Out: } \frac{x^4}{4} + \frac{3 x^2 y^2}{2} *\right)$

In[5]:= u2 = Integrate[N1, y] $\left(* \text{ Out: } \frac{3 x^2 y^2}{2} + \frac{y^4}{4} *\right)$

Hence an implicit solution is $x^4 + 6 x^2 y^2 + y^4 = c$ because

In[6]:= u = x^4 + 6 x^2 y^2 + y^4

In[7]:= Simplify[(D[u, x] dx + D[u, y] dy)/4]

Out[7]= dy y $(3 x^2 + y^2)$ + dx $(x^3 + 3 x y^2)$

Pr.1.11.

In[1]:= ode = y'[x] == 2 (y[x] - 1) Tan[2 x]

In[2]:= sol = DSolve[ode, y[x], x]

In[3]:= sol2 = sol /. x -> 0   (* Out: {{y[0] $\rightarrow 1 + C[1]$}} *)

In[4]:= C0 = Solve[sol2[[1, 1, 2]] == 4, C[1]]

In[5]:= yp = sol /. C[1] -> C0[[1, 1, 2]]   (* Out: {{y[x] $\rightarrow 1 + 3 \operatorname{Sec}[2 x]$}} *)

In[6]:= Plot[yp[[1, 1, 2]], {x, 0, 0.7}]

Pr.1.13.

The answer is $y = (x + c)e^{-kx}$. To obtain it from the integral formula, type

In[1]:= h = Integrate[k, x]

In[2]:= y = Exp[-h] (Integrate[Exp[h] Exp[-k x], x] + c)

DSolve gives the answer if you type

In[3]:= Clear[y]   (* Unassign y, which has just been used. *)

In[4]:= sol = DSolve[y'[x] + k y[x] == Exp[-k x], y[x], x]

Out[4]= {{y[x] $\rightarrow e^{-kx} x + e^{-kx} C[1]$}}

Pr.1.15.

In[1]:= ode = y'[x] - 3 y[x] == -5 y[x]^2

In[2]:= sol = DSolve[ode, y[x], x] $\left(* \text{ Out: } \left\{\left\{y[x] \rightarrow \frac{3 e^{3 x}}{5 e^{3 x} - e^{C[1]}}\right\}\right\} *\right)$

In[3]:= y1 = sol /. Exp[C[1]] -> 2

In[4]:= y2 = sol /. Exp[C[1]] -> 0

In[5]:= y3 = sol /. Exp[C[1]] -> -10

In[6]:= Plot[{y1[[1, 1, 2]], y2[[1, 1, 2]], y3[[1, 1, 2]]}, {x, 0, 1},
        PlotRange -> {0, 1}, AxesLabel -> {Time x, Population}]

Pr.1.17. Type the ODE, solve it by DSolve , then substitute the special values for which you are supposed to plot the solution. Note that the time interval of the transient current is practically very short, due to the rapid decrease of the exponential term. C is protected; write C1 . Also write w for ω .

```
In[1]:= ode = R i'[t] + i[t]/C1 == D[E0 Sin[w t], t]
```

```
In[2]:= sol = DSolve[ode, i[t], t]
```

```
In[3]:= sol2 = Simplify[%]
```

```
In[4]:= ipartic = sol2 /. {R -> 1, C1 -> 1, w -> 1, E0 -> 220, C[1] -> -110}
```

Out[4]= $\{\{i[t] \to -110\,e^{-t} + 110(\text{Cos}\,[t] + \text{Sin}\,[t])\}\}$

```
In[5]:= Plot[ipartic[[1, 1, 2]], {t, 0, 20},
        Ticks -> {Automatic, {0, 50, -50, 100, -100, 150, -150}} ]
```

Pr.1.19.

```
In[1]:= y0 = 0
```

```
In[2]:= pic[1] = y0 + Integrate[1 + (y0 /. x -> t)^2, {t, 0, x}]
```

Out[2]= x

```
In[3]:= pic[2] = y0 + Integrate[1 + (pic[1] /. x -> t)^2, {t, 0, x}]
```

Out[3]= $x + \dfrac{x^3}{3}$

```
In[4]:= pic[3] = y0 + Integrate[1 + (pic[2] /. x -> t)^2, {t, 0, x}]
```

Out[4]= $x + \dfrac{x^3}{3} + \dfrac{2\,x^5}{15} + \dfrac{x^7}{63}$

CHAPTER 2,   page 28

Pr.2.1.

```
In[1]:= ode = y''[x] + 3 y'[x] + 2 y[x] == 0
```

```
In[2]:= sol = DSolve[ode, y[x], x]      (* Out: {{y[x] → e^{-2x} C[1] + e^{-x} C[2]}} *)
```

```
In[3]:= yp = DSolve[{ode, y[0] == 4, y'[0] == -2}, y[x], x]
```

Out[3]= $\{\{y[x] \to e^{-2x}(-2 + 6\,e^x)\}\}$

```
In[4]:= Plot[yp[[1, 1, 2]], {x, 0, 5}]
```

Pr.2.5.   Substitute

```
In[1]:= y = (c1 + c2 x + c3 x^2) Exp[-x]
```

into the left-hand side of the given ODE

```
In[2]:= D[y, x, x, x] + 3 D[y, x, x] + 3 D[y, x] + y
```

Out[2]= $-6\,c3\,e^{-x} + 3\,e^{-x}(c2 + 2\,c3\,x) + 3\,(e^{-x}(c2 + 2\,c3\,x) - e^{-x}(c1 + c2\,x + c3\,x^2)) +$

$\qquad 3\,(2\,c3\,e^{-x} - 2\,e^{-x}(c2 + 2\,c3\,x) + e^{-x}(c1 + c2\,x + c3\,x^2))$

```
In[3]:= Simplify[%]                                        (* Out: 0 *)
```

Pr.2.7.

```
In[1]:= Clear[y]
```

In[2]:= ode = y''[t] + 10 y'[t] + 16 y[t] == 0

In[3]:= sol = DSolve[ode, y[t], t]

Out[3]= $\{\{y[t] \to e^{-8t} C[1] + e^{-2t} C[2]\}\}$

In[4]:= y1 = DSolve[{ode, y[0] == 1, y'[0] == -1}, y[t], t]

Out[4]= $\left\{\left\{y[t] \to e^{-8t}\left(-\dfrac{1}{6} + \dfrac{7 e^{6t}}{6}\right)\right\}\right\}$

In[5]:= y2 = DSolve[{ode, y[0] == 1, y'[0] == 0}, y[t], t]

Out[5]= $\left\{\left\{y[t] \to e^{-8t}\left(-\dfrac{1}{3} + \dfrac{4 e^{6t}}{3}\right)\right\}\right\}$

In[6]:= y3 = DSolve[{ode, y[0] == 1, y'[0] == 1}, y[t], t]

Out[6]= $\left\{\left\{y[t] \to e^{-8t}\left(-\dfrac{1}{2} + \dfrac{3 e^{6t}}{2}\right)\right\}\right\}$

In[7]:= Plot[{y1[[1, 1, 2]], y2[[1, 1, 2]], y3[[1, 1, 2]]}, {t, 0, 2},
        PlotLabel -> Overdamping, AxesLabel -> {t, y}]

Pr.2.9. $L = 1$ m $= 100$ cm, $g = 981 \mathrm{cm/sec^2}$,

In[1]:= L = 100.0;   g = 981.0;

In[2]:= ode = m T''[t] + 9.81 m T[t] == 0

In[3]:= DSolve[{ode, T[0] == 0, T'[0] == 1}, T[t], t]

Out[3]= {{T[t] → 0.319275 Sin[3.13209 t]}}

Since 3.13209 equals about π, the pendulum ticks about once per second.

Pr.2.11.

In[1]:= Clear[y]

In[2]:= ode = y''[x] + y[x] == 0

In[3]:= DSolve[{ode, y[0] == 3, y[Pi] == -3}, y[x], x]

Out[3]= {{y[x] → 3 Cos[x] − C[1] Sin[x]}}

Show that a solution with arbitrary constant c is

In[4]:= y1 = 3 Cos[x] + c Sin[x]

You see that $y_1(0) = 3$ and the right boundary condition is satisfied, too:

In[5]:= y1 /. x -> Pi   (* Out: −3 *)

In[6]:= ode /. {y[x] -> y1, y''[x] -> D[y1, x, x]}   (* Out: True *)

Pr.2.13.   Substitute a general solution into the ODE and determine a and b.

In[1]:= y = x (c1 Cos[Log[x]] + c2 Sin[Log[x]])

In[2]:= ode = x^2 D[y, x, x] + a x D[y, x] + b y == 0

The (rather long) response must hold for any x, say, for $x = 1$, and for any c_1 and c_2.
To get two equations for a and b, take $c_1 = 0$, $c_2 = 1$ and then $c_1 = 1$, $c_2 = 0$.

In[3]:= eq1 = ode /. {c1 -> 0, c2 -> 1, x -> 1}

Out[3]= 1 + a == 0

In[4]:= eq2 = ode /. {c1 -> 1, c2 -> 0, x -> 1}

Out[4]= −1 + a + b == 0

In[5]:= Solve[{eq1, eq2}, {a, b}]　　　　　　　　(* Out: {{a → −1, b → 2}} *)

Hence the ODE is

In[6]:= Clear[y]　　　　　　　　　　　(* This makes y general again. *)

In[7]:= ode2 = x^2 y''[x] − x y'[x] + 2 y[x] == 0

In[8]:= DSolve[ode2, y[x], x]　　　　　　　　　(* Check by solving. *)

Pr.2.15.

In[1]:= Clear[y]　　　　　　　　　　　(* This makes y general again. *)

In[2]:= ode = y''[t] + 4 y[t] == −12 Sin[2 t]

In[3]:= yp = DSolve[{ode, y[0] == 1, y'[0] == 3}, y[t], t]

In[4]:= Plot[yp[[1, 1, 2]], {t, 0, 10}]

Pr.2.17.

In[1]:= ode = y''[t] + 100 y[t] == 36 Cos[8 t]

In[2]:= sol = DSolve[{ode, y[0] == 0, y'[0] == 0}, y[t], t]

Out[2]= {{y[t] → Cos[8 t] − Cos[10 t]}}

This solution can be written as $2 \sin 9t \sin t$. This explains the occurrence of beats.

In[3]:= sol2 = TrigReduce[2 Sin[9 t] Sin[t]]

Out[3]= Cos[8 t] − Cos[10 t]

In[4]:= Plot[sol2, {t, 0, 30}]

Pr.2.19.
Solve the homogeneous ODE to obtain a basis of solutions needed for calculating the Wronskian W.

In[1]:= Clear[y]

In[2]:= ode = y''[x] − 2 y'[x] + y[x] == 0

In[3]:= DSolve[%, y[x], x]　　　　　　(* Out: {{y[x] → e^x C[1] + e^x x C[2]}} *)

In[4]:= y1 = Exp[x];　　　y2 = x Exp[x];

In[5]:= W = y1 D[y2, x] − y2 D[y1, x]

In[6]:= W = Simplify[%]　　　　　　　　　　(* Out: e^{2x} *)

In[7]:= r = 3 x^(3/2) Exp[x]　　　　　　　　(* Out: $3 e^x x^{3/2}$ *)

In[8]:= yp = −y1 Integrate[y2 r/W, x] + y2 Integrate[y1 r/W, x]

Out[8]= $\dfrac{12}{35} e^x x^{7/2}$

CHAPTER 3,   page 40

Pr.3.1. $p = 3$, $q = 2$, $\delta = 1$ gives a node.

In[1]:= sys = {y1'[t] == y1[t], y2'[t] == 2 y2[t]}

In[2]:= DSolve[sys, {y1[t], y2[t]}, t]

Out[2]= {{y1[t] → e^t C[1], y2[t] → e^{2t} C[2]}}

In[3]:= <<Graphics'PlotField'

In[4]:= PlotVectorField[{y1, 2 y2}, {y1, -6, 6}, {y2, -6, 6}]

Pr.3.3. $y_1' = y_2$, $y_2' = -9 y_1$,

In[1]:= Clear[y1, y2]

In[2]:= <<Graphics'PlotField'

In[3]:= eq1r = y2   (* Out: y2 *)

In[4]:= eq2r = -9 y1   (* Out: −9 y1 *)

In[5]:= P0 = PlotVectorField[{eq1r, eq2r}, {y1, -4, 4}, {y2, -4, 4},
        Axes -> True, AxesLabel -> {y1, y2}]

Pr.3.5. Type the system, then the matrix and its eigenvalues and eigenvectors – there is only one – and finally apply DSolve to confirm the form of a general solution derived in AEM.

In[1]:= sys = {y1'[t] == 4 y1[t] + y2[t], y2'[t] == -y1[t] + 2 y2[t]}

In[2]:= A = {{4, 1}, {-1, 2}}   (* Out: {{4, 1}, {−1, 2}} *)

In[3]:= Eigenvectors[A]   (* Out: {{−1, 1}, {0, 0}} *)

In[4]:= DSolve[sys, {y1[t], y2[t]}, t]

Out[4]= {{y1[t] → e^{3t} (C[1] + t C[1] + t C[2]), y2[t] → −e^{3t} (t C[1] − C[2] + t C[2])}}

Pr.3.7. Type the ODE as a system,

In[1]:= sys = {y1'[t] == y2[t], y2'[t] == -y1[t]/9}

In[2]:= <<Graphics'PlotField'

In[3]:= Clear[y]

In[4]:= PlotVectorField[{y2, -y1/9}, {y1, -7, 7}, {y2, -4, 4}, Axes ->   True,
        AxesLabel -> {y, y'}]

Pr.3.9. Type the first equation algebraically solved for $i_1'(t)$. Type the second equation differentiated with respect to t and then algebraically solved for $i_2'(t)$.

In[1]:= Clear[i1, i2],

In[2]:= eq1 = i1'[t] == -4 i1[t] + 4 i2[t] + 12

In[3]:= eq2 = i2'[t] == (1/10) (4 i1'[t] - 4 i2[t])

In[4]:= DSolve[{eq1, eq2, i1[0] == 0, i2[0] == 0}, {i1[t], i2[t]}, t]

Out[4]= {{i1[t] → e^{-2t}($-8+5e^{6t/5}+3e^{2t}$), i2[t] → $4e^{-2t}$($-1+e^{6t/5}$)}}

In[5]:= `sol = FullSimplify[%]`

Out[5]= {{i1[t] → $3-8e^{-2t}+5e^{-4t/5}$, i2[t] → $4e^{-2t}$($-1+e^{6t/5}$)}}

In[6]:= `Plot[{sol[[1, 1, 2]], sol[[1, 2, 2]]}, {t, 0, 10}]`

CHAPTER 4,   page 52

Pr.4.1.

In[1]:= `Series[Tan[x], {x, 0, 9}]`

In[2]:= `Coefficient[%, x^7]`   (* Out: $\frac{17}{315}$ *)

Pr.4.3.

In[1]:= `Clear[a, x, m, s]`

In[2]:= `ser = Sum[a[m] x^m, {m, s - 1, s + 1}]`

Out[2]= x^{-1+s}a[$-1+s$]$+x^s$a[s]$+x^{1+s}$a[$1+s$]

In[3]:= `eq = D[ser, x] - 2 x ser`

Out[3]= ($-1+s$)x^{-2+s}a[$-1+s$]$+sx^{-1+s}$a[s]$+$

　　　($1+s$)x^sa[$1+s$]$-2x$(x^{-1+s}a[$-1+s$]$+x^s$a[s]$+x^{1+s}$a[$1+s$])

In[4]:= `eq2 = Expand[eq]`

Out[4]= $-x^{-2+s}$a[$-1+s$]$+sx^{-2+s}$a[$-1+s$]$-2x^s$a[$-1+s$]$+$

　　　sx^{-1+s}a[s]$-2x^{1+s}$a[s]$+x^s$a[$1+s$]$+sx^s$a[$1+s$]$-2x^{2+s}$a[$1+s$]

In[5]:= `c = Coefficient[eq2, x^s]`   (* Out: -2a[$-1+s$]$+$a[$1+s$]$+s$a[$1+s$] *)

In[6]:= `sol = Solve[c == 0, a[s + 1]]`

Out[6]= $\left\{\left\{a[1+s] \rightarrow \frac{2a[-1+s]}{1+s}\right\}\right\}$

Now rewrite this formula with $s+1$ replaced by s, hence $s-1$ replaced by $s-2$. Then type the beginning of the recursion, namely, $a_{-1}=0$, $a_0=1$. Then you get the desired partial sum as follows.

In[7]:= `Clear[a, s]`

In[8]:= `a[s_] := a[s] = 2 a[s - 2]/s`

The `s_` means that the `s` on the right-hand side is local to the function.

In[9]:= `a[-1] = 0`   (* Out: 0 *)

In[10]:= `a[0] = 1`   (* Out: 1 *)

In[11]:= `Table[a[s], {s, 0, 5}]`   (* Out: $\left\{1,0,1,0,\frac{1}{2},0\right\}$ *)

In[12]:= `Sum[a[s] x^s, {s, 0, 10}]`   (* Out: $1+x^2+\frac{x^4}{2}+\frac{x^6}{6}+\frac{x^8}{24}+\frac{x^{10}}{120}$ *)

Pr.4.5. A basis of solutions is 1 and $\operatorname{arctanh} x = (1/2) \ln\left((1+x)/(1-x)\right)$. Commands:

In[1]:= `Clear[y]`

In[2]:= `ode = (1 - x^2) y''[x] - 2 x y'[x] == 0`

Out[2]= $-2 x\, y'[x] + (1 - x^2)\, y''[x] == 0$

In[3]:= `sol = DSolve[ode, y[x], x]`

Out[3]= $\left\{\left\{y[x] \to C[2] + \dfrac{1}{2} C[1] \operatorname{Log}[-1 + x] - \dfrac{1}{2} C[1] \operatorname{Log}[1 + x]\right\}\right\}$

In[4]:= `ser = Series[sol[[1, 1, 2]], {x, 0, 20}]`

Out[4]= $\left(\dfrac{1}{2} i\,\pi\,C[1] + C[2]\right) - C[1]\, x - \dfrac{1}{3} C[1]\, x^3 - \dfrac{1}{5} C[1]\, x^5 - \dfrac{1}{7} C[1]\, x^7 - \dfrac{1}{9} C[1]\, x^9 -$
$\dfrac{1}{11} C[1]\, x^{11} - \dfrac{1}{13} C[1]\, x^{13} - \dfrac{1}{15} C[1]\, x^{15} - \dfrac{1}{17} C[1]\, x^{17} - \dfrac{1}{19} C[1]\, x^{19} + O[x]^{21}$

Pr.4.7. For the error experiment you may need to use `SetPrecision[...,20]`, etc.

In[1]:= `f = Exp[x]`

In[2]:= `S = Sum[(2 m + 1)/2 Integrate[f LegendreP[m, x], {x, -1, 1}]*`
`"LegendreP"[m, x], {m, 0, 6}]`

In[3]:= `S2 = N[S]`

Out[3]= $1.1752\, \mathtt{LegendreP}[0., x] + 1.10364\, \mathtt{LegendreP}[1., x] + 0.357814\, \mathtt{LegendreP}[2., x] +$
$0.0704556\, \mathtt{LegendreP}[3., x] + 0.00996513\, \mathtt{LegendreP}[4., x] +$
$0.00109959\, \mathtt{LegendreP}[5., x] + 0.0000994543\, \mathtt{LegendreP}[6., x]$

In[4]:= `S3 = N[Sum[(2 m + 1)/2 Integrate[f LegendreP[m, x], {x, -1, 1}]`
`LegendreP[m, x],{m, 0, 6}]]`

In[5]:= `Plot[{S3, f}, {x, -3, 3}]`

The order of the error is about 10^{-n+2}.

Pr.4.9.

In[1]:= `ode = x y''[x] + (1 - 2 x) y'[x] + (x - 1) y[x] == 0`

In[2]:= `ypart = DSolve[{ode, y[1] == 1, y'[1] == 1}, y[x], x]`

In[3]:= `sol = DSolve[ode, y[x], x]`   (* Out: $\{\{y[x] \to e^x C[1] + e^x C[2] \operatorname{Log}[x]\}\}$ *)

In[4]:= `sol /. x -> 0`   (* Out: $\{\{y[0] \to C[1] + C[2]\,(-\infty)\}\}$ *)

Pr.4.11.

In[1]:= `Clear[a, b, c]`

In[2]:= `ode = x (1 - x) y''[x] + (c - (a + b + 1) x) y'[x]- a b y[x] == 0`

In[3]:= `sol = DSolve[ode, y[x], x]`

Out[3]= $\{\{y[x] \to C[1]\, \mathtt{Hypergeometric2F1}[a, b, c, x] +$
$x^{1-c} C[2]\, \mathtt{Hypergeometric2F1}[1 + a - c, 1 + b - c, 2 - c, x]\}\}$

The hypergeometric function $F(a, b, c; x)$ is sometimes also denoted by ${}_2F_1(a, b, c; x)$.

This motivates the Mathematica notation shown here.

In[4]:= `sol /. {a -> -n, c -> b, x -> -t}`

Out[4]= $\{\{y[-t] \rightarrow (1+t)^n C[1] + (-t)^{1-b} C[2] \text{ Hypergeometric2F1}[1-b-n, 1, 2-b, -t]\}\}$

Pr.4.13.

In[1]:= `Clear[y]`

In[2]:= `ode = x (1 - x)y''[x] + (2 - 4 x) y'[x] - 2 y[x] == 0`

In[3]:= `ypartic = DSolve[{ode, y[0] == 1, y'[0] == 1}, y[x], x]`

> Power::"infy": Infinite expression $\frac{1}{0}$ encountered
>
> ...

Out[3]= {}

In[4]:= `sol = DSolve[ode, y[x], x]`

Out[4]= $\left\{\left\{y[x] \rightarrow \dfrac{C[1]}{1-x} + \dfrac{C[2]}{(1-x)\,x}\right\}\right\}$

In[5]:= `solprime = D[sol, x]`

Out[5]= $\left\{\left\{y'[x] \rightarrow \dfrac{C[1]}{(1-x)^2} - \dfrac{C[2]}{(1-x)\,x^2} + \dfrac{C[2]}{(1-x)^2\,x}\right\}\right\}$

Inspection of `sol` and `solprime` shows that both initial conditions are satisfied if you choose $C[1] = 1$, $C[2] = 0$. Obtain the parameter value $c = 2$ by inspection of the given ODE and a and b by typing

In[6]:= `Solve[{a + b + 1 == 4, a b == 2}, {a, b}]`

Out[6]= $\{\{a \rightarrow 1, b \rightarrow 2\}, \{a \rightarrow 2, b \rightarrow 1\}\}$

Pr.4.15.

3.83171 (exact 7S: 3.83171), 7.02156 (exact 7S: 7.01559). Commands:

In[1]:= `s = Series[BesselJ[1, x], {x, 0, 20}]`

In[2]:= `p = Normal[%]`

In[3]:= `NSolve[p == 0, x]`

In[4]:= `FindRoot[BesselJ[1, x] == 0, {x, 3}]`   (* Out: {x → 3.83171} *)

In[5]:= `FindRoot[BesselJ[1, x] == 0, {x, 8}]`   (* Out: {x → 7.01559} *)

Pr.4.17.

In[1]:= `Clear[y]`

In[2]:= `y = BesselY[0, x]`

In[3]:= `x D[y, x, x] + D[y, x] + x y`

In[4]:= `Simplify[%]`   (* Out: $\frac{1}{2}$ (x BesselY[0, x] − 2 BesselY[1, x] + x BesselY[2, x]) *)

In[5]:= `FullSimplify[%]`   (* Out: 0 *)

In[6]:= `y1 = BesselY[1, x]`

In[7]:= `x^2 D[y1, x, x] + x D[y1, x] + (x^2 - 1) y1`

In[8]:= FullSimplify[%]   (* Out: 0 *)

In[9]:= Plot[{BesselY[0, x], BesselY[1, x]}, {x, 0, 10}, AxesLabel -> {x, y}]

Pr.4.19.

In[1]:= ode = y''[x] + x^2 y[x] == 0

In[2]:= sol = DSolve[%, y[x], x]

$$\text{Out[2]}= \left\{\left\{\texttt{y[x]} \to \sqrt{x}\ \texttt{BesselJ}\left[-\frac{1}{4},\frac{x^2}{2}\right]\texttt{C[1]} + \sqrt{x}\ \texttt{BesselJ}\left[\frac{1}{4},\frac{x^2}{2}\right]\texttt{C[2]}\right\}\right\}$$

Hence the transformation is $y = u\sqrt{x}$, $x^2/2 = z$ and gives a Bessel equation for u as a function of z, with parameter $1/4$.

CHAPTER 5,   page 64

Pr.5.1.

In[1]:= Integrate[Exp[-s t] Sin[Pi t], {t, 0, Infinity}]

In[2]:= Simplify[%, Re[s] > 0]

Pr.5.3.

In[1]:= Integrate[Exp[-s t] k, {t, 0, c}] $\left(\texttt{* Out: } k\left(\dfrac{1}{s} - \dfrac{e^{-c s}}{s}\right)\texttt{ *}\right)$

Pr.5.5.

In[1]:= InverseLaplaceTransform[s/(L^2 s^2 + n^2 Pi^2), s, t]

Pr.5.7.   Type the ODE. Obtain the subsidiary equation. Substitute the initial conditions into it. Solve it algebraically. Find the inverse transform of its solution.

In[1]:= ode = y''[t] + 2 y'[t] - 3 y[t] == 6 Exp[-2 t]

In[2]:= subsid = LaplaceTransform[ode, t, s]

In[3]:= subsid2 = subsid /. {y[0] -> 2, y'[0] -> -14}

In[4]:= Solve[subsid2, LaplaceTransform[y[t], t, s]]

$$\text{Out[4]}= \left\{\left\{\texttt{LaplaceTransform[y[t], t, s]} \to \frac{2\,(-7 - 3s + s^2)}{(2 + s)\,(-3 + 2s + s^2)}\right\}\right\}$$

In[5]:= y = Apart[InverseLaplaceTransform[%[[1, 1, 2]], s, t]]

$$\text{Out[5]}= \frac{11\,e^{-3t}}{2} - 2\,e^{-2t} - \frac{3\,e^{t}}{2}$$

Pr.5.9.

In[1]:= InverseLaplaceTransform[3 (1 - Exp[-Pi s])/(s^2 + 9), s, t]

$$\text{Out[1]}= \texttt{Sin[3 t]}\,(\texttt{UnitStep[t]} + \texttt{UnitStep}[-\pi + t])$$

In[2]:= Plot[%, {t, 0, 20}]

Pr.5.11.

In[1]:= eq = R i[t] + 1/C Integrate[i[tau], {tau, 0, t}] ==
           K UnitStep[t - 1] - K UnitStep[t - 3]

In[2]:= subsid = LaplaceTransform[eq, t, s]

See Example 5.6 for the integral. In the next step, J is used because I is protected.

In[3]:= J = Solve[subsid, LaplaceTransform[i[t], t, s]]

In[4]:= j = InverseLaplaceTransform[J[[1, 1, 2]], s, t]

$$\text{Out[4]}= \frac{1}{R}\left(e^{\frac{1-t}{CR}} K\left(-e^{\frac{2}{CR}}\,\text{UnitStep}\,[-3+t]+\text{UnitStep}\,[-1+t]\right)\right)$$

In[5]:= j0 = j /. {K -> 110, R -> 1, C -> 1}

In[6]:= Plot[j0, {t, 0, 10}]

Pr.5.13.

In[1]:= Clear[y, Y]

In[2]:= ode = y''[t] + y[t] == DiracDelta[t - Pi] - DiracDelta[t - 2 Pi]

In[3]:= subsid = LaplaceTransform[ode, t, s]

In[4]:= Y = Solve[subsid, LaplaceTransform[y[t], t, s]]

$$\text{Out[4]}= \left\{\left\{\text{LaplaceTransform}[y[t],t,s] \to -\frac{e^{-2\pi s}-e^{-\pi s}-s\,y[0]-y'[0]}{1+s^2}\right\}\right\}$$

Ignore the messages you will obtain in response to the next command.

In[5]:= sol = InverseLaplaceTransform[Y[[1, 1, 2]], s, t]

In[6]:= yp = sol /. {y[0] -> 0, y'[0] -> 1}

Out[6]= Sin [t] UnitStep [t] − Sin [t] UnitStep [− 2 π + t] − Sin [t] UnitStep [− π + t]

In[7]:= Plot[yp, {t, 0, 20}]

Pr.5.15.   The full-wave rectification of sin t is $|\sin t|$. For $0 \le t \le 5\pi$ this can be typed as

In[1]:= r = Sin[t] (1 - 2 UnitStep[t - Pi] + 2 UnitStep[t - 2 Pi]
           - 2 UnitStep[t - 3 Pi] + 2 UnitStep[t - 4 Pi])

Out[1]= Sin [t] (1 + 2 UnitStep [− 4 π + t] − 2 UnitStep [− 3 π + t] +
           2 UnitStep [− 2 π + t] − 2 UnitStep [− π + t])

In[2]:= Plot[r, {t, 0, 5 Pi}, Ticks -> {{2 Pi, 4 Pi}, Automatic}]

CHAPTER 6,   page 77

Pr.6.1.

In[1]:= A = {{5, -3, 0}, {6, 1, -4}}

In[2]:= B = {{-2, 4, -1}, {1, 1, 5}}

In[3]:= 3 A

In[4]:= A - B

In[5]:= Transpose[2 A - (1/2) B]

Pr.6.3.   Type

In[1]:= A = {{1, 3, 2}, {3, 5, 0}, {2, 0, 4}}

In[2]:= B = {{0, 2, 1}, {-2, 0, -3}, {-1, 3, 0}}

In[3]:= c = {1, 0, -2}

In[4]:= d = {3, 1, 2}

      A is symmetric.

      B is skew-symmetric.

      This explains why you obtain two zero matrices. Further commands:

In[5]:= A.A

In[6]:= %.%

In[7]:= B.B

In[8]:= %.%   (* Not %^2 *)

In[9]:= A.B - B.A   (* Out: {{-16, -2, -12}, {-2, 12, 4}, {-12, 4, 4}} *)

In[10]:= A - Transpose[A]

In[11]:= B + Transpose[B]

In[12]:= Det[A]   (* Out: -36 *)

In[13]:= Det[B]   (* Out: 0 *)

In[14]:= A.c

In[15]:= B.(c - 3 d)   (* Out: {-14, 40, -1} *)

In[16]:= c.d   (* Out: -1 *)

Pr.6.7.

In[1]:= A = {{Cos[t], -Sin[t]}, {Sin[t], Cos[t]}}

In[2]:= A.A.A.A

In[3]:= Simplify[%]   (* Out: {{Cos[4t], -Sin[4t]}, {Sin[4t], Cos[4t]}} *)

Pr.6.9.   It seems that $n_{min} = m + 2$.

Pr.6.11.

In[1]:= A = {{0, -2, -1}, {-2, 3, 2}, {-1, 2, 1}}

In[2]:= F = Inverse[A]   (* Out: {{-1, 0, -1}, {0, -1, 2}, {-1, 2, -4}} *)

In[3]:= B = {{1, 2, 3}, {2, 3, 4}, {3, 4, 6}}

In[4]:= G = Inverse[B]   (* Out: {{-2, 0, 1}, {0, 3, -2}, {1, -2, 1}} *)

In[5]:= Inverse[A.B] - G.F   (* Out: {{0, 0, 0}, {0, 0, 0}, {0, 0, 0}} *)

Pr.6.13.

In[1]:= c = {3, 2, -2, 1, 0};

In[2]:= d = {2, 0, .3, 0, 4};

In[3]:= e = {1, -3, -2,-1, 1};

In[4]:= c.d   (* Out: 0 *)

In[5]:= d.e   (* Out: 0 *)

In[6]:= e.c   (* Out: 0 *)

Pr.6.15.   Type the vectors and then the square root of their dot products $\mathbf{c} \bullet \mathbf{c}$, $\mathbf{d} \bullet \mathbf{d}$, $\mathbf{e} \bullet \mathbf{e}$,

In[1]:= c = {3, 2, -2, 1, 0}

In[2]:= Sqrt[c.c]   (* Out: $3\sqrt{2}$ *)

Pr.6.17. $\det \mathbf{H} = 1/2160, 1/6\,048\,000, 1/266\,716\,800\,000$. The commands are as follows.

In[1]:= n = 3;

In[2]:= Clear[j, k]

In[3]:= f = 1/(j + k - 1);

In[4]:= H = Table [f, {j, 1, n}, {k, 1, n}]

In[5]:= Inverse[H]

In[6]:= Det[H]

In[7]:= <<LinearAlgebra`MatrixManipulation`

In[8]:= HilbertMatrix[3]

Out[8]= $\{\{1, \frac{1}{2}, \frac{1}{3}\}, \{\frac{1}{2}, \frac{1}{3}, \frac{1}{4}\}, \{\frac{1}{3}, \frac{1}{4}, \frac{1}{5}\}\}$

Pr.6.19.   Row reduction of the matrix with the given vectors as row vectors shows linear dependence of these vectors.

In[1]:= a = {0, 16, 0, -24, 0}

In[2]:= b = {1, 0, -1, 0, 2}

In[3]:= c = {0, -14, 0, 21, 0}

In[4]:= A = Join[{a}, {b}, {c}]

In[5]:= RowReduce[A]

Out[5]= $\left\{\{1, 0, -1, 0, 2\}, \{0, 1, 0, -\frac{3}{2}, 0\}, \{0, 0, 0, 0, 0\}\right\}$

CHAPTER 7,   page 86

Pr.7.1.

In[1]:= A = {{-2, 2, -3}, {2, 1, -6}, {-1, -2, 0}}

In[2]:= B = Transpose[A]

In[3]:= SYM = (A + B)/2; MatrixForm[SYM]

In[4]:= SKEW = (A - B)/2

In[5]:= SYM + SKEW   (* For checking *)

Pr.7.3. Show that the inverse equals the transpose.

In[1]:= A = {{Cos[t], Sin[t]}, {-Sin[t], Cos[t]}}

In[2]:= Inverse[A]

In[3]:= Simplify[%]//MatrixForm   (* Out: $\begin{pmatrix} \text{Cos}[t] & -\text{Sin}[t] \\ \text{Sin}[t] & \text{Cos}[t] \end{pmatrix}$ *)

In[4]:= Inverse[A] - Transpose[A]

In[5]:= Simplify[%]   (* 2 × 2 zero matrix *)

Now show that the dot product of vectors **u** and **v** equals the dot product of the vectors **Au** and **Av**, call them **w** and **z**, respectively.

In[6]:= u = {u1, u2};   v = {v1, v2};

In[7]:= Dot[u, v]   (* Out: u1 v1 + u2 v2 *)

In[8]:= w = A.u

Out[8]= {u1 Cos[t] + u2 Sin[t], u2 Cos[t] - u1 Sin[t]}

In[9]:= z = A.v

Out[9]= {v1 Cos[t] + v2 Sin[t], v2 Cos[t] - v1 Sin[t]}

In[10]:= Dot[w, z]

Out[10]= (u2 Cos[t] - u1 Sin[t]) (v2 Cos[t] - v1 Sin[t]) +

    (u1 Cos[t] + u2 Sin[t]) (v1 Cos[t] + v2 Sin[t])

In[11]:= Simplify[%]   (* Out: u1 v1 + u2 v2 *)

Pr.7.7. Note that the spectrum is real, although **A** is not Hermitian.

In[1]:= A = {{ 4 + 12 I, -12 - 12 I, 12 - 12 I},

    {-6 + 6 I,   10 - 6 I,   6 + 6 I},

    { 6 + 6 I,   -6 + 6 I,   -2 - 6 I}}

In[2]:= B = Transpose[Conjugate[A]]

In[3]:= HE = (A + B)/2

In[4]:= SH = (A - B)/2

In[5]:= HE + SH   (* For checking *)

In[6]:= Eigenvalues[A]   (* Out: {-8, 4, 16} *)

In[7]:= Eigenvectors[A]   (* Out: {{i, 0, 1}, {0, -i, 1}, {-i, 1, 0}} *)

Pr.7.9.

In[1]:= F = {{I/2, Sqrt[3]/2},   {Sqrt[3]/2, I/2}}

Out[1]= $\{\{\frac{i}{2}, \frac{\sqrt{3}}{2}\}, \{\frac{\sqrt{3}}{2}, \frac{i}{2}\}\}$

In[2]:= F.F.F.F.F.F   (* Out: {{-1, 0}, {0, -1}} *)

In[3]:= F.F.F.F.F.F.F.F.F.F.F.F   (* Out: {{1, 0}, {0, 1}} *)

In[4]:= e = Eigenvalues[F]

Out[4]= $\left\{\frac{1}{2}\left(i - \sqrt{3}\right), \frac{1}{2}\left(i + \sqrt{3}\right)\right\}$

In[5]:= Abs[e]   (* Out: {1, 1} *)

Show that the inverse of F^2 is unitary.

In[6]:= G = Inverse[F.F]   (* Out: $\left\{\left\{\frac{1}{2}, -\frac{i\sqrt{3}}{2}\right\}, \left\{-\frac{i\sqrt{3}}{2}, \frac{1}{2}\right\}\right\}$ *)

In[7]:= Inverse[G] - Conjugate[Transpose[G]]   (* Out: {{0, 0}, {0, 0}} *)

Show that F satisfies Cayley's theorem.

In[8]:= p = CharacteristicPolynomial[F, λ]   (* Out: $-1 - i\lambda + \lambda^2$ *)

In[9]:= -IdentityMatrix[2] - I F + F.F   (* Out: {{0, 0}, {0, 0}} *)

Pr.7.11. Proceed as in Example 7.5 in this Guide. In particular, obtain the relation between the eigenvectors in the form $\mathbf{y_j} = \mathbf{P}^{-1}\mathbf{x_j}$, $j = 1, 2, 3$.

In[1]:= A = {{-1, -3, 3}, {-6, 2, 6}, {-3, 3, 5}}

In[2]:= P = {{3, -1, 1}, {-15, 6, -5}, {5, -3, 2}}

In[3]:= Q = Inverse[P]

In[4]:= B = Q.A.P; MatrixForm[B]   (* Out: $\begin{pmatrix} -124 & 72 & -48 \\ 267 & -130 & 96 \\ 696 & -372 & 260 \end{pmatrix}$ *)

In[5]:= Eigenvalues[A]   (* Out: {-4, 2, 8} *)

In[6]:= Eigenvalues[B]   (* Out: {-4, 2, 8} *)

In[7]:= ea = Eigenvectors[A]   (* Out: {{1, 1, 0}, {1, 0, 1}, {0, 1, 1}} *)

In[8]:= eb = Eigenvectors[B]   (* Out: {{-4, 6, 19}, {-4, 5, 18}, {-2, 1, 7}} *)

In[9]:= ea.Transpose[Q]   (* Out: {{-4, 6, 19}, {-4, 5, 18}, {-2, 1, 7}} *)

CHAPTER 8,   page 95

Pr.8.1.

In[1]:= P = {1, 2, 3};   Q = {2, 4, 6};

In[2]:= v = Q - P

In[3]:= Sqrt[v.v]   (* Out: $\sqrt{14}$ *)

Pr.8.3.   These forces are in equilibrium, their resultant is **0**. Indeed,

In[1]:= p = {4, -2, -3};   q = {8, 8, 1};   u = {-12, -6, 2};

In[2]:= p + q + u   (* Out: {0,0,0} *)

Pr.8.5.

> In[1]:= b = {2, 0, -5};   c = {4, -2, 1};
>
> In[2]:= (2 b).(5 c)   (* Out: 30 *)
>
> (Try it without parentheses.)
>
> In[3]:= 10 b.c   (* Out: 30 *)

Pr.8.7.   Obtain the components of the displacement vector **d** as differences of corresponding coordinates of the points. Then obtain the work as a dot product.

> In[1]:= p = {2, 6, 6};   A = {3, 4, 0};   B = {5, 8, 0};
>
> In[2]:= p.(B - A)   (* Out: 28 *)

Pr.8.9.

> In[1]:= a = {1, 2, 0};   b = {-3, 2, 0};   c = {2, 3, 4};
>
> In[2]:= Cross[a, b].c   (* Out: 32 *)
>
> In[3]:= a.Cross[b, c]   (* Out: 32 *)

Pr.8.11.   They are linearly dependent because their scalar triple product is zero:

> In[1]:= u = {3, 5, 9};   v = {73, -56, 76};   w = {-4, 7, -1};
>
> In[2]:= <<Calculus'VectorAnalysis'
>
> In[3]:= u.Cross[v, w]   (* Out: 0 *)

Pr.8.13.   Show that $L - R = 0$, where

> In[1]:= a = {a1, a2, a3};   b = {b1, b2, b3};
>
> In[2]:= <<Calculus'VectorAnalysis'
>
> In[3]:= L = Cross[a, b] . Cross[a, b]
>
> Out[3]= $(-a2\,b1 + a1\,b2)^2 + (a3\,b1 - a1\,b3)^2 + (-a3\,b2 + a2\,b3)^2$
>
> In[4]:= R = a.a b.b - (a.b)^2
>
> Out[4]= $-(a1\,b1 + a2\,b2 + a3\,b3)^2 + (a1^2 + a2^2 + a3^2)\,(b1^2 + b2^2 + b3^2)$
>
> In[5]:= L - R
>
> Out[5]= $(-a2\,b1 + a1\,b2)^2 + (a3\,b1 - a1\,b3)^2 + (-a3\,b2 + a2\,b3)^2 + (a1\,b1 + a2\,b2 + a3\,b3)^2$
>
> $\qquad -(a1^2 + a2^2 + a3^2)\,(b1^2 + b2^2 + b3^2)$
>
> In[6]:= Simplify[%]   (* Out: 0 *)

Pr.8.15.

> In[1]:= r = {t, Cosh[t]}
>
> In[2]:= rprime = D[r, t]   (* Out: {1, Sinh[t]} *)
>
> In[3]:= ip = rprime.rprime   (* Out: $1 + Sinh[t]^2$ *)
>
> In[4]:= f1 = Simplify[%]   (* Out: $Cosh[t]^2$ *)
>
> In[5]:= Sqrt[f1]   (* Out: $\sqrt{Cosh[t]^2}$ *)

In[6]:= f2 = Simplify[%, t > 0]   (* Out: Cosh[t] *)

In[7]:= length = Integrate[f2, {t, 0, 1}]   (* Out: Sinh[1] *)

In[8]:= Plot[r[[2]], {t, 0, 1}, PlotRange -> {0, 1.5}]

Pr.8.17.

In[1]:= r = {t, t^2, t^3}

In[2]:= r1 = D[r, t]   (* Out: $\{1, 2\,t, 3\,t^2\}$ *)

In[3]:= r2 = D[r1, t]   (* Out: $\{0, 2, 6\,t\}$ *)

In[4]:= r3 = D[r2, t]   (* Out: $\{0, 0, 6\}$ *)

In[5]:= <<Calculus`VectorAnalysis`

In[6]:= r1.Cross[r2, r3]/((r1.r1)(r2.r2) - (r1.r2)^2)

$$\text{Out}[6]= \frac{12}{-(4\,t + 18\,t^3)^2 + (4 + 36\,t^2)\,(1 + 4\,t^2 + 9\,t^4)}$$

In[7]:= Simplify[%]   (* Out: $\frac{3}{1 + 9\,t^2 + 9\,t^4}$ *)

Pr.8.19.

In[1]:= <<Calculus`VectorAnalysis`

In[2]:= SetCoordinates[Cartesian[x, y, z]]   (* You must carry z along. *)

In[3]:= f = Log[x^2 + y^2]

In[4]:= v = Grad[f]   (* Out: $\left\{\frac{2\,x}{x^2 + y^2}, \frac{2\,y}{x^2 + y^2}, 0\right\}$ *)

In[5]:= % /. {x -> 2, y -> 0}   (* Out: $\{1, 0, 0\}$ *)

In[6]:= <<Graphics`

In[7]:= PlotVectorField[{2 x/(x^2 + y^2), 2 y/(x^2 + y^2)},
            {x, 0, 0.5}, {y, 0, 0.5}]

Pr.8.21.   The cone is the surface $f = 0$, where

In[1]:= f = x^2 + y^2 - z^2   (* Out: $x^2 + y^2 - z^2$ *)

In[2]:= <<Calculus`VectorAnalysis`

In[3]:= SetCoordinates[Cartesian[x, y, z]]

Now type a normal vector **N** of the cone, its length $|\mathbf{N}|$, and the unit normal vector $\mathbf{n} = (1/|\mathbf{N}|)\mathbf{N}$. Write NO since N is protected.

In[4]:= NO = Grad[f]   (* N is protected. *)

In[5]:= lengthNO = Sqrt[NO.NO]

In[6]:= n = NO/lengthNO

$$\text{Out}[6]= \left\{\frac{2\,x}{\sqrt{4\,x^2 + 4\,y^2 + 4\,z^2}}, \frac{2\,y}{\sqrt{4\,x^2 + 4\,y^2 + 4\,z^2}}, -\frac{2\,z}{\sqrt{4\,x^2 + 4\,y^2 + 4\,z^2}}\right\}$$

In[7]:= n /. {x -> 3, y -> 4, z -> 5}   (* Out: $\left\{\frac{3}{5\sqrt{2}}, \frac{2\sqrt{2}}{5}, -\frac{1}{\sqrt{2}}\right\}$ *)

Pr.8.23.

In[1]:= `<<Calculus'VectorAnalysis'`

In[2]:= `SetCoordinates[Cartesian[x, y, z]]`

In[3]:= `v = (x^2 + y^2)^(-1){-y, x, 0}` $(* \text{ Out: } \{-\frac{y}{x^2+y^2}, \frac{x}{x^2+y^2}, 0\} *)$

Out[3]= $\left[-\dfrac{y}{x^2+y^2}, \dfrac{x}{x^2+y^2}\right]$

In[4]:= `Div[v]` $(* \text{ Out: } 0 *)$

CHAPTER 9,   page 107

Pr.9.1.   Integrate from $t = 0$ to 2π.

In[1]:= `F = {2 z, x, -y}`

In[2]:= `r = {Cos[t], Sin[t], 2 t}`

In[3]:= `FC = F /. {x -> r[[1]], y -> r[[2]], z -> r[[3]]}`

Out[3]= $\{4\,t, \text{Cos}[t], -\text{Sin}[t]\}$

In[4]:= `rprime = D[r, t]`

In[5]:= `integrand = FC.rprime`

Out[5]= $\text{Cos}[t]^2 - 2\,\text{Sin}[t] - 4\,t\,\text{Sin}[t]$

In[6]:= `Integrate[integrand, {t, 0, 2 Pi}]` $(* \text{ Out: } 9\pi *)$

Pr.9.3.   Use $\mathbf{F} = [3x^2, 2yz, y^2]$. Show that $\operatorname{curl}\mathbf{F} = 0$. Find f such that $\mathbf{F} = \operatorname{grad} f$. Use $f(B) - f(A)$.

In[1]:= `F = {3 x^2, 2 y z, y^2}`

In[2]:= `<<Calculus'VectorAnalysis'`

In[3]:= `SetCoordinates[Cartesian[x, y, z]]`

In[4]:= `Curl[F]` $(* \text{ Out: } \{0, 0, 0\} *)$

This shows that a potential must exist.

In[5]:= `F[[1]]` $(* \text{ Out: } 3x^2 *)$

In[6]:= `Integrate[F[[1]], x]` $(* \text{ Out: } x^3 *)$

In[7]:= `Integrate[F[[2]], y]` $(* \text{ Out: } y^2 z *)$

In[8]:= `Integrate[F[[3]], z]` $(* \text{ Out: } y^2 z *)$

In[9]:= `f = x^3 + y^2 z` $(* \text{ Out: } x^3 + y^2 z \quad \text{A potential.} *)$

In[10]:= `Grad[f]` $(* \text{ Out: } \{3x^2, 2yz, y^2\} \quad \text{Checking} *)$

In[11]:= `f1 = f /. {x -> 1, y -> -1, z -> 7}` $(* \text{ Out: } 8 *)$

In[12]:= `f2 = f /. {x -> 0, y -> 1, z -> 2}` $(* \text{ Out: } 2 *)$

In[13]:= `f1 - f2` $(* \text{ Out: } 6 *)$

In[14]:= (f /. {x -> 1, y -> -1, z -> 7}) -
          f /. {x -> 0, y -> 1, z -> 2}   (* Out: 6 *)

The parentheses (...) are essential. They tell that the first substitution is finished. Try it without (...).

Pr.9.5.   Use polar coordinates given by $x = r \cos \theta$, $y = r \sin \theta$. The element of area is $r \, dr \, d\theta$. Denote the coordinates of the center of gravity by xbar and ybar . Note that they are equal, for reasons of symmetry of the quarter-disk. Use $dx \, dy = r \, dr \, d\theta$.

In[1]:= Clear[r]

In[2]:= M = Integrate[Integrate[1*r, {r, 0, 1}], {θ, 0, Pi/2}]   (* Total mass *)

In[3]:= xbar = 1/M Integrate[Integrate[r Cos[θ] r, {r, 0, 1}], {θ, 0, Pi/2}]

Out[3]= $\dfrac{4}{3\pi}$

Pr.9.7.   Integrate over y from $1 + x^4$ to 2. Sketch the region of integration to find out that the integration over x extends from -1 to 1. The answer will be 16/5. Thus type

In[1]:= F = {Exp[y]/x, Exp[y] Log[x] + 2 x}

Type the integrand G of the double integral on the left-hand side of the formula,

In[2]:= G = D[F[[2]], x] - D[F[[1]], y]   (* Out: 2 *)

In[3]:= Integrate[Integrate[2, {y, 1 + x^4, 2}], {x, -1, 1}]

Pr.9.9.   The familiar formula πab follows from

In[1]:= r = {a Cos[t], b Sin[t]}

Out[1]= {a Cos [t], b Sin [t]}   (* Ellipse *)

In[2]:= r1 = D[r, t]   (* Out: {-a Sin[t], b Cos[t]} *)

In[3]:= area = 1/2 Integrate[r[[1]] r1[[2]] - r[[2]] r1[[1]], {t, 0, 2 Pi}]

Pr.9.11. $z = 1 - x - y$. Hence a representation r , its partial derivatives ru and rv , and a normal vector N are

In[1]:= r = {u, v, 1 - u - v}   (* Out: {u, v, 1 − u − v} *)

In[2]:= ru = D[r, u]   (* Out: {1, 0, −1} *)

In[3]:= rv = D[r, v]   (* Out: {0, 1, −1} *)

In[4]:= <<Calculus'VectorAnalysis'

In[5]:= SetCoordinates[Cartesian[x, y, z]]

In[6]:= N0 = Cross[ru, rv]   (* Out: {1, 1, 1} *)

In[7]:= F = {x^2, 0, 3 y^2}   (* Out: {x^2, 0, 3 y^2} *)

Let FS denote **F** on the plane S. Denote the integrand by f .

In[8]:= FS = F /. {x -> r[[1]], y -> r[[2]], z -> r[[3]]}

Out[8]= {u^2, 0, 3 v^2}

In[9]:= f = FS.N0   (* Out: $u^2 + 3v^2$ *)

$z = 1 - x - y = 1 - u - v = 0$ gives $u = 1 - v$ in the xy-plane $z = 0$. Hence integrate from $u = 0$ to $1 - v$ and then from $v = 0$ to 1.

In[10]:= Integrate[Integrate[f, {u, 0, 1 - v}], {v, 0, 1}]   (* Out: $\frac{1}{3}$ *)

Pr.9.13. A parametric representation of the sphere is as follows. Then type the partial derivatives, a normal vector N , then F and FS , which is **F** on the sphere. Integrate the dot product of FS and N over u from 0 to $\pi/2$ and over v from 0 to $\pi/2$.

In[1]:= r = {Cos[v] Cos[u], Cos[v] Sin[u], Sin[v]}

Out[1]= {Cos[u] Cos[v], Cos[v] Sin[u], Sin[v]}

In[2]:= ru = D[r, u]   (* Out: {−Cos[v] Sin[u], Cos[u] Cos[v], 0} *)

In[3]:= rv = D[r, v]   (* Out: {−Cos[u] Sin[v], −Sin[u] Sin[v], Cos[v]} *)

In[4]:= <<Calculus'VectorAnalysis'

In[5]:= SetCoordinates[Cartesian[x, y, z]]

In[6]:= N0 = Cross[ru, rv]

Out[6]= {Cos[u] Cos[v]^2, Cos[v]^2 Sin[u],

Cos[u]^2 Cos[v] Sin[v] + Cos[v] Sin[u]^2 Sin[v]}

In[7]:= F = {0, x, 0}   (* Out: {0, x, 0} *)

In[8]:= FS = F /. x -> r[[1]]   (* Out: {0, Cos[u] Cos[v], 0} *)

In[9]:= int = FS.N0   (* Out: Cos[u] Cos[v]^3 Sin[u] *)

In[10]:= Integrate[Integrate[int, {u, 0, Pi/2}], {v, 0, Pi/2}]   (* Out: $\frac{1}{3}$ *)

Pr.9.15. The square of the distance of a point (x, y, z) from the x-axis (with respect to which the moment of inertia is taken) is $y^2 + z^2$. This is the integrand. Use cylindrical coordinates defined by $x = x$, $y = r \cos \theta$, $z = r \sin \theta$. Then that square of the distance is simply r^2 and the volume element is $dx\, r\, dr\, d\theta$. Represent the cylinder as shown, with $0 \le x \le h$, $0 \le r \le a$, $0 \le \theta \le 2\pi$. Integrate accordingly. Answer: $h\, a^4\, \pi/2$.

In[1]:= Clear[r, theta, h]

In[2]:= R = {x, r Cos[theta], r Sin[theta]}   (* Cylinder *)

In[3]:= Integrate[Integrate[Integrate[(r^2) r, {x, 0, h}], {r, 0, a}],
        {theta, 0, 2 Pi}]

Pr.9.17. Type **F** and then its divergence, which will turn out to be constant, so that the problem becomes very simple.

In[1]:= <<Calculus'VectorAnalysis'

In[2]:= SetCoordinates[Cartesian[x, y, z]]

In[3]:= F = {9 x, y Cosh[x]^2, -z Sinh[x]^2}

Out[3]= {9 x, y Cosh[x]^2, −z Sinh[x]^2}

In[4]:= g = Div[F]   (* Out: $9 + \text{Cosh}[x]^2 - \text{Sinh}[x]^2$ *)

In[5]:= g = Simplify[%]   (* Out: 10 *)

Hence the answer is 10 times the volume $(4/3)\pi abc$ of the solid ellipsoid. Here, $a = 3$, $b = 6$, $c = 2$, as can be seen by dividing the formula for S by 36, that is, $x^2/3^2 + y^2/6^2 + z^2/2^2 = 1$. Hence the answer is 480π.

Pr.9.19. Use the volume element $r^2 \sin v \, dr \, du \, dv$, which gives the factor $r^2 \sin v$ in the last command.

$\text{In}[1]:=$ F = {x^3, y^3, z^3}

$\text{In}[2]:=$ <<Calculus'VectorAnalysis'

$\text{In}[3]:=$ SetCoordinates[Cartesian[x, y, z]]

$\text{In}[4]:=$ g = Div[F]

$\text{In}[5]:=$ R = {r Cos[u] Sin[v], r Sin[u] Sin[v], r Cos[v]}

   r is variable!

$\text{In}[6]:=$ g2 = g /. {x -> R[[1]], y -> R[[2]], z -> R[[3]]}

$\text{Out}[6]=$ $3\,r^2\,\text{Cos}\,[v]^2 + 3\,r^2\,\text{Cos}\,[u]^2\,\text{Sin}\,[v]^2 + 3\,r^2\,\text{Sin}\,[u]^2\,\text{Sin}\,[v]^2$

$\text{In}[7]:=$ g2 = Simplify[%]   (* Out: $3\,r^2$ *)

$\text{In}[8]:=$ Integrate[Integrate[Integrate[g2 r^2 Sin[v], {r, 0, 3}],
         {u, 0, 2 Pi}], {v, 0, Pi}]

$\text{Out}[8]=$ $\dfrac{2916\,\pi}{5}$

CHAPTER 10,   page 119

Pr.10.1. $f(x)$ is odd. Hence $a_n = 0$. Instead of integrating from $-\pi$ to π you can integrate from 0 to π and multiply the result by 2.

$\text{In}[1]:=$ bn = 2/Pi Integrate[3 Sin[n x], {x, 0, Pi}]

$\text{Out}[1]=$ $\dfrac{6\left(\dfrac{1}{n} - \dfrac{\text{Cos}\,[n\,\pi]}{n}\right)}{\pi}$

$\text{In}[2]:=$ S = Sum[bn Sin[n x], {n, 1, 20}]

$\text{In}[3]:=$ Plot[S, {x, -Pi, Pi}]

Pr.10.3. The function is neither even nor odd–always watch carefully for what interval a function is given! Note that without the computer the calculations would be involved. The plot shows again the Gibbs phenomenon at the jump.

$\text{In}[1]:=$ a0 = 1/(2 Pi) Integrate[(x/(2 Pi))^4, {x, 0, 2 Pi}]   (* Out: $\frac{1}{5}$ *)

$\text{In}[2]:=$ an = 1/Pi Integrate[(x/(2 Pi))^4 Cos[n x], {x, 0, 2 Pi}]

$\text{In}[3]:=$ bn = 1/Pi Integrate[(x/(2 Pi))^4 Sin[n x], {x, 0, 2 Pi}]

$\text{In}[4]:=$ Plot[a0 + Sum[an Cos[n x] + bn Sin[n x], {n, 1, 10}],
       {x, 0, 4 Pi}, PlotRange -> {-0.1, 1}]

Pr.10.5.   Answer: the terms $\sin t$ and $\sin 2t$ because their coefficients -4π and -2π are the absolutely largest ones.

In[1]:= `an = Simplify[1/Pi Integrate[t^2 Cos[n t], {t, 0, 2 Pi}]]`

In[2]:= `bn = Simplify[1/Pi Integrate[t^2 Sin[n t], {t, 0, 2 Pi}]]`

In[3]:= `Table[{an, bn}, {n, 1, 10}]`

Out[3]= $\left\{\{4, -4\pi\}, \{1, -2\pi\}, \left\{\frac{4}{9}, -\frac{4\pi}{3}\right\}, \left\{\frac{1}{4}, -\pi\right\}, \left\{\frac{4}{25}, -\frac{4\pi}{5}\right\}, \left\{\frac{1}{9}, -\frac{2\pi}{3}\right\},\right.$
$\left. \left\{\frac{4}{49}, -\frac{4\pi}{7}\right\}, \left\{\frac{1}{16}, -\frac{\pi}{2}\right\}, \left\{\frac{4}{81}, -\frac{4\pi}{9}\right\}, \left\{\frac{1}{25}, -\frac{2\pi}{5}\right\}\right\}$

Pr.10.7. $p = 2L = 2, L = 1$. The function is even, hence $b_n = 0$. Instead of integrating from -1 to 1 you could take twice the integral from 0 to 1.

In[1]:= `a0 = 1/(2 1) Integrate[3 x^2, {x, -1, 1}]`   (* Out: 1 *)

In[2]:= `an = 1/1 Integrate[3 x^2 Cos[n Pi x], {x, -1, 1}]`

Out[2]= $3\left(\dfrac{4\,\text{Cos}\,[n\pi]}{n^2\,\pi^2} + \dfrac{2\,(-2 + n^2\,\pi^2)\,\text{Sin}\,[n\pi]}{n^3\,\pi^3}\right)$

In[3]:= `S = a0 + Sum[an Cos[n Pi x], {n, 1, 4}]`

Out[3]= $1 - \dfrac{12\,\text{Cos}\,[\pi x]}{\pi^2} + \dfrac{3\,\text{Cos}\,[2\pi x]}{\pi^2} - \dfrac{4\,\text{Cos}\,[3\pi x]}{3\,\pi^2} + \dfrac{3\,\text{Cos}\,[4\pi x]}{4\,\pi^2}$

In[4]:= `P1 = Plot[S, {x, -3, 3}];`

In[5]:= `P2 = Plot[3 (x + 2)^2, {x, -3, -1}];`

In[6]:= `P3 = Plot[3 x^2, {x, -1, 1}];`

In[7]:= `P4 = Plot[3 (x - 2)^2, {x, 1, 3}];`

In[8]:= `Show[P1, P2, P3]`

Pr.10.9.   `Even` and `Odd` (below) denote the half-range cosine and sine series, respectively.

In[1]:= `a0 = 2/(2 L) Integrate[x, {x, 0, L}]`   (* Out: $\frac{L}{2}$ *)

In[2]:= `an = Simplify[2/L Integrate[x Cos[n Pi x/L], {x, 0, L}]]`

Out[2]= $\dfrac{2L\,(-1 + \text{Cos}\,[n\pi] + n\pi\,\text{Sin}\,[n\pi])}{n^2\,\pi^2}$

In[3]:= `bn = Simplify[2/L Integrate[x Sin[n Pi x/L], {x, 0, L}]]`

Out[3]= $\dfrac{2L\,(-n\pi\,\text{Cos}\,[n\pi] + \text{Sin}\,[n\pi])}{n^2\,\pi^2}$

In[4]:= `Even = a0 + Sum[an Cos[n Pi x/L], {n, 1, 5}]`

Out[4]= $\dfrac{L}{2} - \dfrac{4L\,\text{Cos}\,\left[\frac{\pi x}{L}\right]}{\pi^2} - \dfrac{4L\,\text{Cos}\,\left[\frac{3\pi x}{L}\right]}{9\,\pi^2} - \dfrac{4L\,\text{Cos}\,\left[\frac{5\pi x}{L}\right]}{25\,\pi^2}$

In[5]:= `Odd = Sum[bn Sin[n Pi x/L], {n, 1, 5}]`

Out[5]= $\dfrac{2L\,\text{Sin}\,\left[\frac{\pi x}{L}\right]}{\pi} - \dfrac{L\,\text{Sin}\,\left[\frac{2\pi x}{L}\right]}{\pi} + \dfrac{2L\,\text{Sin}\,\left[\frac{3\pi x}{L}\right]}{3\pi} - \dfrac{L\,\text{Sin}\,\left[\frac{4\pi x}{L}\right]}{2\pi} + \dfrac{2L\,\text{Sin}\,\left[\frac{5\pi x}{L}\right]}{5\pi}$

Pr.10.11. It is typical that the error is oscillating with x. Of course, you can expect the error to increase as x comes close to the jumps.

In[1]:= bn = 1/Pi Integrate[x^3 Sin[n x], {x, -Pi, Pi}]

In[2]:= Error = x^3 - Sum[bn Sin[n x], {n, 1, 5}]

In[3]:= Plot[Error, {x, -Pi, Pi}]

Pr.10.13. The present function is discontinuous, so that you should expect large values of the minimum square error.

In[1]:= bn = 1/Pi Integrate[x Sin[n x], {x, -Pi, Pi}]

In[2]:= SN = N[Table[Integrate[x^2, {x, -Pi, Pi}] - Pi Sum[bn^2, {n, 1, N0}],
        {N0, 1, 10}]]

Out[2]= {8.10448, 4.96289, 3.56662, 2.78123, 2.27857, 1.92951, 1.67305, 1.4767,

        1.32156, 1.1959}

In[3]:= N[Integrate[x^2, {x, -Pi, Pi}] - Pi Sum[bn^2, {n, 1, 100}]]

Out[3]= 0.125037

CHAPTER 11,   page 131

Pr.11.1. Note the nodes (points that do not move). S[[1]] , .., S[[4]] call the solutions individually. Read and follow the instructions on animation in Example 11.1 in this Guide.

In[1]:= <<Graphics'Animation'

In[2]:= S = Table[Sin[n x] Sin[n t], {n, 1, 4}]

Out[2]= Sin [t] Sin [x], Sin [2 t] Sin [2 x], Sin [3 t] Sin [3 x], Sin [4 t] Sin [4 x]

In[3]:= S[[1]]   (* Out: Sin [t] Sin [x] *)

In[4]:= Animate[Plot[S[[3]], {x, 0, 2 Pi}], {t, 0, Pi}, PlotRange -> {-1, 1}]

Pr.11.3. You need the inverse of the given transformation, which you obtain by Solve .

In[1]:= Clear[f, g]

In[2]:= u = f[x + y] + g[2 x - y]

In[3]:= D[u, x, x] + D[u, x, y] - 2 D[u, y, y]

Out[3]= 2 f''[x + y] + 2 g''[2 x - y] - 2 (f''[x + y] + g''[2 x - y])

In[4]:= Simplify[%]   (* Out: 0 *)

Pr.11.5. Yes. Only trivially by $-k$.

In[1]:= Clear[u, u2, k, K]

In[2]:= u = (c1 Exp[k x] + c2 Exp[-k x]) (c3 Exp[k y] + c4 Exp[-k y])

In[3]:= D[u, x, x] - D[u, y, y]

In[4]:= Simplify[%]   (* Out: 0 *)

In[5]:= u2 = (c1 Exp[k x] + c2 Exp[-k x]) (c3 Exp[K y] + c4 Exp[-K y])

In[6]:= D[u2, x, x] - D[u2, y, y]

In[7]:= sol = Simplify[%]   (* No reduction to zero *)

In[8]:= sol /. K -> k   (* Out: 0 *)

In[9]:= sol /. K -> -k   (* Out: 0 *)

Pr.11.7.

In[1]:= Clear[u, F, G]

In[2]:= u[x,t] = F[x] G[t]

In[3]:= pde = D[u[x,t], t, t] == c^2 D[u[x,t], x, x, x, x]

In[4]:= eq = pde[[1]]/(c^2 u[x,t]) == pde[[2]]/(c^2 u[x,t])

$$\text{Out[4]}= \frac{G''[t]}{c^2 G[t]} == \frac{F^{(4)}[x]}{F[x]}$$

In[5]:= ode1 = eq[[2]] F[x] == k^2 F[x]   (* Out: $F^{(4)}[x] == k^2 F[x]$ *)

In[6]:= sol1 = DSolve[ode1, F[x], x]

$$\text{Out[6]}= \left\{\left\{F[x] \rightarrow e^{-\sqrt{k}\,x} C[1] + e^{-i\sqrt{k}\,x} C[2] + e^{i\sqrt{k}\,x} C[3] + e^{\sqrt{k}\,x} C[4]\right\}\right\}$$

In[7]:= ode2 = eq[[1]] == k^2   (* Out: $\frac{G''[t]}{c^2 G[t]} == k^2$ *)

In[8]:= sol2 = DSolve[ode2, G[t], t]   (* Out: $\{\{G[t] \rightarrow e^{-ckt} C[1] + e^{ckt} C[2]\}\}$ *)

In[9]:= answer = sol1[[1, 1, 2]] sol2[[1, 1, 2]]

$$\text{Out[9]}= (e^{-ckt} C[1] + e^{ckt} C[2])\left(e^{-\sqrt{k}\,x} C[1] + e^{-i\sqrt{k}\,x} C[2] + e^{i\sqrt{k}\,x} C[3] + e^{\sqrt{k}\,x} C[4]\right)$$

Of course, the arbitrary constants should all be numbered differently, say, C[1] , ...,
C[6] .

Pr.11.9.   Recall that the ends of the bar at $x = 0$ and $x = \pi$ are kept at temperature 0.

In[1]:= Clear[F, G, c]

In[2]:= u[x,t] = F[x] G[t]

In[3]:= pde = D[u[x, t], t] == c^2 D[u[x, t], x, x]

Out[3]= F[x] G'[t] == c^2 G[t] F''[x]

In[4]:= eq = pde[[1]]/(c^2 u[x,t]) == pde[[2]]/(c^2 u[x,t]) == -k^2

$$\text{Out[4]}= \frac{G'[t]}{c^2 G[t]} == \frac{F''[x]}{F[x]} == -k^2$$

In[5]:= sol1 = DSolve[eq[[2]] == -k^2, F[x], x]

Out[5]= {{F[x] → C[2] Cos [k x] + C[1] Sin [k x]}}

In[6]:= (sol1[[1, 1, 2]] /. x -> 0) == 0   (* Out: C[2] == 0 *)

In[7]:= (sol1[[1, 1, 2]] /. {C[1] -> 1, C[2] -> 0, x -> Pi}) == 0

Out[7]= Sin [k π] == 0

(In this command the parentheses $(\ldots)$ are important. Try without.) Hence $k = 1, 2, \ldots$. ($k = -1, -2, \ldots$ would give the same functions since $\sin(-a) = -\sin a$.) This gives $F1 = \sin nx$, $n = 1, 2, \ldots$. Furthermore, since now $-k^2 = -c^2$ in eq, type

In[8]:= sol2 = DSolve[eq[[1]] == -n^2, G[t], t]

Out[8]= $\{\{$G[t] $\rightarrow$ e$^{-c^2 n^2 t + C[1]}\}\}$

Here Mathematica reuses C[1] for a different arbitrary constant.

In[9]:= answer = Sin[n x] sol2[[1, 1, 2]]   (* Out: e$^{-c^2 n^2 t + C[1]}$ Sin[n x] *)

Pr.11.11.   Because of the boundary condition the series solution reduces to a single term,

In[1]:= An = 2/(Pi Sinh[n Pi/2]) Integrate[Sin[x] Sin[n x], {x, 0, Pi}]

In[2]:= Table[An, {n, 1, 10}]

        Power:: infy: Infinite expression $\frac{1}{0}$ encountered

Out[2]= Indeterminate, 0, 0, 0, 0, 0, 0, 0, 0, 0

In[3]:= A1 = 2/(Pi Sinh[Pi/2]) Integrate[Sin[x] Sin[x], {x, 0, Pi}]

In[4]:= u1 = A1 Sin[x] Sinh[y]   (* Out: Csch$\left[\frac{\pi}{2}\right]$ Sin[x] Sinh[y] *)

In[5]:= Plot3D[u1, {x, 0, Pi}, {y, 0, Pi/2}]

Pr.11.13. $\cos \lambda_{22} t \sin 0.5\pi x \sin \pi y$, $\lambda_{22} = 5\pi\sqrt{0.25 + 1}$. Hence type

In[1]:= <<Graphics'Animation'

In[2]:= Animate[Plot3D[Sin[Pi x/2] Sin[Pi y] Cos[5 Pi Sqrt[5/4] t],
      {x, 0, 4}, {y, 0, 2}], {t, 0, 1}]

To slow down the motion, replace t by, say, $0.1\,t$.

Pr.11.15.   You need the zero

In[1]:= k = FindRoot[BesselJ[1, x], {x, 3}]   (* Out: {x- > 3.83171} *)

In[2]:= <<Graphics'Animation'

In[3]:= Animate[ParametricPlot3D[{r Cos[Theta], r Sin[Theta],
      BesselJ[1, k[[1, 2]] r] Cos[Theta] Cos[k[[1, 2]] t]},
      {r, 0, 1}, {Theta, 0, 2 Pi}], {t, 0, 2 Pi}]

To slow down the motion, replace the t in Cos[k[[1, 2]] t] by, say, 0.1 t

CHAPTER 12,   page 147

Pr.12.1.

In[1]:= z1 = 8 + 3 I   (* Out: $8 + 3i$ *)

In[2]:= z2 = 9 - 2 I   (* Out: $9 - 2i$ *)

In[3]:= z1 + z2   (* Out: $17 + i$ *)

In[4]:= z1 - z2   (* Out: $-1 + 5i$ *)

In[5]:= z1 z2   (* Out: $78 + 11\,i$ *)

In[6]:= z1/z2   (* Out: $\frac{66}{85} + \frac{43\,i}{85}$ *)

In[7]:= N[z1/z2]   (* Out: $0.776471 + 0.505882\,i$ *)

In[8]:= Abs[z1/z2]   (* Out: $\sqrt{\frac{73}{85}}$ *)

In[9]:= Abs[z1]/Abs[z2]   (* Out: $\sqrt{\frac{73}{85}}$ *)

In[10]:= Re[z1]   (* Out: 8 *)

In[11]:= Im[z1^2]   (* Out: 48 *)

In[12]:= Arg[z1]   (* Out: $\mathrm{ArcTan}\left[\frac{3}{8}\right]$ *)

Pr.12.3.   Start from the right-hand sides.

In[1]:= z = x + I y   (* Space between I and [y]! *)

In[2]:= zbar = Conjugate[z]   (* Out: Conjugate[$x + iy$] *)

In[3]:= zbar = ComplexExpand[%]   (* Out: $x - iy$ *)

By the trick in Example 12.1 you get the same,

In[4]:= zbar = z /. Complex[x_, y_] -> Complex[x, -y]   (* Out: $x - iy$ *)

In[5]:= (z + zbar)/2   (* Out: x *)

In[6]:= (z - zbar)/(2 I)   (* Out: y *)

Pr.12.5.   You obtain the the real sequence S2 of 41 pairs of the real and imaginary parts from the complex sequence S. Observe the two different ranges of n. Be careful with the various brackets in S2; each of them is needed. P2 plots the unit circle.

In[1]:= zn = (0.9 + 0.4 I)^n

In[2]:= S = Table[{Re[zn], Im[zn]}, {n, -20, 20}]

In[3]:= P1 = ListPlot[S, Prolog -> AbsolutePointSize[4],
            AspectRatio -> Automatic]

In[4]:= P2 = ParametricPlot[{Cos[t], Sin[t]}, {t, 0, 2 Pi}]

In[5]:= Show[P1, P2]

Pr.12.7.

In[1]:= Clear[S]

In[2]:= Roots[z^3 - 1 - I == 0, z]

Out[2]= $z == (-1)^{2/3}\,(1+i)^{1/3}\,||\,z == -(-1)^{1/3}\,(1+i)^{1/3}\,||\,z == (1+i)^{1/3}$

In[3]:= S = Table[2^(1/6) {Cos[Pi/12 + j 2 Pi/3], Sin[Pi/12 + j 2 Pi/3]},
            {j, 0, 2}]

Out[3]= $\left\{\left\{\dfrac{1+\sqrt{3}}{2\,2^{1/3}}, \dfrac{-1+\sqrt{3}}{2\,2^{1/3}}\right\}, \left\{-\dfrac{1}{2^{1/3}}, \dfrac{1}{2^{1/3}}\right\}, \left\{-\dfrac{-1+\sqrt{3}}{2\,2^{1/3}}, -\dfrac{1+\sqrt{3}}{2\,2^{1/3}}\right\}\right\}$

In[4]:= P1 = ListPlot[S, Prolog -> AbsolutePointSize[4]]

In[5]:= P2 = ParametricPlot[{2^(1/6) Cos[t], 2^(1/6) Sin[t]}, {t, 0, 2 Pi}]

In[6]:= Show[P1, P2, AspectRatio -> Automatic]

Pr.12.9. You can plot the two circles as follows,

In[1]:= P1 = ParametricPlot[{2 + Sqrt[8] Cos[t], 2 + Sqrt[8] Sin[t]},
        {t, 0, 2 Pi}]

In[2]:= P2 = ParametricPlot[{2 + 2 Cos[t], 2 + 2 Sin[t]}, {t, 0, 2 Pi}]

In[3]:= Show[P1, P2, AspectRatio -> Automatic]

Pr.12.11. No. Neither of the Cauchy-Riemann equations is satisfied.

In[1]:= z = x + I y

In[2]:= zbar = ComplexExpand[Conjugate[z]]   (* Out: $x - i\,y$ *)

In[3]:= u = ComplexExpand[Re[z z zbar]]   (* Out: $x^3 + x\,y^2$ *)

In[4]:= v = ComplexExpand[Im[z z zbar]]   (* Out: $x^2\,y + y^3$ *)

In[5]:= D[u, x] - D[v, y]   (* Out: $2x^2 - 2y^2$ *)

In[6]:= D[u, y] + D[v, x]   (* Out: $4\,x\,y$ *)

Pr.12.15. The interior of the circle is mapped onto the exterior and conversely. If you don't see the correspondence, map the square portion by portion, e.g., $0.5 \le x \le 0.6$, then $0.6 \le x \le 0.7$, etc. Similarly in the y-direction if necessary.

In[1]:= <<Graphics'ComplexMap'

In[2]:= P1 = CartesianMap[Identity, {1/2, 3/2}, {1/2, 3/2}]

In[3]:= Clear[z]

In[4]:= w[z_] = 1/z

In[5]:= P2 = CartesianMap[w, {1/2, 3/2}, {1/2, 3/2}]

In[6]:= P3 = ParametricPlot[{Cos[t], Sin[t]}, {t, 0, 2 Pi}]

In[7]:= Show[P1, P2, P3]

Pr.12.17. The images coincide on the real axis. The elliptical ring opens up when you shorten.

In[1]:= <<Graphics'ComplexMap'

In[2]:= w[z_] = Cos[z]

In[3]:= CartesianMap[w, {0, 2 Pi}, {1/2, 1}]

Pr.12.19. Those values suggest that Ln z is discontinuous along the negative real axis.

In[1]:= Clear[z]

In[2]:= f[z_] = Log[z]   (* Out: Log[z] *)

In[3]:= f[-5]   (* Out: $i\pi + $ Log[5] *)

In[4]:= ComplexExpand[f[-12 - 16 I]]   (* Out: $i\left(-\pi + \text{ArcTan}\left[\frac{4}{3}\right]\right) + $ Log[20] *)

In[5]:= ComplexExpand[f[1 + I]]   (* Out: $\frac{i\pi}{4} + \frac{\text{Log}[2]}{2}$ *)

In[6]:= ComplexExpand[f[1 - I]]   (* Out: $-\frac{i\pi}{4} + \frac{\text{Log}[2]}{2}$ *)

In[7]:= f[-10 + 0.1 I]   (* Out: $2.30264 + 3.13159\,i$ *)

In[8]:= f[-10 - 0.1 I]   (* Out: $2.30264 - 3.13159\,i$ *)

CHAPTER 13,   page 153

Pr.13.1.

In[1]:= z = 1 + I + (2 + I) t   (* Out: $(1+i) + (2+i)\,t$ *)

In[2]:= zdot = D[z, t]   (* Out: $2+i$ *)

In[3]:= Integrate[Re[z] zdot, {t, 0, 1}]   (* Out: $4+2i$ *)

Pr.13.3. You can leave the two integrals separate or take them together as shown here. Answer $4+i$

In[1]:= z1 = 1 + I + I t (* Out: $(1+i) + it$ Type a space between I and t! *)

In[2]:= z1dot = D[z1, t]   (* Out: i *)

In[3]:= z2 = 1 + 2 I + 2 t   (* Out: $(1+2i) + 2t$ *)

In[4]:= z2dot = D[z2, t]   (* Out: 2 *)

In[5]:= Integrate[Re[z1] z1dot + Re[z2] z2dot, {t, 0, 1}]   (* Out: $4+i$ *)

Pr.13.5. Answer $12\pi i$. Commands:

In[1]:= Clear[z]

In[2]:= f = (4 z^2 + 17 z - 68)/(z^3 - 12 z + 16)   (* Out: $\frac{-68 + 17z + 4z^2}{16 - 12z + z^3}$ *)

In[3]:= Apart[%]   (* Out: $-\frac{3}{(-2+z)^2} + \frac{6}{-2+z} - \frac{2}{4+z}$ *)

No contribution from the first fraction for which $f(z) = -3$ and thus $f'(z) = 0$ in (1) and (2) in Example 13.3 in this Guide. From the second fraction, $2\pi i \times 6$ by Cauchy's formula. No contribution from the third fraction since -4 lies outside the contour. To confirm this, you may type

In[4]:= z = 3 Exp[I t]   (* Out: $3\,e^{it}$ *)

In[5]:= zdot = D[z, t]   (* Out: $3\,i\,e^{it}$ *)

In[6]:= Integrate[f zdot, {t, 0, 2 Pi}]

Out[6]= $3\,i\left(\frac{2}{15}\left(9\,i + 15\pi - 15\,i\,\text{Log}[5] - 5\,i\,\text{Log}[7]\right) + \frac{2}{15}\left(-9\,i + 15\pi + 15\,i\,\text{Log}[5] + 5\,i\,\text{Log}[7]\right)\right)$

In[7]:= Simplify[%]   (* Out: $12\,i\,\pi$ *)

Pr.13.7. Denote the straight segment by $z1$ and the parabolic arc by z2. Accordingly, type

In[1]:= z1 = (1 + I) t   (* Out: $(1+i)\,t$ *)

In[2]:= z1bar = ComplexExpand[Conjugate[z1]]   (* Out: $(1-i)\,t$ *)

ComplexExpand regards t as real, as it should be.

In[3]:= z1dot = D[z1, t]   (* Out: $1+i$ *)

In[4]:= Integrate[z1bar z1dot, {t, 0, 1}]   (* Out: 1 *)

In[5]:= z2 = t + I t^2   (* Out: $t+it^2$ Space after I ! *)

In[6]:= z2bar = ComplexExpand[Conjugate[z2]]   (* Out: $t-it^2$ *)

In[7]:= z2dot = D[z2, t]   (* Out: $1+2it$ *)

t is again regarded as real.

In[8]:= Integrate[z2bar z2dot, {t, 0, 1}]   (* Out: $1+\frac{i}{3}$ *)

This shows path dependence of the integral of the nonanalytic function $\bar{z}$.

Pr.13.9. On the four sides, starting from the origin and going around clockwise, you have the following, from which you can read the integrands $(-y^2)i$, etc. of the four integrals, as shown in the command. Answer $-1-i$.

$z =$	iy	$i+x$	$1+i-iy$	$1-x$
$\dot{z} =$	i	1	$-i$	-1
$\mathrm{Re}\, z^2 =$	$-y^2$	$-1+x^2$	$2y-y^2$	$(1-x)^2$
Range	$y = 0..1$	$x = 0..1$	$y = 0..1$	$x = 0..1$

In[1]:= Integrate[-y^2 I, {y, 0, 1}] + Integrate[-1 + x^2, {x, 0, 1}]
    + Integrate[(2 y - y^2) (-I), {y, 0, 1}]
    + Integrate[(1 - x)^2 (-1), {x, 0, 1}]

CHAPTER 14,   page 160

Pr.14.1. Type the commands. S is the complex sequence. For plotting obtain the real sequence S2 from it (see Example 14.1 in this Guide).

In[1]:= zn = 1 - 1/n^2 + I (2 + 4/n)

In[2]:= S = Table[zn, {n, 1, 10}]

In[3]:= S2 = Table[{Re[S[[n]]], Im[S[[n]]]}, {n, 1, 10}]

In[4]:= ListPlot[S2, Prolog -> AbsolutePointSize[4], PlotRange -> {0, 4}]

Pr.14.3. The series converges. Commands:

In[1]:= zn = (20 + 30 I)^n/n!

In[2]:= zn1 = zn /. n -> n + 1

In[3]:= FullSimplify[zn1/zn]   (* Out: $\frac{20+30i}{1+n}$ *)

In[4]:= Limit[%, n -> Infinity]   (* Out: 0 *)

Pr.14.5.

In[1]:= zn = (3 I)^n n!/n^n   (* Out: $(3i)^n n^{-n} n!$ *)

In[2]:= zn1 = zn /. n -> n + 1   (* Out: $i^{1+n} 3^{1+n} (1+n)^{-1-n} (1+n)!$ *)

In[3]:= `FullSimplify[zn1/zn]`   (* Out: $3i\left(\frac{n}{1+n}\right)^n$ *)

In[4]:= `ComplexExpand[Abs[%]]`   (* Out: $3\left(\frac{\sqrt{n^2}}{\sqrt{(1+n)^2}}\right)^n$ *)

In[5]:= `Limit[%, n -> Infinity]`   (* Out: $\frac{3}{e}$ *)

Since $3/e > 1$, the series diverges. To really understand what is going on, do the simplification of the quotient $|z_{n+1}/z_n|$ by hand, using $(n+1)!/n! = n+1$ and remembering from calculus that the limit of $(1+1/n)^n$ as $n \to \infty$ equals e.

Pr.14.7.

In[1]:= `an = (3 n)!/(2^n (n!)^3)`   (* Out: $\frac{2^{-n}(3n)!}{n!^3}$ *)

In[2]:= `an1 = an /. n -> n + 1`   (* Out: $\frac{2^{-1-n}(3(1+n))!}{(1+n)!^3}$ *)

In[3]:= `FullSimplify[an/an1]`   (* Out: $\frac{2(1+n)^2}{3(1+3n)(2+3n)}$ *)

In[4]:= `Limit[%, n -> Infinity]`   (* Out: $\frac{2}{27}$ *)

Pr.14.9.   Type `Log`, not `Ln`. The commands are

In[1]:= `an = D[Log[z], {z, n}]/n!`   (* Out: $\frac{\text{Log}^{(n)}[z]}{n!}$ *)

In[2]:= `an1 = an /. z -> 1`   (* Out: $\frac{\text{Log}^{(n)}[1]}{n!}$ *)

In[3]:= `S = Table[an1, {n, 1, 5}]`   (* Out: $\left\{1, -\frac{1}{2}, \frac{1}{3}, -\frac{1}{4}, \frac{1}{5}\right\}$ *)

In[4]:= `Sum[S[[n]] (z - 1)^n, {n, 1, 5}]`

Out[4]= $-1 - \frac{1}{2}(-1+z)^2 + \frac{1}{3}(-1+z)^3 - \frac{1}{4}(-1+z)^4 + \frac{1}{5}(-1+z)^5 + z$

Note that Mathematica has put z (missing in the first term) at the very end.

Pr.14.11.   Integrate the Maclaurin series of $1/(1+z^2)$. Since $\arctan 0 = 0$, the constant of integration is 0. Commands:

In[1]:= `S = Series[1/(1 + z^2), {z, 0, 10}]`

In[2]:= `Integrate[S, z]`

In[3]:= `Series[ArcTan[z], {z, 0, 11}]`

Pr.14.13.   The number of terms needed is found by trial and error. The command `Coefficient` gives the coefficients.

In[1]:= `S = Series[z/(Exp[z] - 1), {z, 0, 14}]`

In[2]:= `Table[n! Coefficient[S, z^n], {n, 1, 14}]`

Out[2]= $\left\{-\frac{1}{2}, \frac{1}{6}, 0, -\frac{1}{30}, 0, \frac{1}{42}, 0, -\frac{1}{30}, 0, \frac{5}{66}, 0, -\frac{691}{2730}, 0, \frac{7}{6}\right\}$

Pr.14.15. S is the sequence of partial sums to be plotted.

In[1]:= S = Table[x^4 Sum[1/(1 + x^4)^m, {m, 1, 2^n}], {n, 1, 3}];

In[2]:= Plot[{S[[1]], S[[2]], S[[3]]}, {x, -2, 2}]

CHAPTER 15,   page 169

Pr.15.1.

In[1]:= f = Cosh[z]/(z - Pi I)^2

In[2]:= FullSimplify[Series[f, {z, Pi I, 4}]]

$$\text{Out[2]}= -\frac{1}{(z-i\pi)^2} - \frac{1}{2} - \frac{1}{24}(z-i\pi)^2 - \frac{1}{720}(z-i\pi)^4 + 0[z-i\pi]^5$$

Pr.15.3. S shows that $f(z)$ has an essential singularity at $z = 0$ and the residue is 0 (there is no term in $1/z$).

In[1]:= f = Exp[-1/z^2]/z^2   (* Out: $\dfrac{e^{-\frac{1}{z^2}}}{z^2}$ *)

In[2]:= Series[f, {1/z, 0, 10}]

General:: ivar: $\frac{1}{z}$ is not a valid variable

$$\text{Out[2]}= \text{Series}\left[\frac{e^{-\frac{1}{z}}}{z^2}, \left\{\frac{1}{z}, 0, 10\right\}\right]$$

In[3]:= g = f /. z -> 1/w   (* Out: $e^{-w^2} w^2$ *)

In[4]:= Series[g, {w, 0, 14}]

$$\text{Out[4]}= w^2 - w^4 + \frac{w^6}{2} - \frac{w^8}{6} + \frac{w^{10}}{24} - \frac{w^{12}}{120} + \frac{w^{14}}{720} + 0[w]^{15}$$

In[5]:= S = % /. w -> 1/z

$$\text{Out[5]}= \left(\frac{1}{z}\right)^2 - \left(\frac{1}{z}\right)^4 + \frac{1}{2}\left(\frac{1}{z}\right)^6 - \frac{1}{6}\left(\frac{1}{z}\right)^8 + \frac{1}{24}\left(\frac{1}{z}\right)^{10} - \frac{1}{120}\left(\frac{1}{z}\right)^{12} + \frac{1}{720}\left(\frac{1}{z}\right)^{14} + 0\left[\frac{1}{z}\right]^{15}$$

Pr.15.5. The partial fraction expansion of $f(z)$ shows that $f(z)$ has a simple pole with residue $-1/32$ at 0 and a pole of fifth order with residue $1/32$ at 2. The last two commands shown are just for confirmation.

In[1]:= f = (3 z^4 -18 z^3 + 36 z^2 - 24 z + 1)/

         (z^6 - 10 z^5 + 40 z^4 - 80 z^3 + 80 z^2 - 32 z)

In[2]:= Apart[%]

$$\text{Out[2]}= \frac{1}{2(-2+z)^5} - \frac{1}{4(-2+z)^4} + \frac{1}{8(-2+z)^3} + \frac{47}{16(-2+z)^2} + \frac{1}{32(-2+z)} - \frac{1}{32z}$$

In[3]:= Residue[f, {z, 0}]   (* Out: $-\frac{1}{32}$ *)

In[4]:= Residue[f, {z, 2}]   (* Out: $\frac{1}{32}$ *)

Pr.15.7.

In[1]:= f = (-z^2 - 22 z + 8)/(z^3 - 5 z^2 + 4 z)

In[2]:= Solve[Denominator[f] == 0, z]   (* Out: $\{\{z \to 0\}, \{z \to 1\}, \{z \to 4\}\}$ *)

In[3]:= Residue[f, {z, 0}]   (* Out: 2 *)

In[4]:= Residue[f, {z, 1}]   (* Out: 5 *)

In[5]:= Residue[f, {z, 4}]   (* Out: −8 *)

In[6]:= Apart[f]   (* This confirms the values of the residues. *)

Out[6]= $-\dfrac{8}{-4+z} + \dfrac{5}{-1+z} + \dfrac{2}{z}$

Pr.15.9. $1 - e^z = 1 - 1$ at $z = 0$. Indeed,

In[1]:= f = 1/(1 - Exp[z])   (* Out: $\dfrac{1}{1-e^z}$ *)

In[2]:= Solve[1 - Exp[z] == 0, z]

Solve:: ifun: Inverse functions are being used by Solve, so some solutions may not be found

Out[2]= $\{\{z \to 0\}\}$

The following series shows that the singularity of f at $z = 0$ is a simple pole with residue -1. The command Residue confirms this value.

In[3]:= Series[f, {z, 0, 5}]   (* Out: $-\dfrac{1}{z} + \dfrac{1}{2} - \dfrac{z}{12} + \dfrac{z^3}{720} - \dfrac{z^5}{30240} + 0[z]^6$ *)

In[4]:= Residue[f, {z, 0}]   (* Out: −1 *)

Since e^z is periodic with period $2\pi i$, further zeros (poles of f) are at $2n\pi i$, the residues of f at those points being the same because of periodicity. For confirmation of some of them, type

In[5]:= Table[Residue[f, {z, 2 n Pi I}], {n, -5, 5}]

Pr.15.11. $f(z)$ has four simple poles, all inside the contour.

In[1]:= f = z Cosh[Pi z]/(z^4 + 13 z^2 + 36)   (* Out: $\dfrac{z\,\mathrm{Cosh}[\pi z]}{36 + 13z^2 + z^4}$ *)

In[2]:= Solve[Denominator[f] == 0, z]

Out[2]= $\{\{z \to -2i\}, \{z \to 2i\}, \{z \to -3i\}, \{z \to 3i\}\}$

In[3]:= 2Pi I (Residue[f, {z, -2 I}] + Residue[f, {z, 2 I}] +
         Residue[f, {z, -3 I}] + Residue[f, {z, 3 I}])   (* Out: $\frac{4i\pi}{5}$ *)

Pr.15.13.   Proceed as in Example 15.4 in this Guide.

In[1]:= t = I Log[z]

In[2]:= c = Cos[t]   (* Out: $\dfrac{1+z^2}{2z}$ *)

In[3]:= c2 = Cos[2 t]   (* Out: Cosh[2 Log[z]] *)

In[4]:= d = D[t, z]   (* Out: $\dfrac{i}{z}$ *)

In[5]:= `int = d c/(13 - 12 c2)`   (* Out: $\dfrac{i\,(1+z^2)}{2\,z^2\,(13-12\,\mathrm{Cosh}[2\,\mathrm{Log}\,[z]])}$ *)

In[6]:= `int2 = Simplify[%]`   (* Out: $-\dfrac{i\,(1+z^2)}{2\,(6-13\,z^2+6\,z^4)}$ *)

In[7]:= `zeros = Solve[Denominator[int2] == 0, z]`

Out[7]= $\left\{\left\{z\to-\sqrt{\tfrac{2}{3}}\right\},\left\{z\to\sqrt{\tfrac{2}{3}}\right\},\left\{z\to-\sqrt{\tfrac{3}{2}}\right\},\left\{z\to\sqrt{\tfrac{3}{2}}\right\}\right\}$

These zeros are simple (why?). The last two lie outside the contour. The first two lie inside. The residues at the latter are obtained from (2) in Example 15.3,

In[8]:= `Res = Numerator[int2]/D[Denominator[int2], z]`

Out[8]= $-\dfrac{i\,(1+z^2)}{2\,(-26\,z+24\,z^3)}$

In[9]:= `Res /. z -> zeros[[1, 1, 2]]`   (* Out: $-\dfrac{i}{4\sqrt{6}}$ *)

In[10]:= `Res /. z -> zeros[[2, 1, 2]]`   (* Out: $\dfrac{i}{4\sqrt{6}}$ *)

The sum of these two residues is 0, and so is the integral. Confirmation:

In[11]:= `Clear[t]`

In[12]:= `Integrate[Cos[t]/(13 - 12 Cos[2 t]), {t, 0, 2 Pi}]`   (* Out: 0 *)

Pr.15.15.   Simple poles at 2 and $-1+i\sqrt{3}$. (Third pole in the lower half-plane.) Residues from (2) in Example 15.3 in this Guide. Commands:

In[1]:= `f = z/(8 - z^3)`

In[2]:= `zeros = Solve[Denominator[f] == 0, z]`

Out[2]= $\{\{z\to 2\},\{z\to-2\,(-1)^{1/3}\},\{z\to 2\,(-1)^{2/3}\}\}$

In[3]:= `N[%]`   (* Out: $\{\{z\to 2.\},\{z\to-1.-1.73205\,i\},\{z\to-1.+1.73205\,i\}\}$ *)

In[4]:= `res = Numerator[f]/D[Denominator[f], z]`   (* Out: $-\dfrac{1}{3\,z}$ *)

In[5]:= `res1 = res /. z -> zeros[[1, 1, 2]]`   (* Out: $-\dfrac{1}{6}$ *)

In[6]:= `res2 = res /. z -> zeros[[3, 1, 2]]`   (* Out: $\tfrac{1}{6}\,(-1)^{1/3}$ *)

In[7]:= `ComplexExpand[Pi I res1 + 2 Pi I res2]`   (* Out: $-\dfrac{\pi}{2\sqrt{3}}$ *)

In[8]:= `N[%]`   (* Out: −0.9069 Try to get the answer by using Integrate *)

CHAPTER 16,   page 177

Pr.16.1.   `AspectRatio -> Automatic` (at the end) means equal scales on both axes, so that you get circles.

In[1]:= `Clear[z]`

In[2]:= `F = z^2`   (* Out: z^2 *)

In[3]:= `z = x + I y`   (* Out: $x+i\,y$ *)

In[4]:= Phi = ComplexExpand[Re[F]]   (* Out: $x^2 - y^2$ *)

In[5]:= y0 = Solve[Phi == k, y]   (* Out: $\{\{y \to -\sqrt{-k+x^2}\}, \{y \to \sqrt{-k+x^2}\}\}$ *)

In[6]:= S1 = Table[y0[[1, 1, 2]], {k, -10, 10}]

In[7]:= S2 = Table[y0[[2, 1, 2]], {k, -10, 10}]

In[8]:= Plot[Evaluate[{S1, S2}], {x, -2, 2}, AspectRatio -> Automatic]

    Plot::plnr : $\sqrt{-10+x^2}$ is not a machine-size real number
    at x = -0.834043'*^-7

(And several similar messages due to the square root of negative numbers.)

Pr.16.3. Orthogonality fails at integer multiples of π on the real axis, the points where $(\cos z)' = -\sin z = 0$.

In[1]:= z = x + I y   (* Out: $x + iy$ *)

In[2]:= F = Cos[z]   (* Out: Cos$[x + iy]$ *)

In[3]:= Phi = ComplexExpand[Re[F]]   (* Out: Cos$[x]$ Cosh$[y]$ *)

In[4]:= Psi = ComplexExpand[Im[F]]   (* Out: $-$Sin$[x]$ Sinh$[y]$ *)

In[5]:= sol1 = Solve[Phi == k1, y]

    Solve::ifun : Inverse functions are being used by Solve, so some
        solutions may not be found

Out[5]= $\{\{y \to -$ArcCosh$[k1$ Sec $[x]]\}, \{y \to$ ArcCosh$[k1$ Sec $[x]]\}\}$

In[6]:= sol2 = Solve[Psi == k2, y]

    Solve::ifun : Inverse functions are being used by Solve, so some
        solutions may not be found

Out[6]= $\{\{y \to -$ArcSinh$[k2$ Csc $[x]]\}\}$

In[7]:= S1 = Table[sol1[[1, 1, 2]], {k1, -5, 5}];

In[8]:= S2 = Table[-sol1[[1, 1, 2]], {k1, -5, 5}];

In[9]:= S3 = Table[sol2[[1, 1, 2]], {k2, -5, 5}];

In[10]:= Plot[Evaluate[{S1, S2, S3}], {x, -4, 4}, PlotRange -> {-4, 4}]

    Plot::plnr : -ArcCosh[-5 Sec[x]] is not a machine-size real number
        at x = -1.28742

(And several similar messages due to the square root of negative numbers.)

In[11]:= Plot3D[Phi, {x, -4, 4}, {y, -4, 4}]

Pr.16.5. Proceed as in Example 16.2 in this Guide, that is, obtain the inverse of $w = F(z)$ by typing

In[1]:= Clear[z]

In[2]:= <<Graphics'ComplexMap'

In[3]:= w = Log[-(z + 1)/(z - 1)]   (* Out: Log$\left[\frac{-1-z}{-1+z}\right]$ *)

In[4]:= z0 = Solve[(-1 - z)/(-1 + z) == W, z, InverseFunctions -> True]

Out[4]= $\left\{\left\{z \to \frac{-1+W}{1+W}\right\}\right\}$

In this function $z = z(W)$ divide numerator and denominator by exp $(W/2)$, obtaining $z = -\tanh(W/2)$. Interchange notations (and use again w instead of W) to get the inverse function $w = -\tanh(z/2)$, which you now use in conformal mapping. Experiment with the choice of the rectangle to be mapped.

In[5]:= Clear[w]

In[6]:= w[z_] = -Tanh[z/2]

In[7]:= CartesianMap[w, {-2, 2}, {-2, 2}]

Pr.16.7.

In[1]:= z = x + I y   (* Out: $x + i\,y$ *)

In[2]:= F = z^3   (* Out: $(x + i\,y)^3$ *)

In[3]:= Fbar = ComplexExpand[Conjugate[F]]

Out[3]= $x^3 - 3x y^2 + i(-3x^2 y + y^3)$

In[4]:= Phi = Simplify[(F + Fbar)/2]

Out[4]= $x^3 - 3x y^2$

In[5]:= y1 = Solve[Phi == k, y]

Out[5]= $\left\{\left\{y \to -\dfrac{i\sqrt{k - x^3}}{\sqrt{3}\,\sqrt{x}}\right\}, \left\{y \to \dfrac{i\sqrt{k - x^3}}{\sqrt{3}\,\sqrt{x}}\right\}\right\}$

In[6]:= S1 = Table[y1[[1, 1, 2]], {k, -5, 5}];

In[7]:= S2 = Table[y1[[2, 1, 2]], {k, -5, 5}];

In[8]:= Plot[Evaluate[{S1, S2}], {x, -4, 4}, AspectRatio -> Automatic,
           PlotRange -> {-4, 4}, AxesLabel -> {x, y}]

Plot::plnr : $-\dfrac{i\sqrt{-5-x^3}}{\sqrt{3}}$ is not a machine-size real number at x = -1.6535

(And several similar messages due to the square root of negative numbers.)

Pr.16.9.   At $x = 3$, $y = -3$ you have $u = (x - 1)(y - 1) = 2 \times (-4) = -8$. The same value is obtained as the mean over the circle, as follows. Type a parametric representation of the circle C. Then you automatically obtain u on C. Finally integrate.

In[1]:= x = 3 + Cos[t]   (* Out: $3 + \text{Cos}[t]$ *)

In[2]:= y = -3 + Sin[t]   (* Out: $-3 + \text{Sin}[t]$ *)

In[3]:= Phi = (x - 1) (y - 1)   (* Out: $(2 + \text{Cos}[t])(-4 + \text{Sin}[t])$ *)

In[4]:= mean = 1/(2 Pi) Integrate[Phi, {t, 0, 2 Pi}]   (* Out: -8 *)

CHAPTER 17,   page 189

Pr.17.1.

In[1]:= sol = Solve[x^2 - 30 x + 1 == 0, x]

Out[1]= $\left\{\left\{x \to \dfrac{1}{15 + 4\sqrt{14}}\right\}, \left\{x \to 15 + 4\sqrt{14}\right\}\right\}$

Here you see that Mathematica uses the improved formula.

In[2]:= s2 = N[sol[[2, 1, 2]]]   (* Out: 29.9666 *)

In[3]:= s1 = 30 - 29.9666   (* Out: 0.0334   Loss of 3 significant digits *)

No misunderstandings: Mathematica would fill this up to 6S (see below), but this is not the point of Example 17.1 and of this problem!

In[4]:= s1b = 10/s2   (* Out: 0.0333705 *)

In[5]:= N[sol[[1, 1, 2]]]   (* Out: 0.0333705 *)

Pr.17.3. $x = g(x) = (x + 0.12)^{1/4}$. Substituting $x = x_{n-1}$ into $(x + 0.12)^{1/4}$ gives x_n.

In[1]:= Clear[x]

In[2]:= x[0] = 1   (* Out: 1 *)

In[3]:= Do[x[n] = N[(x[n-1] + 0.12)^(1/4)], {n, 1, 10}]

In[4]:= Table[x[n], {n, 0, 10}]

In[5]:= x[10]^4 - x[10]   (* Out: 0.12 *)

Pr.17.5.   For the formula see Example 17.4 in this Guide.

In[1]:= Clear[x]

In[2]:= f[x_] = 1 - x^2/4 + x^4/64 - x^6/2304

In[3]:= x[0] = 2;   x[1] = 2.5;

In[4]:= Do[x[n+1] = x[n] - N[f[x[n]](x[n] - x[n-1])/(f[x[n]] - f[x[n-1]])],
        {n, 1, 6}]

In[5]:= Table[x[n], {n, 0, 6}]

Out[5]= {2, 2.5, 2.39635, 2.39157, 2.39165, 2.39165, 2.39165}

In[6]:= FindRoot[BesselJ[0, x], {x, 2}]   (* Out: {x → 2.40483} *)

Pr.17.7.   Proceed as in Example 17.6 in this Guide.

In[1]:= data = {{1.0, 1.0}, {1.02, 0.9888}, {1.04, 0.9784}}

Out[1]= {{1., 1.}, {1.02, 0.9888}, {1.04, 0.9784}}

In[2]:= InterpolatingPolynomial[{data[[1]], data[[2]], data[[3]]}, x]

Out[2]= 1. + (- 0.56 + 1. (- 1.02 + x)) (- 1. + x)

In[3]:= p2 = Expand[%]   (* Out: 2.58 - 2.58 x + 1. x^2 *)

In[4]:= p2 /. x -> 1.01   (* Out: 0.9943 *)

In[5]:= p2 /. x -> 1.03   (* Out: 0.9835 *)

The following way of obtaining the interpolation polynomial is simpler (in particular, in the case of larger data sets and higher-order interpolation polynomials.

In[6]:= u = {1.0, 1.02, 1.04};   v = {1.0, 0.9888, 0.9784};

In[7]:= Seq = Table[{u[[j]], v[[j]]}, {j, 1, 3}]

Out[7]= {{1., 1.}, {1.02, 0.9888}, {1.04, 0.9784}}

In[8]:= InterpolatingPolynomial[Seq, x]

Out[8]= $1. + (-0.56 + 1. (-1.02 + x)) (-1. + x)$

Pr.17.9.   Note the large oscillations of the polynomial between the nodes.

In[1]:= w = Table[{j, 0}, {j, -5, 5}];

In[2]:= w[[6]] = {0, 1}

In[3]:= w

Out[3]= {{−5, 0}, {−4, 0}, {−3, 0}, {−2, 0}, {−1, 0}, {0, 1}, {1, 0}, {2, 0}, {3, 0}, {4, 0}, {5, 0}}

In[4]:= InterpolatingPolynomial[w, x]

Out[4]= $(1 + x) (2 + x) (3 + x) (4 + x) (5 + x) \left(\frac{1}{120} + \left(-\frac{1}{120} + \left(\frac{1}{240} + \left(-\frac{1}{720} + \left(\frac{1}{2880} + \frac{4 - x}{14400}\right) (-3 + x)\right) (-2 + x)\right) (-1 + x)\right) x\right)$

In[5]:= p10 = Expand[%]

Out[5]= $1 - \frac{5269\,x^2}{3600} + \frac{1529\,x^4}{2880} - \frac{341\,x^6}{4800} + \frac{11\,x^8}{2880} - \frac{x^{10}}{14400}$

In[6]:= N[%]

Out[6]= $1. - 1.46361\,x^2 + 0.530903\,x^4 - 0.0710417\,x^6 + 0.00381944\,x^8 - 0.0000694444\,x^{10}$

In[7]:= Plot[p10, {x, -5, 5}]

Pr.17.11.

In[1]:= <<NumericalMath`SplineFit`

In[2]:= Seq = Table[{j, 0}, {j, -5, 5}]

Out[2]= {{−5, 0}, {−4, 0}, {−3, 0}, {−2, 0}, {−1, 0}, {0, 0}, {1, 0}, {2, 0}, {3, 0}, {4, 0}, {5, 0}}

In[3]:= Seq[[6]] = {0, 1}   (* This changes {0, 0} to {0, 1} *)

In[4]:= Seq

Out[4]= {{−5, 0}, {−4, 0}, {−3, 0}, {−2, 0}, {−1, 0}, {0, 1}, {1, 0}, {2, 0}, {3, 0}, {4, 0}, {5, 0}}

In[5]:= spl = SplineFit[Seq, Cubic]

Out[5]= SplineFunction[Cubic, {0., 10.}, <>]

In[6]:= spl[0]   (* Out: {−5, 0} *)

In[7]:= spl[0.5]   (* Out: {−4.5, 0.00310773} *)

In[8]:= spl[6]   (* Out: {1, 0} *)

In[9]:= ParametricPlot[spl[u], {u, 0, 10}, PlotRange -> All,
            Compiled -> False]

Pr.17.13.   You need about 250, 250, and 16 subintervals, respectively.

In[1]:= Clear[f, x, b]

In[2]:= f[x_] = Cos[x^2];   b = Sqrt[Pi/2];

In[3]:= SetPrecision[N[Integrate[f[x], {x, 0, b}]], 10] (* Out: 0.9774514243 *)

In[4]:= M = 250;   h = SetPrecision[N[b/M], 10]   (* Out: 0.005013256549 *)

In[5]:= RectRule = SetPrecision[N[h (Sum[f[x + h/2],
        {x, 0, b - h/2, h}])], 10]   (* Out: 0.9774540492 *)

In[6]:= M = 250;   h = SetPrecision[N[b/M], 10]   (* Out: 0.005013256549 *)

In[7]:= TrapezRule = SetPrecision[N[h (f[0]/2 + f[b]/2 + Sum[f[x],
        {x, h, b - h, h}])], 10]   (* Out: 0.9774461744 *)

In[8]:= M = 16;   h = SetPrecision[N[b/M], 10]   (* Out: 0.07833213358 *)

In[9]:= SimpsonRule = SetPrecision[h/3 (f[0] + 4 Sum[f[x],
        {x, h, b - h, 2 h}] + 2 Sum[f[x], {x, 2 h, b - 2 h, 2 h}] + f[b]), 10]

Out[9]= 0.9774547101

Pr.17.15.   For the nodes and coefficients see Example 17.8 in this Guide.

In[1]:= f = Sin[x]/x

In[2]:= Integrate[f, {x, 0, 1}]   (* Out: SinIntegral[1] *)

In[3]:= SetPrecision[%, 10]   (* Out: 0.9460830704 *)

In[4]:= x = (1 + t)/2;   (* Thus $dx = dt/2$. *)

In[5]:= f1 = f /. t -> -Sqrt[3/5]

In[6]:= f2 = f /. t -> 0

In[7]:= f3 = f /. t -> Sqrt[3/5]

In[8]:= 1/2 (5/9 f1 + 8/9 f2 + 5/9 f3)

Out[8]= $\dfrac{1}{2} \left(\dfrac{16}{9} \mathrm{Sin} \left[\dfrac{1}{2} \right] + \dfrac{10 \, \mathrm{Sin} \left[\frac{1}{2} \left(1 - \sqrt{\frac{3}{5}} \right) \right]}{9 \left(1 - \sqrt{\frac{3}{5}} \right)} + \dfrac{10 \, \mathrm{Sin} \left[\frac{1}{2} \left(1 + \sqrt{\frac{3}{5}} \right) \right]}{9 \left(1 + \sqrt{\frac{3}{5}} \right)} \right)$

In[9]:= SetPrecision[%, 10]   (* Out: 0.9460831341   This is 7D accuracy. *)

CHAPTER 18,   page 210

Pr.18.1.

In[1]:= <<LinearAlgebra'MatrixManipulation'

In[2]:= A = {{0, 6, 13}, {6, 0, -8}, {13, -8, 0}};

In[3]:= b = {61, -38, 79};

In[4]:= x = LinearSolve[A, b]   (* Out: {3, −5, 7} *)

In[5]:= A1 = Transpose[Join[Transpose[A], {b}]]

In[6]:= B1 = RowReduce[A1]; MatrixForm[B1]

In[7]:= ReplacePart[A1, A1[[3]], 1]

Out[7]= {{13, −8, 0, 79}, {6, 0, −8, −38}, {13, −8, 0, 79}}

In[8]:= A2 = ReplacePart[%, A1[[1]], 3]; MatrixForm[A2]

Out[8]//MatrixForm=

$$\begin{pmatrix} 13 & -8 & 0 & 79 \\ 6 & 0 & -8 & -38 \\ 0 & 6 & 13 & 61 \end{pmatrix}$$

In[9]:= A3 = ReplacePart[A2, A2[[2]] - 6/13 A2[[1]], 2]

Out[9]= $\{\{13, -8, 0, 79\}, \{0, \frac{48}{13}, -8, -\frac{968}{13}\}, \{0, 6, 13, 61\}\}$

In[10]:= A3 = ReplacePart[A3, A3[[3]] -6/(48/13) A3[[2]], 3]

Out[10]= $\{\{13, -8, 0, 79\}, \{0, \frac{48}{13}, -8, -\frac{968}{13}\}, \{0, 0, 26, 182\}\}$

In[11]:= Clear[x]

In[12]:= x[3]= 1/A3[[3,3]]   A3[[3,4]]   (* Out: 7 *)

In[13]:= x[2] = 1/A3[[2,2]] (A3[[2,4]] - A3[[2,3]] x[3])   (* Out: -5 *)

In[14]:= x[1] = 1/A3[[1,1]](A3[[1,4]] - A3[[1,3]] x[3] - A3[[1,2]] x[2])

Out[14]= 3

Pr.18.3.

In[1]:= <<LinearAlgebra`MatrixManipulation`

In[2]:= A = {{5, 4, 1}, {10, 9, 4}, {10, 13, 15}}

In[3]:= b = {3.4, 8.8, 19.2}

In[4]:= B = LUDecomposition[A]   (* No pivoting was done. *)

Out[4]= {{{5, 4, 1}, {2, 1, 2}, {2, 5, 3}}, {1, 2, 3}, 1}

In[5]:= LUBackSubstitution[%, b]   (* Out: {0.2, 0.4, 0.8} *)

To actually see what is going on, type

In[6]:= M = LUMatrices[B[[1]]]

Out[6]= {{{1, 0, 0}, {2, 1, 0}, {2, 5, 1}}, {{5, 4, 1}, {0, 1, 2}, {0, 0, 3}}}

In[7]:= M[[1]].M[[2]]   (* Product LU gives A. *)

Out[7]= {{5, 4, 1}, {10, 9, 4}, {10, 13, 15}}

In[8]:= y = LinearSolve[M[[1]], b]   (* Out: {3.4, 2., 2.4} *)

In[9]:= x = LinearSolve[M[[2]], y]   (* Out: {0.2, 0.4, 0.8} *)

Pr.18.5.

In[1]:= <<LinearAlgebra`Cholesky`

In[2]:= A = {{9, 6, 12}, {6, 13, 11}, {12, 11, 26}};

In[3]:= b = {17.4, 23.6, 30.8}

In[4]:= U = CholeskyDecomposition[A]; MatrixForm[U]

Out[4]//MatrixForm=

$$\begin{pmatrix} 3 & 2 & 4 \\ 0 & 3 & 1 \\ 0 & 0 & 3 \end{pmatrix}$$

In[5]:= y = LinearSolve[Transpose[U], b]   (* Out: {5.8, 4., 1.2} *)

In[6]:= x = LinearSolve[U, y]   (* Out: {0.6, 1.2, 0.4} *)

In[7]:= A.x   (* Out: {17.4, 23.6, 30.8}   For checking *)

Pr.18.7.

In[1]:= <<LinearAlgebra'MatrixManipulation'

In[2]:= A = {{9, 6, 12}, {6, 13, 11}, {12, 11, 26}};

In[3]:= B = Transpose[Join[A, IdentityMatrix[3]]];

In[4]:= Clear[F]

In[5]:= F = RowReduce[B]

$$\text{Out[5]}= \left\{\left\{1, 0, 0, \frac{217}{729}, -\frac{8}{243}, -\frac{10}{81}\right\}, \left\{0, 1, 0, -\frac{8}{243}, \frac{10}{81}, -\frac{1}{27}\right\}, \left\{0, 0, 1, -\frac{10}{81}, -\frac{1}{27}, \frac{1}{9}\right\}\right\}$$

In[6]:= Inv = TakeColumns[F, -3]   (* Take the last three columns of F. *)

or find a SubMatrix, where the {1, 4} means that you pick the submatrix starting at Row 1 and Column 4, and {3, 3} means that you take 3 rows and 3 columns – that is, the submatrix consisting of the last three columns of F.

In[7]:= Inv = SubMatrix[F, {1, 4},   {3, 3}]; MatrixForm[Inv]

Out[7]//MatrixForm=

$$\begin{pmatrix} \dfrac{217}{729} & -\dfrac{8}{243} & -\dfrac{10}{81} \\ -\dfrac{8}{243} & \dfrac{10}{81} & -\dfrac{1}{27} \\ -\dfrac{10}{81} & -\dfrac{1}{27} & \dfrac{1}{9} \end{pmatrix}$$

In[8]:= A.Inv   (* Out: {{1, 0, 0}, {0, 1, 0}, {0, 0, 1}}   For checking *)

Pr.18.9.

In[1]:= <<LinearAlgebra'MatrixManipulation'

In[2]:= x = {7, -12, 5, 0}   (* Out: {7, −12, 5, 0} *)

In[3]:= y = 10 x   (* Out: {70, −120, 50, 0} *)

In[4]:= VectorNorm[x, 1]

   VectorNorm::prec : VectorNorm has received a vector with infinite precision.

Out[4]= VectorNorm[{7, −12, 5, 0}, 1]

In[5]:= x = {7., -12, 5, 0}   (* Out: {7., −12, 5, 0} *)

In[6]:= VectorNorm[x, 1]   (* Out: 24. *)

In[7]:= VectorNorm[x, 2]   (* Out: 14.7648 *)

In[8]:= VectorNorm[x]   (* Out: 12 *)

In[9]:= y = 10 x   (* Out: {70., −120, 50, 0} *)

In[10]:= VectorNorm[y, 1]   (* Out: 240. *)

In[11]:= VectorNorm[y, 2]   (* Out: 147.648 *)

In[12]:= VectorNorm[y]   (* Out: 120. *)

Pr.18.11.

In[1]:= <<LinearAlgebra`MatrixManipulation`

In[2]:= A = {{9, 6, 12}, {6, 13, 11}, {12, 11, 26}}

In[3]:= MatrixConditionNumber[A, 1]

> MatrixConditionNumber::prec : MatrixConditionNumber has received a matrix with infinite precision.

Out[3]= MatrixConditionNumber[{{9, 6, 12}, {6, 13, 11}, {12, 11, 26}}, 1]

In[4]:= A = {{9., 6, 12}, {6, 13, 11}, {12, 11, 26}};

In[5]:= MatrixConditionNumber[A, 1]   (* Out: 22.2483 *)

In[6]:= MatrixConditionNumber[A]   (* Out: 22.2483 *)

In[7]:= B = Inverse[A]

In[8]:= FrobA = Sqrt[Sum[Sum[A[[j,k]]^2, {k, 1, 3}], {j, 1, 3}]]

Out[8]= 39.0896

In[9]:= FrobB = Sqrt[Sum[Sum[B[[j,k]]^2, {k, 1, 3}], {j, 1, 3}]]

Out[9]= 0.389343

In[10]:= KappaFrob = FrobA FrobB

Out[10]= 15.2193

Pr.18.13.

In[1]:= <<NumericalMath`PolynomialFit`

In[2]:= data = {{0, 2}, {1, 2}, {2, 1}, {3, 3}, {5, 2}, {7, 3}, {9, 3}, {10, 4}, {11, 3}}

In[3]:= f = PolynomialFit[data, 1]   (* Out: FittingPolynomial[<>, 1] *)

In[4]:= f[4]   (* Out: 2.34328   This evaluates f at a given point, 4. *)

In[5]:= Clear[x]

In[6]:= p = Expand[f[x]]   (* Out: 1.70647 + 0.159204 x *)

In[7]:= P1 = Plot[p, {x, -1, 12}]

In[8]:= P2 = ListPlot[data]

In[9]:= Show[P1, P2, Prolog -> AbsolutePointSize[4]]

Pr.18.15.

```
In[1]:= <<NumericalMath`PolynomialFit`
In[2]:= data = {{2, 0}, {3, 3}, {5, 4}, {6, 3}, {7, 1}}
In[3]:= Clear[f]
In[4]:= f = PolynomialFit[data, 2]
In[5]:= f[5]                    (* Out: 4.14286   This evaluates f at a given point, 4. *)
In[6]:= Clear[x]
In[7]:= p = Expand[f[x]]           (* Out: -8.35714 + 5.44643 x - 0.589286 x^2 *)
In[8]:= P1 = Plot[p, {x, 1, 8}]
In[9]:= P2 = ListPlot[data]
In[10]:= Show[P1, P2, Prolog -> AbsolutePointSize[4]]
```

Pr.18.17.   The command Eigenvalues gives the eigenvalues. The command Eigenvectors gives eigenvectors of Euclidean norm 1. The command Eigensystem gives the eigenvalues and their independent eigenvectors. For more details, click on "Help", then on "Master Index", and then type "Eigensystem".

```
In[1]:= A = {{0.49, 0.02, 0.22}, {0.02, 0.28, 0.2}, {0.22, 0.2, 0.4}}
In[2]:= Eigenvalues[A]                       (* Out: {0.72, 0.36, 0.09} *)
In[3]:= Eigenvectors[A]
In[4]:= Eigensystem[A]
```

Pr.18.19.

```
In[1]:= A = {{3, 2, 3}, {2, 6, 6}, {3, 6, 3}}; MatrixForm[A]
In[2]:= x = {1, 1, 1};   N0 = 10;                       (* N is protected. *)
```

Now use the program in Example 18.9 in this Guide. The response to Steps 1, 2, 10 is

```
In[3]:= Do [
        y = N[A.x];
             Print["y = ", y];
        q = N[x.y/x.x];
             Print["q = ", q];
        delta = N[Sqrt[y.y/x.x - q^2]];
             Print["delta = ", delta];
        z = Sort[y];
        x = N[y Sign[z[[1]] + z[[3]]]/Max[Abs[y[[1]]], Abs[y[[3]]]]];
             Print["x = ", x],
        {j, 1, N0}
        ]                                          (* N is protected. *)
```

y = {8, 14, 12}
q = 11.3333
delta = 2.49444
x = {0.666667, 1.16667, 1.}
y = {7.33333, 14.3333, 12.}
q = 11.9802
delta = 0.444553

      ⋮

x = {0.6, 1.2, 1.}
y = {7.2, 14.4, 12.}
q = 12.
delta = 2.38419×10^{-7}
x = {0.6, 1.2, 1.}

CHAPTER 19,   page 228

Pr.19.1. $f(x, y) = -0.1y$. Use the program as well as the plotting commands in Example 19.1.

In[1]:= Clear[x, y]

In[2]:= f[x_, y_] = -0.1 y   (* Out: -0.1y *)

In[3]:= h = 0.1;   M = 10;   x[0] = 0; y[0] = 2;   (* N is protected! *)

In[4]:= Do [y[n + 1] = N[y[n] + h f[x[n], y[n]]];
        x[n + 1] = N[x[n] + h],
        {n, 0, M}]

In[5]:= Table[{x[n], y[n], 2 Exp[-0.1 x[n]] - y[n]}, {n, 0, M}]

The two pairs of braces in table are essential. Try without.

Pr.19.3.   For $h = 0.01$ the error of $y(1)$ is about 0.00009.

In[1]:= f[x_, y_] = -0.1 y   (* Out: -0.1y *)

In[2]:= h = 0.01;   M = 100;   x[0] = 0; y[0] = 2;   (* N is protected! *)

In[3]:= Do [y[n + 1] = N[y[n] + h f[x[n], y[n]]];
        x[n + 1] = N[x[n] + h],
        {n, 0, M}]

In[4]:= Table[{x[n], y[n], 2 Exp[-0.1 x[n]] - y[n]}, {n, 0, M}]

Pr.19.5.   The global error of the method is of order h^2, so that halving h should reduce the error to about 1/4 of its original value. The numerical values obtained confirm this; e.g., for $h = 0.1$ and 0.05 and $x = 1.3$ you get the errors 0.10 and 0.03, for $x = 1.4$ you get 0.41 and 0.14, approximately. For $x = 1.5$ you are too close to $\pi/2$ and you get 3.6 and 1.6. For $x = 1.55$ and $h = 0.05$ you get the approximate value 27, the true value being 48, approximately.

In[1]:= f[x_, y_] = 1 + y^2

In[2]:= h = 0.05;   M = 31;   x[0] = 0;   y[0] = 0;

In[3]:= Do [

ystar = y[n] + h f[x[n], y[n]];   (* Auxiliary value *)

y[n + 1] = y[n] + h/2 (f[x[n], y[n]] + f[x[n] + h, ystar]);

x[n + 1] = x[n] + h,

{n, 0, M}

]

In[4]:= Table[{x[n], y[n], Tan[x[n]] - y[n]}, {n, 0, M}]

Out[4]= {{0, 0, 0}, {0.05, 0.0500625, −0.0000207916}, {0.1, 0.100376, −0.0000414242},

$\vdots$

{1.55, 26.6722, 21.4063}}

Pr.19.7. Use the module in Example 19.3 in this Guide (if you have not saved the module in that example you will need to do it now). Type the data as shown. The errors differ by a factor $1/2^4$, approximately.

In[1]:= <<RK.m

In[2]:= f[x_, y_] = -0.2 x y

In[3]:= x[0] = 0;   y[0] = 1;   h = 0.2;   M = 50;

In[4]:= RK[f, x, y, h, M]

In[5]:= Table[{x[n], y[n], Exp[-0.1 x[n]^2] - y[n]}, {n, 0, 50}]

Out[5]= {{0, 1, 0}, {0.2, 0.996008, 1.06581×10^{-11}},

$\vdots$

{10., 0.0000454429, -4.29335×10^{-8}}}

In[6]:= x[0] = 0;   y[0] = 1;   h = 0.1;   M = 100;

In[7]:= RK[f, x, y, h, M]

In[8]:= Table[{x[n], y[n], Exp[-0.1 x[n]^2] - y[n]}, {n, 0, 100}]

Out[8]= {{0, 1, 0}, {0.1, 0.999, 4.17444×10^{-14}}, {0.2, 0.996008, 4.14224×10^{-13}},

$\vdots$

{10., 0.0000454023, -2.32563×10^{-9}}}

Pr.19.9. Remember that Runge-Kutta (needed as a starter in Adams-Moulton) has been saved under RK in Example 19.3 in this Guide, and the Adams-Moulton module has been saved under AMRK in Example 19.4. Load by typing

In[1]:= <<RK.m

In[2]:= <<AMRK.m

Then call the module and use Table to obtain the answer.

In[3]:= f[x_, y_] = x + y   (* Out: x + y *)

In[4]:= x[0] = 0;   y[0] = 0;   h = 0.1;   M = 10;

In[5]:= AdamsMoultonRK[f, x, y, h, M]

In[6]:= Table[{x[n], y[n], Exp[x[n]] - x[n] - 1 - y[n]}, {n, 0, 10}]

Out[6]= {{0, 0, 0}, {0.1, 0.00517083, 8.47423×10^{-8}},

    {0.2, 0.0214026, 1.87309×10^{-7}},

    {0.3, 0.0498585, 3.10513×10^{-7}}, {0.4, 0.0918245, 1.57286×10^{-7}},

    {0.5, 0.148721, -3.76386×10^{-8}}, {0.6, 0.222119, -2.73068×10^{-7}},

    {0.7, 0.313753, -5.57891×10^{-7}}, {0.8, 0.425542, -8.99635×10^{-7}},

    {0.9, 0.559604, -1.30709×10^{-6}}, {1., 0.718284, -1.79029×10^{-6}}}

Pr.19.11.   Load the module (see Example 19.5 in this Guide) by typing

In[1]:= Clear[f, x, y, h, M]

In[2]:= <<RKS.m

Type the data. Call the module. Use Table for obtaining the computed values, with the errors computed by using the exact solution as obtained by DSolve. The response consists of x, y_1, y_2, the error of y_1, and the error of y_2.

In[3]:= f[x_, y_] := {2 y[[1]] - 4 y[[2]], y[[1]] - 3 y[[2]]}   (* Colon! *)

In[4]:= x[0] = 0;   y[0] = {3, 0};   h = 0.1;   M = 5;

In[5]:= RKS[f, x, y, h, M]

In[6]:= Table[{x[n], y[n][[1]], y[n][[2]],
            4 Exp[x[n]] - Exp[-2 x[n]] - y[n][[1]],
            Exp[x[n]] - Exp[-2 x[n]] - y[n][[2]]}, {n, 0, M}]

Out[6]= {{0, 3, 0, 0, 0}, {0.1, 3.60195, 0.286437, 2.91922×10^{-6}, 2.665×10^{-6}},

    {0.2, 4.21529, 0.551078, 4.97431×10^{-6}, 4.41238×10^{-6}},

    {0.3, 4.85062, 0.801042, 6.43086×10^{-6}, 5.49932×10^{-6}},

    {0.4, 5.51796, 1.04249, 7.49457×10^{-6}, 6.12188×10^{-6}},

    {0.5, 6.227, 1.28084, 8.32537×10^{-6}, 6.42906×10^{-6}}}

In[7]:= sys = {D[z1[t], t] == 2 z1[t] - 4 z2[t],
            D[z2[t], t] == z1[t] - 3 z2[t]}

In[8]:= sol = DSolve[{sys[[1]], sys[[2]], z1[0] == 3, z2[0] == 0},
            {z1[t], z2[t]}, t]

Out[8]= {{z1[t] → e^{-2t} ($-1 + 4 e^{2t}$), z2[t] → e^{-2t} ($-1 + e^{2t}$)}}

In[9]:= Plot[{sol[[1, 1, 2]], sol[[1, 2, 2]]}, {t, 0, 5}]

Pr.19.13.   Type the coefficient matrix **A** as in Example 19.7 and the vector **b** of the system **Ax** = **b** resulting from the boundary values.

```
In[1]:= A = Table[ Switch[j - k, 0, -4, 1, 1, -1, 1, p, 1, -p, 1, _, 0],
                  {j, p^2}, {k, p^2}]; MatrixForm[A]
In[2]:= A[[p, p + 1]] = 0; A[[p + 1, p]] = 0; A[[2 p, 2 p + 1]] = 0;
            A[[2 p + 1, 2 p]] = 0; MatrixForm[A]
In[3]:= b = -{110, 0, 110, 110, 0, 110, 330,  220, 330}
In[4]:= N[LinearSolve[A, b]]
```

Out[4]= {70.7143, 62.8571, 70.7143, 110., 110., 110., 149.286, 157.143, 149.286}

```
In[5]:= X = Partition[%, p]; MatrixForm[X]
```

Out[5]//MatrixForm=

$$\begin{pmatrix} 70.7143 & 62.8571 & 70.7143 \\ 110. & 110. & 110. \\ 149.286 & 157.143 & 149.286 \end{pmatrix}$$

```
In[6]:= Answer = {X[[3]], X[[2]], X[[1]]}; MatrixForm[Answer]
```

Out[6]//MatrixForm=

$$\begin{pmatrix} 149.286 & 157.143 & 149.286 \\ 110. & 110. & 110. \\ 70.7143 & 62.8571 & 70.7143 \end{pmatrix}$$

```
In[7]:= P3 = ContourPlot[x, {x, 0, 12}, {y, 0, 12}, Contours -> 3,
           ContourShading -> False]
In[8]:= P4 = ContourPlot[y, {x, 0, 12}, {y, 0, 12}, Contours -> 3,
           ContourShading -> False]
In[9]:= Show[P3, P4]
```

Pr.19.15.   125 steps will give $t = 125 \times 0.0016 = 0.2$. Accordingly, type

```
In[1]:= Clear[n, r, h, k]
In[2]:= n = 24;    r = 1;    h = 1/25;    M = 125;
In[3]:= A = Table[Switch[j - k, 0, 4, 1, -1, -1, -1, _, 0], {j, n}, {k, n}];
In[4]:= Clear[u]
In[5]:= Do[u[k] = N[Sin[Pi k h]], {k, 0, n + 1}]
In[6]:= Table[u[k], {k, 0, n + 1}]
In[7]:= Do [
            b = Table[0, {i, 1, n}];
            Do[b[[k]] = u[k - 1] + u[k + 1], {k, 1, n}];
            v = N[LinearSolve[A, b]];
                Print[v];
            Do[u[k] = v[[k]], {k, 1, n}],
            {j, 1, M}
            ]
```

All the new values are somewhat smaller than those in Example 19.8. They are (old values in parentheses)

$$
\begin{array}{llll}
t = 0.04: & 0.396267 & (0.399274), & 0.641174 & (0.646039) \\
t = 0.08: & 0.267151 & (0.271221), & 0.43226 & (0.438844) \\
t = 0.12: & 0.180105 & (0.184236), & 0.291417 & (0.2981) \\
t = 0.16: & 0.121422 & (0.125149), & 0.196464 & (0.202495) \\
t = 0.20: & 0.0818588 & (0.0850118), & 0.13245 & (0.137552)
\end{array}
$$

CHAPTER 20,   page 234

Pr.20.1.   7 steps. Use the module from Example 20.1 in this Guide. Type

In[1]:= Clear[x, y, z, t, S, T, f]

In[2]:= <<Calculus'VectorAnalysis'

In[3]:= SetCoordinates[Cartesian[x,y,u]]

In[4]:= << SD.m

In[5]:= f[x_, y_] = x^2 + 10 y^2   (* Out: $x^2 + 10 y^2$ *)

In[6]:= X[0] = 3; Y[0] = -2; T[0] = 0;   M = 7;

In[7]:= SD[f, X, Y, M]

In[8]:= S = Table[N[{X[j], Y[j]}], {j, 0, 6}]

In[9]:= L = Line[S];

In[10]:= Show[Graphics[L], Axes -> True, AxesLabel -> {x, y}]

In[11]:= Table[N[{X[j], Y[j], f[X[j], Y[j]]}], {j, 0, M}]

In[12]:= f[X[7], Y[7]]   (* Out: 0.0000778187 *)

Pr.20.3.   The path is spiraling around the origin. Use the module from Example 20.1.

In[1]:= Clear[x, y, z, t, S, T, f]

In[2]:= <<Calculus'VectorAnalysis'

In[3]:= SetCoordinates[Cartesian[x,y,u]]

In[4]:= << SD.m

In[5]:= f[x_, y_] = 5 x^2 - y^2   (* Out: $5 x^2 - y^2$ *)

In[6]:= X[0] = 1/2; Y[0] = 1/2; M = 6;   T[0] = 0;

In[7]:= SD[f, X, Y, M]

In[8]:= S = Table[N[{X[j], Y[j]}], {j, 0, 6}]

In[9]:= L = Line[S];

In[10]:= Show[Graphics[L], Axes -> True, AxesLabel -> {x,y}]

Pr.20.5.   Maximize $z = x_1 + x_2$. The matrix $\mathbf{T}_2$ shows that $z = 80/4 + 60/2 = 50$ is the maximum.

```
In[1]:= T0 = {{1, -1, -1,  0,  0,  0,  0},
              {0,  2,  3,  1,  0,  0, 130},
              {0,  3,  8,  0,  1,  0, 300},
              {0,  4,  2,  0,  0,  1, 140}}
In[2]:= T1 = {T0[[1]] + 1/4 T0[[4]], T0[[2]] - 2/4 T0[[4]],
              T0[[3]] - 3/4 T0[[4]], T0[[4]]}; MatrixForm[T1]
In[3]:= T2 = {T1[[1]] + 1/4 T1[[2]], T1[[2]], T1[[3]] - 13/4 T1[[2]],
              T1[[4]] - T1[[2]]}; MatrixForm[T2]
```

Out[3]//MatrixForm=

$$
\begin{pmatrix}
1 & 0 & 0 & \dfrac{1}{4} & 0 & \dfrac{1}{8} & 50 \\[2mm]
0 & 0 & 2 & 1 & 0 & -\dfrac{1}{2} & 60 \\[2mm]
0 & 0 & 0 & -\dfrac{13}{4} & 1 & \dfrac{7}{8} & 0 \\[2mm]
0 & 4 & 0 & -1 & 0 & \dfrac{3}{2} & 80
\end{pmatrix}
$$

CHAPTER 22,   page 243

Pr.22.1.

```
In[1]:= <<Statistics'DescriptiveStatistics'
In[2]:= S = {203, 199, 198, 201, 200, 201, 201};
In[3]:= N[Mean[S]]                                      (* Out: 200.429 *)
In[4]:= N[Variance[S]]                                  (* Out: 2.61905 *)
```

Pr.22.3.

```
In[1]:= Clear[x, n, p]
In[2]:= <<Statistics'DescriptiveStatistics'
In[3]:= BinDist = BinomialDistribution[10, 0.5]
In[4]:= Table[{x, PDF[BinDist, x], CDF[BinDist, x]}, {x, 0, 10}]
```

Out[4]= {{0, 0.000976563, 0.000976563}, {1, 0.00976563, 0.0107422},

{2, 0.0439453, 0.0546875}, {3, 0.117188, 0.171875},

{4, 0.205078, 0.376953}, {5, 0.246094, 0.623047},

{6, 0.205078, 0.828125}, {7, 0.117188, 0.945313},

{8, 0.0439453, 0.989258}, {9, 0.00976563, 0.999023},

{10, 0.000976563, 1}}

```
In[5]:= mu = Sum[x PDF[BinDist, x], {x, 0, 10}]            (* Out: 5. *)
In[6]:= var = Sum[(x - mu)^2 PDF[BinDist, x], {x, 0, 10}]  (* Out: 2.5 *)
```

In[7]:= Mean[BinDist]   (* Out: 5. *)

In[8]:= Variance[BinDist]   (* Out: 2.5 *)

Pr.22.7.   In 8! = 40320 ways. Type 8! .

Pr.22.11.

In[1]:= <<Statistics'DiscreteDistributions'

In[2]:= hyp = HypergeometricDistribution[10, 50, 100]

In[3]:= S = Table[N[PDF[hyp, x]], {x, 0, 10}]

In[4]:= <<Graphics'

In[5]:= BarChart[S, PlotRange -> {0, 1}]

Pr.22.13.

In[1]:= mu = Integrate[x/(b - a), {x, a, b}]

In[2]:= Simplify[%]   (* Out: $\frac{a+b}{2}$ *)

In[3]:= var = Integrate[(x - mu)^2/(b - a), {x, a, b}]

In[4]:= Simplify[%]   (* Out: $\frac{1}{12}(a-b)^2$ *)

Pr.22.15.   1084 hours (practically 100 hours).

In[1]:= <<Statistics'ContinuousDistributions'

In[2]:= ND = NormalDistribution[1000, 100]

In[3]:= Quantile[ND, 0.8]   (* Out: 1084.16 *)

CHAPTER 23,   page 256

Pr.23.1.   For instance, a do-loop may be used as shown and will save unnecessary typing.

In[1]:= rn = Table[Random[Integer, {1, 50}], {j, 1, 5}]

In[2]:= SeedRandom[6]

In[3]:= <<Statistics'DescriptiveStatistics'

In[4]:= Do[mean[k] = Mean[Table[Random[Integer, {1, 50}], {j, 1, 5}]],
        {k , 1, 100}]

In[5]:= S = Table[N[mean[k]], {k, 1, 100}]

In[6]:= Sort[S]

In[7]:= << Graphics'Graphics'

In[8]:= Histogram[S]

Pr.23.3.

In[1]:= <<Statistics'ContinuousDistributions'

In[2]:= c = Quantile[NormalDistribution[0, 1], 0.975]   (* Out: 1.95996 *)

```
In[3]:= Clear[n]
In[4]:= length = 2 c/Sqrt[n]        (* 'length' with l, not L. Try with L. *)
In[5]:= Plot[length, {n, 0, 500}, PlotRange -> {0, 1}]
```

Pr.23.5.

```
In[1]:= <<Statistics'DescriptiveStatistics'
In[2]:= <<Statistics'ContinuousDistributions'
In[3]:= S = {17.3, 17.8, 18.0, 17.7, 18.2, 17.4, 17.6, 18.1}
In[4]:= n = Length[S];
In[5]:= nvar = (n - 1) Variance[S]                        (* Out: 0.73875 *)
In[6]:= c1 = Quantile[ChiSquareDistribution[7], 0.025]    (* Out: 1.68987 *)
In[7]:= c2 = Quantile[ChiSquareDistribution[7], 0.975]    (* Out: 16.0128 *)
In[8]:= conf1 = nvar/c2                                  (* Out: 0.0461351 *)
In[9]:= conf2 = nvar/c1                                  (* Out: 0.437164 *)
```

Pr.23.7. Hypothesis: Not better. Alternative: Better. If the hypothesis is true, the random variable $X = $ *Number of patients cured among 400* is approximately normal with mean $np = 300$ and variance $npq = 75$ (mean and variance of the binomial distribution with $p = 75\%$). The test is right-sided. 310 is less than the critical 314. Accept the hypothesis and assert that the better result is due to randomness.

```
In[1]:= <<Statistics'ContinuousDistributions'
In[2]:= c = Quantile[NormalDistribution[300, Sqrt[75]], 0.95]
Out[2]= 314.245
```

Pr.23.9. The test is right-sided. Accept the hypothesis because $y < c$. Indeed,

```
In[1]:= <<Statistics'ContinuousDistributions'
In[2]:= n = 20;    alpha = 0.05;
In[3]:= y = (n - 1) 1.^2/0.8^2                           (* Out: 29.6875 *)
In[4]:= c  = Quantile[ChiSquareDistribution[19], 1 - alpha]
Out[4]= 30.1435
```

Pr.23.11. Two-sided test. Use the t-distribution with 7 degrees of freedom. Accept the hypothesis because t_0 lies between $-c$ and c.

```
In[1]:= <<Statistics'DescriptiveStatistics'
In[2]:= <<Statistics'ContinuousDistributions'
In[3]:= Sa = {0.4, -0.6, 0.2, 0.0, 1.0, 1.4, 0.4, 1.6}
In[4]:= n = Length[Sa]                                   (* Out: 8 *)
In[5]:= m = Mean[Sa]                                     (* Out: 0.55 *)
In[6]:= var = Variance[Sa]                               (* Out: 0.545714 *)
```

In[7]:= t0 = (m - 0)/Sqrt[var/n]   (* Out: 2.10584 *)

In[8]:= c = Quantile[StudentTDistribution[n - 1], 0.975]   (* Out: 2.36462 *)

Pr.23.13. Use the chi-square distribution with $n - 1 = 2$ degrees of freedom. Reject the hypothesis that the traffic on the three lanes is the same.

In[1]:= <<Statistics`ContinuousDistributions`

In[2]:= n = 3;

In[3]:= e = N[(920 + 870 + 750)/3]   (* Out: 846.667 *)

In[4]:= b = {920, 870, 750}

In[5]:= Sum[(b[[j]] - e)^2/e, {j, 1, 3}]   (* Out: 18.0315 *)

In[6]:= Quantile[ChiSquareDistribution[n - 1], 0.95]   (* Out: 5.99146 *)

Pr.23.15. In the command Fit you have to type 1, x, x^2 because you want a quadratic polynomial.

In[1]:= <<Statistics`DescriptiveStatistics`

In[2]:= xSa = {1, 2, 4, 5, 7, 8}

In[3]:= ySa = {7, 5, 2, 1, 2, 4}

In[4]:= data = Table[{xSa[[j]], ySa[[j]]}, {j, 1, 6}]

In[5]:= f = Fit[data, {1, x, x^2}, x]

Out[5]= $10.3867 - 3.49333\,x + 0.333333\,x^2$

In[6]:= P1 = Plot[f, {x, 0, 10}]

In[7]:= P2 = ListPlot[data]

In[8]:= Show[{P1, P2}, Prolog -> AbsolutePointSize[4]]

INDEX
OF MATHEMATICA COMMANDS
AND KEYWORDS